LATIN-AMERICAN SCHOOL OF PHYSICS XXXI ELAF

LATIN-AMERICAN SCHOOL OF PHYSICS XXXI ELAF

New Perspectives on Quantum Mechanics

Mexico City, Mexico   July-August 1998

EDITORS
Shahen Hacyan
Rocío Jáuregui
Instituto de Física, UNAM, México

Ramón López-Peña
ICN, UNAM, México

AIP CONFERENCE PROCEEDINGS 464

American Institute of Physics   **Woodbury, New York**

Editors:

Shahen Hacyan
Instituto de Física
Universidad Nacional Autónoma de México
Apartado Postal 20-364
01000 México, D.F.
México
E-mail: hacyan@fenix.ifisicacu.unam.mx

Rocío Jáuregui
Instituto de Física
Universidad Nacional Autónoma de México
Apartado Postal 20-364
01000 México, D.F.
México
E-mail: rocio@fenix.ifisicacu.unam.mx

Ramón López-Peña
Instituto de Ciencias Nucleares
Universidad Nacional Autónoma de México
Apartado Postal 70-543
04510 México, D.F.
México
E-mail: lopez@nuclecu.unam.mx

We do not claim copyright for the reprint of the paper published in *Phys. Rev. D*, **28**, 2477–2484 (1983), pp. 230–237 in this book.

Authorization to photocopy items for internal or personal use, beyond the free copying permitted under the 1978 U.S. Copyright Law (see statement below), is granted by the American Institute of Physics for users registered with the Copyright Clearance Center (CCC) Transactional Reporting Service, provided that the base fee of $15.00 per copy is paid directly to CCC, 222 Rosewood Drive, Danvers, MA 01923. For those organizations that have been granted a photocopy license by CCC, a separate system of payment has been arranged. The fee code for users of the Transactional Reporting Service is: 1-56396-856-8/ 99 /$15.00.

© 1999 American Institute of Physics

Individual readers of this volume and nonprofit libraries, acting for them, are permitted to make fair use of the material in it, such as copying an article for use in teaching or research. Permission is granted to quote from this volume in scientific work with the customary acknowledgment of the source. To reprint a figure, table, or other excerpt requires the consent of one of the original authors and notification to AIP. Republication or systematic or multiple reproduction of any material in this volume is permitted only under license from AIP. Address inquiries to Office of Rights and Permissions, 500 Sunnyside Boulevard, Woodbury, NY 11797-2999; phone: 516-576-2268; fax: 516-576-2499; e-mail: rights@aip.org.

L.C. Catalog Card No. 99-60216
ISBN 1-56396-856-8
ISSN 0094-243X
DOE CONF- 980781

Printed in the United States of America

Contents

Preface vii
Sponsors ix
Lecturers xi
Organizing Committee xiii

I. QUANTUM OPTICS AND TRAPPED PARTICLES

Quantum Optics in Cavities and the Classical Limit of
Quantum Mechanics 3
    L. Davidovich
Cavity Quantum Electrodynamics: A Review of Rydberg Atom-Microwave
Experiments on Entanglement and Decoherence 45
    S. Haroche
Laser Cooling and Trapping of Neutral Atoms 67
    L. A. Orozco
Magnetic Trapping, Evaporative Cooling, and Bose-Einstein Condensation 91
    S. L. Rolston
Quantum Optics with Trapped Ions 111
    R. C. Thompson

II. FOUNDATIONS OF QUANTUM MECHANICS

Linear Stochastic Electrodynamics: Looking for the Physics behind
Quantum Theory 151
    L. de la Peña and A. M. Cetto
Conventional Quantum Mechanics Without Wave Function and
Density Matrix 191
    V. I. Man'ko
A Tutorial on Quantum Distribution Functions for Spin-1/2 Systems
and Einstein-Podolsky-Rosen Correlations 221
    M. O. Scully, H. Lee, E. Gómez, and R. Ortega-Martínez

III. TOPICS IN QUANTUM MECHANICS AND QUANTUM CHAOS

Spectral Properties of Classically Chaotic Systems 253
    F. Leyvraz
Interference Phenomena in Electronic Transport Through Chaotic Cavities:
An Information-Theoretic Approach 281
    P. A. Mello and H. U. Baranger
Topics in Quantum Mechanics 335
    M. Moshinsky

List of Participants... 355
Author Index... 367

PREFACE

The XXXI Latin American School of Physics (Escuela Latinoamericana de Física, ELAF) took place in Mexico City from July 27 to August 14, 1998. The subject of this School was "New Perspectives on Quantum Mechanics." The aim of the XXXI ELAF was to present a general overview of the current state of the art on the foundations of quantum physics, both from a theoretical and an experimental perspective, and with particular emphasis on the most advanced research in this field and on future technological applications. For this purpose, we tried to get a balance between theoretical and experimental physics, selecting the lecturers accordingly. Thirteen lectures of five hours each were given by leading physicists working actively on the frontiers of quantum physics. To these invited lectures were added about fifteen plenary lectures by other distinguished researchers. Altogether, about one hundred fifty students and researchers attended this school, many of them from Mexico, but also from Brazil, Argentina, Venezuela and Cuba, as well as from some European countries.

The ELAF took place over a period of three weeks. In the first week, the lectures were given by Professors L. Davidovich (Universidad Federal Rio de Janeiro), G. Gabrielse (Harvard University), S. Haroche (Ecole Normale Supérieure), L. Orozco (SUNY, Stony Brook), and M. Scully (Texas A & M University). The topics were related to quantum optics, modern techniques for atom cooling by means of lasers, atomic traps, and the manipulation of atoms in microcavities. The emphasis was on experimental techniques related to fundamental phenomena predicted by quantum mechanics. The second week had a more theoretical focus, with the courses being given by V.I. Man'ko (Lebedev Institute of Physics), P. A. Mello (Universidad Nacional Autónoma de México, UNAM), M. Moshinsky (UNAM), and L. de la Peña (UNAM). They discussed new theoretical methods for the study of quantum phenomena; in particular mesoscopic systems, tomographic methods, several mathematical techniques, and alternative interpretations of quantum mechanics. Finally, the third week was both theoretical and experimental. The courses were given by F. Leyvraz (UNAM), S. Rolston (National Institute of Standards and Technology), J. Sánchez Mondragón (Instituto Nacional de Astrofísica, Óptica y Electrónica) and R. Thompson (Imperial College), on topics of quantum optics of two-level systems, ion traps, Bose-Einstein condensation, and quantum chaos.

On this occasion, the organizers of the school dedicated the ELAF to Professor Marcos Moshinky, a pioneer of physics in Latin-America and one of the founders of ELAF, who has delivered lectures in each of the ELAF sessions organized in Mexico—and the present one was no exception!—.

As organizers, we feel that the ELAF fulfilled all its purposes, which were concerned with the presentation of courses at a very high level, given by distinguished experts working actively at the frontiers of modern physics. We also had many interesting lectures given by postgraduate students and active researchers. Last, but not least, we had the excellent organizing support of the "Colegio Nacional" (the Mexican academy for sciences and arts), as well as the economic support of several Mexican universities and international organizations, which were essential to the success of ELAF 98.

Shahen Hacyan
Chairman, Organizing Committee

SPONSORS

UNIVERSIDAD NACIONAL AUTÓNOMA DE MÉXICO (UNAM)
    Coordinación de la Investigación Científica:
        Instituto de Ciencias Nucleares
        Instituto de Física
    Dirección General de Estudios de Posgrado

UNIVERSIDAD AUTÓNOMA METROPPOLITANA, UNIDAD IZTAPALAPA (UAM-I)
    Departamento de Física

EL COLEGIO NACIONAL

CENTRO INTERNACIONAL DE CIENCIAS A. C., MÉXICO (CIC)

ORGANIZACIÓN DE ESTADOS AMERICANOS (OEA)

CENTRO LATINOAMERICANO DE FÍSICA, BRASIL (CLAF)

CENTRO LATINOAMERICANO DE FÍSICA, MÉXICO (CLAF-MÉXICO)

INTERNATIONAL CENTER FOR THEORETICAL PHYSICS AT TRIESTE (ICTP)

CENTRO DE INVESTIGACIÓN Y DE ESTUDIOS AVANZADOS DEL INSTITUTO POLITÉCNICO NACIONAL (CINVESTAV)
    Departamento de Física

CONSEJO NACIONAL DE CIENCIA Y TECNOLOGÍA (CONACYT)

LECTURERS

DAVIDOVICH, LUIZ
Inst. de Física, U. F. do Rio de Janeiro
Cx. P. 68528, 21945-970 Rio de Janeiro

ldavid@if.ufrj.br
Brazil

DE LA PEÑA, LUIS
Inst. de Física-UNAM
A. Postal 20-364 C.P. 01000

luis@fenix.ifisicacu.unam.mx
México

GABRIELSE, GERALD
Lyman Lab., Dept. Physics
Harvard University
Cambridge, MA 02138

gabrielse@physics.harvard.edu
U.S.A.

HAROCHE, SERGE
Dept. de Physique de l'Ecole Normale
Supérieure, 24 rue Lhomond
75231 Paris Cedex 05

haroche@physique.ens.fr
France

LEYVRAZ, FRANÇOIS
Centro de Ciencias Físicas, UNAM
Av. Universidad 2001 C.P. 62210
Cuernavaca, Mor.

leyvraz@ce.ifisicacu.unam.mx
México

MAN'KO, VLADIMIR
Lebedev Physical Institute
Leninskii Pr. 53, Moscow 117924

manko@na.infn.it
Russia

MELLO, PIER A.
Inst. de Física-UNAM
A. Postal 20-364 C.P. 01000

mello@fenix.ifisicacu.unam.mx
México

MOSHINSKY, MARCOS
Inst. de Física-UNAM
A. Postal 20-364 C.P. 01000

moshi@fenix.ifisicacu.unam.mx
México

OROZCO, LUIS A.   luisorozco@sunysb.edu
Dept. of Physics and Astronomy   U.S.A.
SUNY at Stony Brook, NY 11794-3800

ROLSTON, STEVEN L.   steven.rolston@nist.gov
National Inst. of Standards and   U.S.A.
Technology, PHY A167
Gaithersburg, MD 20899

SANCHEZ MONDRAGON, JAVIER J.   jsanchez@uaem.mx7jsanchez@inaoep.mx
Fac. Ciencias, UAEM   México
Av. Universidad 2001 C.P. 62210
Cuernavaca, Mor.

SCULLY, MARLAN O.   U.S.A.
Department of Physics, Texas A&M Univ
College Station, Texas 77843-4242

THOMPSON, RICHARD C.   r.c.thompson@ic.ac.uk
Blackett Lab., Imperial College   U.K.
Prince Consort Rd., London SW7 2BZ

ORGANIZING COMMITTEE

Hacyan, Shahen (Chairman)   hacyan@fenix.ifisicacu.unam.mx
Inst. de Física-UNAM   México
A. Postal 20-364 C.P. 01000

Castaños, Octavio   ocasta@servidor.unam.mx
Inst. de Ciencias Nucleares-UNAM   México
A. Postal 70-543 C.P. 04510

Cruz, Salvador   México
UAM-Iztapalapa
Av. Michaoacán y Purísma s/n
Col. Vicentina

Hirsch, Jorge G.   hirsch@nuclecu.unam.mx
Inst. de Ciencias Nucleares-UNAM   México
A. Postal 70-543 C.P. 04510

Jáuregui, Rocío (IF-UNAM)   rocio@fenix.ifisicacu.unam.mx
Inst. de Física-UNAM   México
A. Postal 20-364 C.P. 01000

López-Peña, Ramón   lopez@nuclecu.unam.mx
Inst. de Ciencias Nucleares-UNAM   México
A. Postal 70-543 C.P. 04510

Recamier, José   pepe@pepe.ifisicam.unam.mx
Centro de Ciencias Físicas-UNAM   México
Cuernavaca, Morelos

I. QUANTUM OPTICS AND TRAPPED PARTICLES

Quantum Optics in Cavities and the Classical Limit of Quantum Mechanics

Luiz Davidovich

Instituto de Física, Universidade Federal do Rio de Janeiro
Cx. P. 68528, 21945-970 Rio de Janeiro, Rio de Janeiro, Brazil

Abstract. These lectures review some basic techniques of quantum optics, related to the description of the quantized electromagnetic field in phase space and of the interaction between atoms and photons in cavities. The Wigner function is introduced, and some of the methods for measuring this distribution are reviewed. The combination of phase space methods with cavity QED techniques is shown to lead to experiments which are closely connected to fundamental problems regarding the classical limit of quantum mechanics and the quantum theory of measurement.

I INTRODUCTION

One of the most subtle problems in the physics of this century is the relation between the macroscopic world, described by classical physics, and the microscopic world, ruled by the laws of quantum physics. Among the several questions involved in the quantum-classical transition, one stands out in a striking way. As pointed out by Einstein in a letter to Max Born in 1954 [1], it concerns "the inexistence at the classical level of the majority of states allowed by quantum mechanics," namely coherent superpositions of classically distinct states. Indeed, while in the quantum world one frequently comes across coherent superpositions of states (like in Young's two-slit interference experiment, in which each photon is considered to be in a coherent superposition of two wave packets, centered around the classical paths which stem out of each slit), one does not see macroscopic objects in coherent superpositions of two distinct classical states, like a stone which could be at two places at the same time. There is an important difference between a state of this kind and one which would involve just a classical alternative: the existence of quantum coherence between the two localized states would allow in principle the realization of an interference experiment, complementary to the simple observation of the position of the stone. We know all this already from Young's experiment: the observation of the photon path (that is, a measurement which is able to distinguish through which slit the photon has passed) unavoidably destroys the interference fringes.

If one assumes that the usual rules of quantum dynamics are valid up to the macroscopic level, then the existence of quantum interference at the microscopic level necessarily implies that the same phenomenon should occur between distinguishable macroscopic states. This was emphasized by Schrödinger in his famous "cat paradox" [2]. An important role is played by this fact also in quantum measurement theory, as pointed out by Von Neumann [3]. Indeed, let us assume for instance that a microscopic two-level system (states $|+\rangle$ and $|-\rangle$) interacts with a macroscopic measuring apparatus, in such a way that the pointer of the apparatus points to a different (and classically distinguishable!) position for each of the two states, that is, the interaction transforms the joint atom-apparatus initial state into

$$|+\rangle|\uparrow\rangle \to |+\rangle'|\nearrow\rangle,$$
$$|-\rangle|\uparrow\rangle \to |-\rangle'|\nwarrow\rangle.$$

The linearity of quantum mechanics implies that, if the quantum system is prepared in a coherent superposition of the two states, say $|\psi\rangle = (|+\rangle + |-\rangle)/\sqrt{2}$, the final state of the complete system should be a coherent superposition of two correlated states, each of which corresponding to a different position of the pointer:

$$(1/\sqrt{2})(|+\rangle + |-\rangle)|\uparrow\rangle$$
$$\to (1/\sqrt{2})(|+\rangle'|\nearrow\rangle + |-\rangle'|\nwarrow\rangle) = (1/\sqrt{2})(|\nearrow\rangle' + |\nwarrow\rangle'), \qquad (1)$$

where in the last step it was assumed that the two-level system is incorporated into the measurement apparatus after their interaction (for instance, an atom which gets stuck to the detector). One gets, therefore, as a result of the interaction between the microscopic and the macroscopic system, a coherent superposition of two classically distinct states of the macroscopic apparatus. This is actually the situation in Schrödinger's cat paradox: the cat can be viewed as a measuring apparatus of the state of a decaying atom, the state of life or death of the cat being equivalent to the two positions of the pointer. This would imply that one should be able in principle to get interference between the two states of the pointer: it is precisely the lack of evidence of such phenomena in the macroscopic world which motivated Einstein's concern.

Faced with this problem, Von Neumann introduced through his collapse postulate [3] two distinct types of evolution in quantum mechanics: the deterministic and unitary evolution associated to the Schrödinger equation, which describes the establishment of a correlation between states of the microscopic system being measured and distinguishable classical states (for instance, distinct positions of a pointer) of the macroscopic measurement apparatus; and the probabilistic and irreversible process associated with measurement, which transforms the correlated state into a statistical mixture. This separation of the whole process into two steps has been the object of much debate [4–6]; indeed, it would not only imply an intrinsic limitation of quantum mechanics to deal with classical objects, but it would also pose the problem of drawing the line between the microscopic and the macroscopic world.

Several possibilities have been explored as solutions to this paradox, including the proposal that a small non-linear term in the Schrödinger equation, although unnoticeable for microscopic phenomena, could eliminate the coherence between macroscopic states, thus transforming the quantum superpositions into statistical mixtures [4]. The non-observability of the coherence between the two positions of the pointer has been attributed both to the lack of non-local observables with matrix elements between the two corresponding states [7] as well as to the fast decoherence due to dissipation [8–10]. This last approach has been emphasized in recent years: decoherence follows from the irreversible coupling of the observed system to a reservoir [8,9]. In this process, the quantum superposition is turned into a statistical mixture, for which all the information on the system can be described in classical terms, so our usual perception of the world is recovered. Furthermore, for macroscopic superpositions quantum coherence decays much faster than the physical observables of the system, its decay time being given by the dissipation time divided by a dimensionless number measuring the "separation" between the two parts. The statement that these two parts are macroscopically separated implies that this separation is an extremely large number. Such is the case for biological systems like "cats" made of huge number of molecules. In the simple case mentioned by Einstein [1], of a particle split into two spatially separated wave packets by a distance d, the dimensionless measure of the separation is $(d/\lambda_{dB})^2$, where λ_{dB} is the particle de Broglie wavelength. For a particle with mass equal to 1 g at a temperature of 300 K, and $d = 1$ cm, this number is about 10^{40}, and the decoherence is for all purposes instantaneous. This would provide an answer to Einstein's concern: decoherence of macroscopic states would be too fast to be observed.

In these lectures, it will be shown that the study of the interaction between atoms and electromagnetic fields in cavities can help us understand some aspects of this problem. In fact, many recent contributions in the field of quantum optics have led not only to the investigation of the subtle frontier between the quantum and the classical world, but also of hitherto unsuspected quantum mechanical processes like teleportation. Research on quantum optics is therefore intimately entangled with fundamental problems of quantum mechanics.

The whole area of "cavity quantum electrodynamics" is a very recent one. It concerns the interactions between atoms and discrete modes of the electromagnetic field in a cavity, under conditions such that losses due to dissipation and atomic spontaneous emission are very small. Usually, one deals with atomic beams crossing cavities with a high quality factor Q (defined as the product of the angular frequency of the mode and its lifetime, $Q = \omega\tau$). The atoms, prepared in special states and detected after interacting with the field, serve two purposes: they are used to manipulate the field in the cavity, so as to produce the desired states, and also to measure the field.

Several factors contributed to the development of this area. The production of superconducting Niobium cavities, with extremely high quality factors, up to the order of 10^{10}, allows one to keep a photon in the cavity for a time of the order of one second. New techniques of atomic excitation (alkaline atoms, like Rubidium and

Cesium, are frequently used for this purpose) to highly excited levels (principal quantum numbers of the order of 50), and with maximum angular momentum ($\ell = n - 1$) – the so-called planetary Rydberg atoms – have led to the production of atomic beams that interact strongly even with very weak fields, of the order of one photon, due to the large magnitude of the relevant electric dipoles. Besides, the lifetime of these states is large – of the order of the millisecond – which may be understood semiclassically, from the correspondence principle (which should be valid for $n \sim 50$): the electron is always very far away from the nucleus, and therefore its acceleration is small, implying weak radiation and a long lifetime. One should also mention the new techniques of atomic velocity control, which allow the production of approximately monokinetic atomic beams, leading to a precise control of the interaction time between atom and field. For a review of some of the main problems and results in this field, see Ref. [11].

Looking into the problem of the classical limit of quantum mechanics will actually provide us with a useful thread, a "leitmotif" which will lead us to many important techniques of quantum optics. We start therefore with a review of the basic ingredients of quantum optics.

II THE QUANTIZED ELECTROMAGNETIC FIELD

The free-field Hamiltonian for a mode of the electromagnetic field is given by the harmonic oscillator expression

$$\hat{H} = \hbar\omega \left(\hat{N} + \tfrac{1}{2}\right), \tag{2}$$

where $\hat{N} = \hat{a}^\dagger \hat{a}$ is the *number operator*, and $\hat{a}$ and $\hat{a}^\dagger$ satisfy the commutation relation

$$[\hat{a}, \hat{a}^\dagger] = 1. \tag{3}$$

The eigenstates of $\hat{H}$ are denoted by $|n\rangle$, and satisfy the equation

$$\hat{N}|n\rangle = n|n\rangle, \tag{4}$$

while the eigenenergies are given by $E_n = (n + 1/2)\hbar\omega$. It is easy to show that

$$\hat{a}^\dagger|n\rangle = \sqrt{n+1}|n+1\rangle, \quad \hat{a}|n\rangle = \sqrt{n}|n-1\rangle. \tag{5}$$

The eigenvalue n of the number operator $\hat{N}$ is interpreted as the number of photons in the field, while, in view of (5), $\hat{a}$ and $\hat{a}^\dagger$ are the photon *annihilation* and *creation* operators.

The states $|n\rangle$ are the so-called *Fock states*, and have a well-defined number of photons. The state corresponding to $n = 0$ is the *vacuum state*. It is easy to show from the above relations that

$$|n\rangle = \frac{\hat{a}^{\dagger n}}{\sqrt{n!}}|0\rangle.$$

The electric field is expressed in terms of the annihilation and creation operators by

$$\vec{E}(r) = E_w[\hat{a} u(\vec{r})\, \vec{\epsilon} + \hat{a}^\dagger u^*(\vec{r})\, \vec{\epsilon}^*]\,, \tag{6}$$

where $u(\vec{r})$ is a function which describes the spatial dependence of the field mode, $\vec{\epsilon}$ is the polarization vector, and $E_w = \sqrt{\hbar\omega/V}$ is the *field per photon*. Here $V = \int |u(\vec{r})|^2 d^3r$ is the effective volume of the mode, defined so that the expectation value of the electromagnetic energy in the vacuum state, $(1/4\pi)\int \langle 0|[\vec{E}(\vec{r})]^2|0\rangle d^3r$, is equal to the zero-point energy $\hbar\omega/2$. One should note that $\langle n|\hat{E}|n\rangle = 0$, that is, the average electric field is zero in a Fock state.

A special role will be played in the following by the *phase displacement operator*:

$$\hat{U}(\theta) = \exp(-i\theta \hat{N})\,. \tag{7}$$

It follows from the commutation relations that

$$\hat{U}^\dagger(\theta)\hat{a}\hat{U}(\theta) = \hat{a}\exp(-i\theta)\,. \tag{8}$$

For $\theta = \omega t$, the phase displacement operator coincides, up to a factor $\exp(-i\omega t/2)$ coming from the zero-point energy, with the evolution operator corresponding to the Hamiltonian (2), and (8) yields the time evolution of the Heisenberg operator associated with $\hat{a}$.

II.A Quadratures of the electromagnetic field

The quadratures of the electromagnetic field correspond to the position and momentum of a harmonic oscillator:

$$\hat{q} = \frac{1}{\sqrt{2}}\left(\hat{a} + \hat{a}^\dagger\right),\quad \hat{p} = \frac{i}{\sqrt{2}}\left(\hat{a}^\dagger - \hat{a}\right),\quad [\hat{q},\hat{p}] = i\,. \tag{9}$$

This commutation relation implies the Heisenberg inequality $\Delta q \Delta p \geq 1/2$.

From (8) and (9), we see that, for $\theta = \pi$,

$$\hat{U}^\dagger(\pi)\hat{q}\hat{U}(\pi) = -\hat{q}\,,\qquad \hat{U}^\dagger(\pi)\hat{p}\hat{U}(\pi) = -\hat{p}\,,$$

so that $\hat{U}(\pi)$ is the *parity operator*.

Setting $u(\vec{r}) = |u(\vec{r})|\exp[-i\phi(\vec{r})]$ in Eq. (6), we have, in the Heisenberg picture, for the electric field operator in terms of these quadratures (for a real polarization vector):

$$\vec{E}(r,t) = E_w |u(\vec{r})|\sqrt{2}[\hat{q}\cos(\omega t + \phi) - \hat{p}\sin(\omega t + \phi)]\,\vec{\epsilon}\,.$$

This expression is analogous to the one which yields the position of a harmonic oscillator at time t in terms of its initial position and momentum:

$$x(t) = x(t_0) \cos \omega(t - t_0) + \frac{p(t_0)}{m\omega} \sin \omega(t - t_0). \qquad (10)$$

The quadrature eigenstates (which correspond to states with well-defined position and momentum for the harmonic oscillator) will be denoted by

$$\hat{q}|q\rangle = q|q\rangle, \qquad \hat{p}|p\rangle = p|p\rangle.$$

As for the position and momentum eigenstates, these states provide two non-normalizable bases. The corresponding quadrature wave functions are given by

$$\psi(q) = \langle q|\psi\rangle, \ \tilde{\psi}(p) = \langle p|\tilde{\psi}\rangle.$$

Using the phase displacement operator given by (7), it is possible to define *generalized quadratures*:

$$\hat{q}_\theta = \hat{U}^\dagger(\theta)\hat{q}\hat{U}(\theta) = (1/\sqrt{2})\left(\hat{a}e^{-i\theta} + \hat{a}^\dagger e^{i\theta}\right) = \hat{q}\cos\theta + \hat{p}\sin\theta, \qquad (11)$$

$$\hat{p}_\theta = \hat{U}^\dagger(\theta)\hat{p}\hat{U}(\theta) = -\hat{q}\sin\theta + \hat{p}\cos\theta. \qquad (12)$$

It is clear from these expressions that $\hat{U}(\theta)$ is the rotation operator in phase space. For a harmonic oscillator, and with $\theta = \omega t$, (11) and (12) correspond respectively to the position and the momentum of the oscillator at time t, expressed in terms of the position $\hat{q}$ and momentum $\hat{p}$ at time $t = 0$.

II.B Coherent states

An important role is played in quantum optics, and also in the understanding of the classical limit of quantum mechanics, by the *coherent states*, defined as eigenstates of the annihilation operator [12]:

$$\hat{a}|\alpha\rangle = \alpha|\alpha\rangle. \qquad (13)$$

From this definition, it is easy to show that the average number of photons in a coherent state $|\alpha\rangle$ is given by

$$\langle n \rangle = \langle \alpha | \hat{a}^\dagger \hat{a} | \alpha \rangle = |\alpha|^2,$$

while the average value of the electric field coincides with the classical expression for an electromagnetic field with complex amplitude α:

$$\langle \vec{E}(\vec{r}) \rangle = \sqrt{\hbar\omega/V} u(\vec{r}) \vec{\epsilon} \alpha + c.c..$$

It also follows from the definition (13) and the commutation relations that, for a coherent state,

$$\Delta q = \Delta p = 1/\sqrt{2},$$

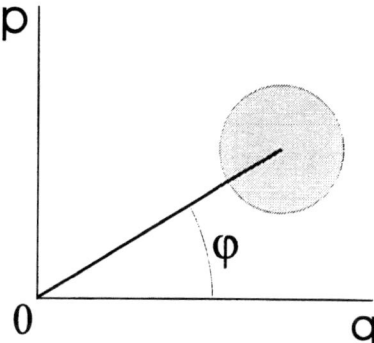

FIGURE 1. Pictorial representation of a coherent state in phase space.

or, more generally, $\Delta q_\theta = 1/2$. Therefore, coherent states are minimum uncertainty states. This property can be pictorially depicted by drawing in phase space a circle, with a radius equal to the uncertainty in Δq or Δp, as shown in Fig. 1

In terms of Fock states, coherent states can be expressed in the following way:

$$|\alpha\rangle = e^{-|\alpha|^2/2} \sum_{n=0}^{\infty} \frac{\alpha^n}{\sqrt{n!}} |n\rangle = e^{-|\alpha|^2/2} \sum_{n=0}^{\infty} \frac{(\alpha \hat{a}^\dagger)^n}{n!} |0\rangle, \qquad (14)$$

corresponding to the photon number distribution

$$p(n) = |\langle n | \alpha \rangle|^2 = \exp(-|\alpha|^2) \frac{|\alpha|^{2n}}{n!} = \exp(-\langle n \rangle) \frac{\langle n \rangle^n}{n!}. \qquad (15)$$

This is a Poisson distribution, with the variance $(\Delta n)^2 = \langle n \rangle$.

The *displacement operator* is defined by

$$\hat{D}(\alpha, \alpha^*) = \exp\left(\alpha \hat{a}^\dagger - \alpha^* \hat{a}\right). \qquad (16)$$

Note that the right-hand side of (14) implies that

$$|\alpha\rangle = e^{-|\alpha|^2/2} e^{\alpha \hat{a}^\dagger} |0\rangle = e^{-|\alpha|^2/2} e^{\alpha \hat{a}^\dagger} e^{-\alpha \hat{a}} |0\rangle = \exp\left(\alpha \hat{a}^\dagger - \alpha^* \hat{a}\right) |0\rangle = \hat{D}(\alpha, \alpha^*) |0\rangle, \qquad (17)$$

where in the last step the Baker-Hausdorff transformation has been used to entangle the annihilation and creation operators in the exponent. Therefore, a coherent state $|\alpha\rangle$ can be obtained by applying the displacement operator $\hat{D}(\alpha, \alpha^*)$ to the vacuum state. The displacement operator is closely connected to the evolution operator corresponding to the interaction of the electromagnetic field with a classical current. Indeed, this interaction is described by the Hamiltonian

$$H_{\text{int}} = \int \vec{J} \cdot \vec{A} \, d^3 r,$$

which can be written in the form $H_{\text{int}} = i(\alpha \hat{a}^\dagger - \alpha^* \hat{a})$. The evolution operator corresponding to this interaction coincides, up to a phase, with $D(\alpha, \alpha^*)$. Therefore, Eq. (17) implies that classical currents generate coherent states from the vacuum.

From the expansion of the coherent states in terms of the Fock states, one easily derives the following orthonormality and completeness relations:

$$|\langle \alpha | \alpha' \rangle|^2 = e^{-|\alpha - \alpha'|^2},$$

$$\frac{1}{\pi} \int d^2\alpha \, |\alpha\rangle\langle\alpha| = \mathcal{I},$$

where $d^2\alpha \equiv d(\Re e\alpha)d(\Im m\alpha)$.

In terms of the quadrature eigenstates $|q\rangle$, one may write:

$$\langle q|\alpha_0\rangle = (1/\pi)^{1/4} e^{-iq_0 p_0/2} e^{ip_0 q} e^{-(q-q_0)^2/2},$$

with $\alpha_0 = (q_0 + ip_0)/\sqrt{2}$.

Therefore, the probability of finding a quadrature $\hat{q}$ of the field with a value q, for a coherent state $|\alpha_0\rangle$, is given by a Gaussian (for the vacuum, $q_0 = p_0 = 0$):

$$P(q) = (1/\pi)^{1/2} \exp[-(q-q_0)^2].$$

II.C Squeezed states

The most general family of minimum uncertainty states, as shown by Pauli [13], corresponds to the following wave function in the q-quadrature representation:

$$\psi(q) = \langle q|\psi\rangle = (2\pi\Delta^2 q)^{-1/4} \exp(ip_0 q) \exp\left(\frac{-(q-q_0)^2}{4\Delta^2 q}\right).$$

The variance $\Delta^2 q$ is not necessarily equal to $1/2$, as it is for coherent states. In the q_θ-representation, one would get the same expression for $\psi(q_\theta)$, with q, q_0, p_0, and $\Delta^2 q$ replaced by q_θ, $q_{\theta 0}$, $p_{\theta 0} = q_{\theta+\pi/2,0}$, and $\Delta^2 q_\theta$, respectively. If $\Delta^2 q_\theta < 1/2$ for some region of values of θ, we say that the state is squeezed.

It is clear that this family of minimum uncertainty squeezed states can be obtained from coherent states through a scale transformation, which compresses say the q-axis and at the same time dilates the p-axis. Thus, the wave function of the squeezed vacuum would be obtained from the one corresponding to a coherent state by the scale transformation [14]:

$$\phi_0(q,\xi) = e^{\xi/2}\psi_0\left(e^\xi q\right), \qquad \tilde{\phi}_0(p,\xi) = e^{-\xi/2}\tilde{\psi}_0\left(e^{-\xi}p\right),$$

where

$$\psi_0(q) = (1/\pi)^{1/4} e^{-q^2/2}.$$

Differentiating $\phi_0(q,\xi) = \langle q|\phi_0(\xi)\rangle$ with respect to ξ, one gets:

$$\frac{\partial}{\partial \xi}\langle q|\phi_0(\xi)\rangle = \frac{\partial}{\partial \xi}\phi_0(q,\xi) = \frac{1}{2}\phi_0 + q\frac{\partial \phi_0}{\partial q} = \frac{1}{2}\left(q\frac{\partial}{\partial q} + \frac{\partial}{\partial q}q\right)\phi_0$$
$$= \frac{i}{2}\langle q|(\hat{q}\hat{p} + \hat{p}\hat{q})|\phi_0(\xi)\rangle. \tag{18}$$

Therefore,
$$\frac{\partial}{\partial \xi}|\phi_0(\xi)\rangle = \frac{i}{2}(\hat{q}\hat{p} + \hat{p}\hat{q})|\phi_0(\xi)\rangle = \frac{1}{2}\left(\hat{a}^2 - \hat{a}^{\dagger 2}\right)|\phi_0(\xi)\rangle,$$

and
$$|0,\xi\rangle = \hat{S}(\xi)|0\rangle,$$
where $\hat{S}(\xi) = \exp\left[(\xi/2)\left(\hat{a}^2 - \hat{a}^{\dagger 2}\right)\right]$ is the *squeezing operator*.

More generally, one could set $\xi = re^{i\theta}$, implying a combination of squeezing and rotation, which would correspond to compressing the quadrature $q_{\theta/2}$ and dilating the quadrature $q_{\theta/2+\pi/2}$.

A more general family of minimum uncertainty squeezed states is obtained by displacing the squeezed vacuum, with the displacement operators introduced before:
$$|\alpha,\xi\rangle = \hat{D}(\alpha)\hat{S}(\xi)|0\rangle. \tag{19}$$

An equivalent result is obtained by applying the squeezing transformation to the coherent state $|\alpha\rangle$.

One should note that the squeezing operator can be interpreted as the evolution operator corresponding to the Hamiltonian
$$H_{\text{int}} \propto \hat{a}^2 - \hat{a}^{\dagger 2}.$$

Hamiltonians of this form occur in non-linear optics, where they describe degenerate parametric amplifiers. A realization of such an amplifier is provided by a KTP crystal (potassium titanium phosphate) pumped by a laser with frequency twice as large as the mode of interest.

The average number of photons in a squeezed state can be immediately obtained from (19):
$$\langle \alpha,\xi|a^{\dagger}a|\alpha,\xi\rangle = |\alpha|^2 + \sinh^2 r.$$

This equation shows that the average number of photons in the squeezed vacuum is different from zero: energy is required to squeeze the vacuum!

II.D Measurement of quadratures

Several methods have been proposed to measure quadratures of the electromagnetic field (for a review, see for instance Ref. [15]). The general idea consists in mixing the signal to be detected with an intense coherent signal, called *local oscillator*, before detection [16].

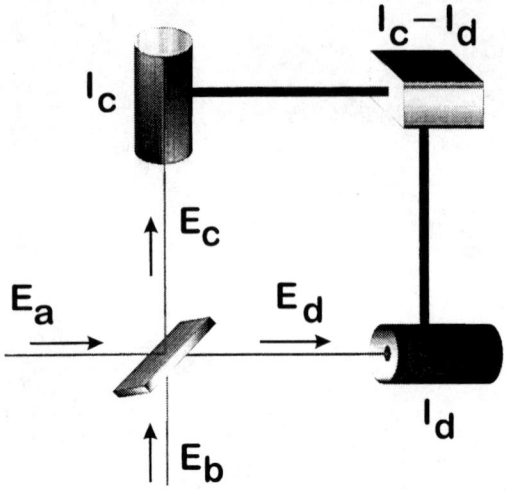

FIGURE 2. Method of balanced homodyne detection

The discussion here is restricted to the method of *balanced homodyne detection*, sketched in Fig. 2. The field to be measured (complex amplitude E_a) is sent on a beam splitter, together with a coherent field (complex amplitude E_b) with the same frequency. One measures then the difference of intensity of the two beams emerging from the beam splitter (complex amplitudes E_c and E_d). The detection is said to be balanced when the mirror transmits 50% of the incident light.

Let r and t be the reflection and transmission coefficients of the mirror, respectively. Let us set:

$$E_c = rE_a - tE_b, \tag{20a}$$
$$E_d = tE_a + rE_b, \tag{20b}$$

or yet, in matrix form,

$$\begin{pmatrix} E_c \\ E_d \end{pmatrix} = \begin{pmatrix} r & -t \\ t & r \end{pmatrix} \begin{pmatrix} E_a \\ E_b \end{pmatrix}. \tag{21}$$

Energy conservation (assuming that losses are negligible) implies that

$$|E_c|^2 + |E_d|^2 = |E_a|^2 + |E_b|^2. \tag{22}$$

From (21) and (22), one gets:

$$|r|^2 + |t|^2 = 1, \tag{23a}$$
$$r^*t - rt^* = 0. \tag{23b}$$

If one takes r real and equal to $\sqrt{\eta}$ (one should note that phases in r and t can be removed by redefining the phases of the incoming and outgoing fields), it follows from (23) that $t = \pm(1-\eta)^{1/2}$. Choosing the positive sign, one gets then

$$\begin{pmatrix} E_c \\ E_d \end{pmatrix} = \begin{pmatrix} \sqrt{\eta} & -\sqrt{1-\eta} \\ \sqrt{1-\eta} & \sqrt{\eta} \end{pmatrix} \begin{pmatrix} E_a \\ E_b \end{pmatrix}. \tag{24}$$

Normalizing the intensity to the photon number, and introducing the annihilation operators through

$$E_a \to \hat{a}, \quad E_b \to \hat{b},$$
$$E_c \to \hat{c}, \quad E_d \to \hat{d},$$

one gets, from (24),

$$\hat{c} = \sqrt{\eta}\,\hat{a} - \sqrt{1-\eta}\,\hat{b}, \qquad \hat{d} = \sqrt{1-\eta}\,\hat{a} + \sqrt{\eta}\,\hat{b}. \tag{25}$$

For balanced detection, $\eta = 1/2$, so that

$$\hat{c} = \frac{1}{\sqrt{2}}(\hat{a}-\hat{b}), \qquad \hat{d} = \frac{1}{\sqrt{2}}(\hat{a}+\hat{b}). \tag{26}$$

Note that conditions (23) imply that the transformation between the field operators corresponding to (21) is unitary (this is the requirement for operators which corresponds to energy conservation for the classical fields).

The difference between the intensities of the fields E_d and E_c is given then by

$$I = \hat{d}^\dagger \hat{d} - \hat{c}^\dagger \hat{c} = \hat{a}^\dagger \hat{b} + \hat{b}^\dagger \hat{a}. \tag{27}$$

Assuming that the field E_b may be described classically (this would be the case for a coherent state with large average photon number), one replaces $\hat{b}$ by $\beta = B\,e^{-i(\omega t - \theta)}$, so that (27) gets transformed into

$$I = B\left[\hat{a}e^{i(\omega t - \theta)} + \hat{a}^\dagger e^{-i(\omega t - \theta)}\right]. \tag{28}$$

Since $\hat{a} = \hat{a}_0 e^{-i\omega t}$ (all fields are taken in the Heisenberg picture), one gets finally,

$$I = B\left(\hat{a}_0\,e^{-i\theta} + \hat{a}_0^\dagger\,e^{i\theta}\right). \tag{29}$$

This equation shows that the difference of intensities, measured by the method of homodyne detection, is directly proportional to the quadrature $\hat{X}(\theta)$ of the field E_a, defined by

$$\hat{X}(\theta) = \frac{1}{\sqrt{2}}\left(\hat{a}_0 e^{-i\theta} + \hat{a}_0^\dagger e^{i\theta}\right). \tag{30}$$

Therefore, by detecting the difference of intensities, as the phase of the local oscillator E_b is changed, one may measure an arbitrary quadrature of the field E_a. In actual experiments, the phase of the local oscillator is adjusted to yield the maximum possible quadrature squeezing. The shot-noise level is determined by blocking the signal field, so that only the local oscillator field reaches the detector. The results of the measurements are spectrally analyzed, leading to the determination of the amount of squeezing as a function of the frequency. In practice, one deals with a continuum of modes, and the above analysis applies to the situation when the frequency window of the detector is much smaller than the linewidth of the light which is being measured.

III REPRESENTATIONS IN PHASE SPACE

III.A The density operator

We present in this section a brief review of the concept of density operator. For a pure state $|\psi\rangle$, the density operator is defined by

$$\hat{\rho} = |\psi\rangle\langle\psi|.$$

If instead one is uncertain about the state of the system, and we know that there is a probability P_ψ for the system to be in state $|\psi\rangle$, the density operator is defined by

$$\hat{\rho} = \sum_\psi P_\psi |\psi\rangle\langle\psi|. \tag{31}$$

The utility of the above definitions can be grasped by writing down, in terms of $\hat{\rho}$, the average value of an observable $\hat{A}$:

$$\langle \hat{A} \rangle = \sum_\psi P_\psi \langle\psi|\hat{A}|\psi\rangle = \text{Tr}(\hat{\rho}\hat{A}), \tag{32}$$

which represents a unified way of expressing the average value, valid both for a pure state and a statistical mixture.

From the definition it follows immediately that $\text{Tr}\hat{\rho} = \sum_\psi P_\psi = 1$.

Also, it is easy to see that $\hat{\rho}$ is Hermitian, and therefore can be diagonalized. If $|\phi_i\rangle$ are the eigenstates of $\hat{\rho}$, then

$$\hat{\rho} = \sum_i P_i |\phi_i\rangle\langle\phi_i|, \quad \langle\phi_i|\phi_j\rangle = \delta_{ij},$$

which implies that

$$\hat{\rho}^2 = \sum_i P_i^2 |\phi_i\rangle\langle\phi_i| \Rightarrow \text{Tr}\hat{\rho}^2 = \sum_i P_i^2 \leq 1.$$

For a pure state, $\text{Tr}(\hat{\rho}^2) = 1$, while for a mixture $\text{Tr}(\hat{\rho}^2) < 1$.

In terms of the Fock basis,

$$\hat{\rho} = \sum_{nm} \rho_{nm} |n\rangle\langle m|.$$

Finally, let us recall that if one has two systems (interacting or not), let us say an atom and an electromagnetic field, a basis for the combined system can be obtained by forming the tensor product of the bases corresponding to each of the two systems. The tensor product of two states $|\psi_A\rangle$ and $|\psi_F\rangle$ corresponding respectively to the systems A and F is written as

$$|\psi_A\rangle \otimes |\psi_F\rangle,$$

corresponding to the density operator

$$\hat{\rho} = \hat{\rho}_A \otimes \hat{\rho}_F.$$

The average of expressions involving products of operators acting on A and on F separately can be written as

$$\langle \hat{A}\hat{F} \rangle = \text{Tr}(\hat{\rho}_A \hat{A})\text{Tr}(\hat{\rho}_F \hat{F}).$$

Of course, a general state of the combined system will not have the form of a tensor product, but can be expressed as a linear combination of tensor product states. A state which cannot be factorized is called an *entangled state*.

States may also be represented by phase space distributions, which allow quantum-mechanical averages of operators to be expressed as classical-like integrations over phase space of c-numbers corresponding to the operators. In these lectures, we will pay special attention to the Wigner distribution.

III.B The Wigner distribution

Phase space probability distributions are very useful in classical statistical physics. Averages of relevant functions of the positions and momenta of the particles can be obtained by integrating these functions with those probability weights.

In quantum mechanics, similar averages are calculated through Eq. (32). Heisenberg's inequality forbids the existence in phase space of *bona fide* probability distributions, since one cannot determine simultaneously the position and the momentum of a particle. In spite of this, phase space distributions may still play a useful role in quantum mechanics, allowing the calculation of the average of operator-valued functions of the position and momentum operators as classical-like integrals of c-number functions. These functions are associated to those operators through correspondence rules, which depend on a previously defined operator ordering.

From all phase space representations, the Wigner distribution is the more natural one, when one looks for a quantum-mechanical analog of a classical probability distribution in phase space. This is a consequence of a beautiful result demonstrated by Bertrand and Bertrand [17], which will be presented here. The following account stays close to the ones in Refs. [14] and [17].

Let us look for a representation for which the *marginal distributions* coincide with the quadrature probability distributions:

$$\int dp\, W(q,p) = \langle q|\hat{\rho}|q\rangle\,, \int dq\, W(q,p) = \langle p|\hat{\rho}|p\rangle\,. \tag{33}$$

One should note that, for a pure state, $\langle q|\hat{\rho}|q\rangle = |\psi(q)|^2$, $\langle p|\hat{\rho}|p\rangle = |\tilde{\psi}(p)|^2$. One should also note that from (33) it follows immediately the normalization property:

$$\int dp\,dq\, W(q,p) = 1\,. \tag{34}$$

Properties (33) must remain true if one rotates the axes in phase space, so that

$$\hat{q}_\theta = \hat{U}^\dagger(\theta)\hat{q}\hat{U}(\theta) = \hat{q}\cos\theta + \hat{p}\sin\theta\,, \tag{35}$$

$$\hat{p}_\theta = \hat{U}^\dagger(\theta)\hat{p}\hat{U}(\theta) = -\hat{q}\sin\theta + \hat{p}\cos\theta\,, \tag{36}$$

or, inversely,

$$\hat{q} = \hat{q}_\theta \cos\theta - \hat{p}_\theta \sin\theta\,, \tag{37}$$

$$\hat{p} = \hat{q}_\theta \sin\theta + \hat{p}_\theta \cos\theta\,. \tag{38}$$

Thus:

$$P(q_\theta) = \int W(q_\theta \cos\theta - p_\theta \sin\theta, q_\theta \sin\theta + p_\theta \cos\theta)dp_\theta\,, \tag{39}$$

where now

$$P(q_\theta) = \langle q|\hat{U}(\theta)\hat{\rho}\hat{U}^\dagger(\theta)|q\rangle\,. \tag{40}$$

Expression (39), which yields the probability distribution for q_θ in terms of the function $W(q,p)$, is called a *Radon transform*. It was investigated in 1917 by the mathematician Johan Radon [18], who showed that, if one knows $P(q_\theta)$ for all angles θ, then one can uniquely recover $W(q,p)$, through the so-called *Radon inverse transform*. If one now identifies $P(q_\theta)$, given by the Radon transform (39), with the quantum expression (40), then it follows that (39) and (40) uniquely determine the function $W(q,p)$, in terms of the density operator $\hat{\rho}$ of the system. The function $W(q,p)$ is in this case precisely the Wigner function of the system.

Before demonstrating this result, let us note that Radon's result is the mathematical basis of tomography. In fact, application of this procedure to medicine (see Fig. 3) has brought the Nobel prize in Medicine to the medical doctors Cormack and Hounsfield in 1979.

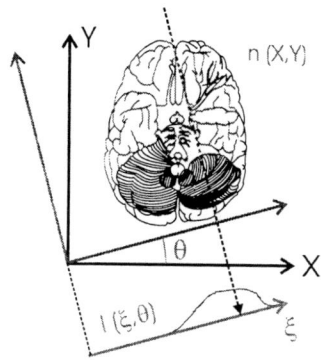

FIGURE 3. Medical tomography. Measurement of the X-ray absorption for all angles along a plane allows one to reconstruct the absorptive part of the refraction index for a slice of the organ under investigation.

III.C Reconstruction of the Wigner function

It is shown now that the distribution $W(q,p)$ may be uniquely determined from the knowledge of $P(q_\theta)$. For this end, let us introduce the *characteristic function* corresponding to $W(q,p)$, which is just the Fourier transform of this distribution:

$$\tilde{W}(u,v) = \int\int W(q,p)\exp(-iuq - ivp)\,dq\,dp. \tag{41}$$

The characteristic function corresponding to $P(q_\theta)$ is introduced in a similar way:

$$\tilde{p}(\xi,\theta) = \int P(q_\theta)\exp(-i\xi q_\theta)\,dq_\theta. \tag{42}$$

Inserting in (42) the expression for $P(q_\theta)$ as a function of W, given by (39), one gets:

$$\tilde{p}(\xi,\theta) = \int\int W(q,p)\exp(-i\xi q_\theta)dq_\theta\,dp_\theta,$$

where $q = q_\theta\cos\theta - p_\theta\sin\theta$ and $p = q_\theta\sin\theta + p_\theta\cos\theta$, and therefore $q_\theta = q\cos\theta + p\sin\theta$.

Changing the integration variables in $\tilde{p}(\xi,\theta)$, so that $(q_\theta,p_\theta)\to(q,p)$, one gets:

$$\tilde{p}(\xi,\theta) = \int_{-\infty}^{+\infty}\int_{-\infty}^{+\infty} W(q,p)\exp[-i\xi(q\cos\theta + p\sin\theta)]dq\,dp. \tag{43}$$

Therefore, $\tilde{p}(\xi,\theta)$ is the Fourier transform of $W(q,p)$ in polar coordinates:

$$\tilde{p}(\xi,\theta) = \tilde{W}(\xi\cos\theta,\xi\sin\theta).$$

This implies that, from $P(q_\theta)$, one can calculate $\tilde{p}(\xi,\theta)$, and from this function one can calculate $W(q,p)$.

This demonstrates the tomographic reconstruction of $W(q,p)$. In order to connect this distribution to the density operator of the system, one uses Eq. (40):

$$\tilde{p}(\xi,\theta) = \int_{-\infty}^{+\infty} \langle q_\theta|\hat{\rho}|q_\theta\rangle e^{-i\xi q_\theta} dq_\theta = \int_{-\infty}^{+\infty} \langle q_\theta|\hat{\rho} e^{-i\xi \hat{q}_\theta}|q_\theta\rangle dq_\theta = \text{Tr}\left[\hat{\rho} e^{-i\xi \hat{q}_\theta}\right]. \quad (44)$$

But $\xi \hat{q}_\theta = \hat{q}\xi\cos\theta + \hat{p}\xi\sin\theta$. Therefore, setting $u = \xi\cos\theta$, $v = \xi\sin\theta$, one gets

$$\tilde{W}(u,v) = \text{Tr}\left[\hat{\rho} e^{-iu\hat{q}-iv\hat{p}}\right].$$

Another form for the $W(q,p)$ can be obtained in the following way. We rewrite the characteristic function $\tilde{W}(u,v)$ as:

$$\tilde{W}(u,v) = \int_{-\infty}^{+\infty} \langle q|\hat{\rho} e^{-iu\hat{q}-iv\hat{p}}|q\rangle\, dq = e^{iuv/2} \int_{-\infty}^{+\infty} \langle q|\hat{\rho} e^{-iu\hat{q}}|q+v\rangle\, dq$$

$$= \int_{-\infty}^{+\infty} e^{-iux}\left\langle x - \frac{v}{2}\middle|\hat{\rho}\middle| x + \frac{v}{2}\right\rangle dx, \quad (45)$$

where we have used that

$$\exp(-iu\hat{q} - iv\hat{p}) = \exp(-iuv/2)\exp(-iu\hat{q})\exp(-iv\hat{p})$$

and we have set $q = x - v/2$. Taking the Fourier transform of $\tilde{W}(u,v)$ given by (45), one gets the following expression for the distribution $W(q,p)$:

$$W(q,p) = \frac{1}{2\pi} \int_{-\infty}^{+\infty} e^{ipx}\left\langle q - \frac{x}{2}\middle|\hat{\rho}\middle| q + \frac{x}{2}\right\rangle dx, \quad (46)$$

which, except for a normalization constant, is the famous expression written down by Wigner [19] in his article *"On the Quantum Correction for Thermodynamic Equilibrium"*, published in 1932. Wigner used this quasi-probability distribution in phase space as a convenient way of calculating quantum corrections to classical statistical mechanics. He wrote in his paper that (46) "was chosen from all possible expressions, because it seems to be the simplest." He added a quite intriguing footnote: "This expression was found by L. Szilard and the author some years ago for another purpose." One has shown here that the Wigner distribution has in fact a quite distinctive feature: it is the only distribution in phase space which yields the correct marginal distributions for any quadrature!

The tomographic procedure has a simple interpretation for a harmonic oscillator. From (10), it is clear that in this case measuring the quadratures for all angles is equivalent to measuring the position of the harmonic oscillator for all times from 0 to $2\pi/\omega$. This implies that the measurement of $|\psi(x,t)|^2$ for $0 < t \leq 2\pi/\omega$ allows one to reconstruct the state $\psi(x,t)$ of the harmonic oscillator.

The question about what is the minimum set of measurements needed to reconstruct the state of a system is actually a very old problem in quantum mechanics. In his article on quantum mechanics in the *Handbuch der Physik* in 1933 [13], Pauli stated that "the mathematical problem, as to whether for given functions $W(x)$ and $\tilde{W}(p)$ [probability distributions in position and momentum space], the wave function ψ, if such a function exists, is always uniquely determined has still not been investigated in all its generality." One knows now the answer to this question: the probability distributions $W(x)$ and $\tilde{W}(p)$ do not form a complete set in the tomographic sense, and therefore are not sufficient to determine uniquely the quantum state of the system.

III.D Expression of the Wigner function in terms of $\hat{a}$ and $\hat{a}^\dagger$

Setting

$$\hat{q} = \frac{1}{\sqrt{2}}\left(\hat{a} + \hat{a}^\dagger\right), \quad \hat{p} = \frac{i}{\sqrt{2}}\left(\hat{a}^\dagger - \hat{a}\right), \quad \lambda = \frac{1}{\sqrt{2}}(u + iv), \qquad (47)$$

one gets

$$\tilde{W}(\lambda, \lambda^*) = \text{Tr}\left(\hat{\rho} e^{-i\lambda^* \hat{a} - i\lambda \hat{a}^\dagger}\right). \qquad (48)$$

The Fourier transform of this expression yields the Wigner function, expressed in terms of $\hat{a}$ and $\hat{a}^\dagger$:

$$W(\alpha, \alpha^*) = \frac{1}{\pi}\int d^2\lambda \; e^{\alpha\lambda^* - \alpha^*\lambda}\tilde{W}(\lambda, \lambda^*). \qquad (49)$$

As shown by Cahill and Glauber [20], this expression may be written in the following way:

$$W(\alpha, \alpha^*) = 2\text{Tr}\left[\hat{\rho}\hat{D}(\alpha, \alpha^*)e^{i\pi\hat{a}^\dagger\hat{a}}\hat{D}^{-1}(\alpha, \alpha^*)\right], \qquad (50)$$

where $\hat{D}(\alpha, \alpha^*) = \exp(\alpha\hat{a}^\dagger - \alpha^*\hat{a})$ is the displacement operator. Since $\exp(i\pi a^\dagger a)$ is the parity operator, this expression shows that the Wigner function is proportional to the average of the displaced parity operator.

The Wigner function given by (50) involves actually a different normalization with respect to the one defined before: one must set $W \to 2\pi W$, so that

$$\int (d^2\alpha/\pi) W(\alpha, \alpha^*) = 1. \qquad (51)$$

III.E Properties of the Wigner distribution

Thorough discussions of properties of the Wigner distribution can be found in Refs. [14] and [21]. Here only some of them are summarized.

It is easy to show that the Wigner function is real and bounded. If one adopts the normalization of (46), so that (34) holds, then Schwarz's inequality implies that $|W(q,p)| \leq 1/\pi$. If the Cahill-Glauber normalization (50) is adopted instead, so that (51) is satisfied, one has $|W(\alpha, \alpha^*)| \leq 2$.

Furthermore, let $W_\psi(q,p)$ be the Wigner function corresponding to the state $\psi(q)$, and $W_\phi(q,p)$ the Wigner function corresponding to the state $\phi(q)$, as given by (46). Then,

$$\left| \int dq \, \psi^*(q)\phi(q) \right|^2 = 2\pi \int dq \int dp \, W_\psi W_\phi \, .$$

This relation has several consequences. First, setting $\psi(q) = \phi(q)$, one gets $\int dq \int dp \, [W(q,p)]^2 = 1/2\pi$. More generally, it is easy to show that

$$\mathrm{Tr}\left(\hat{\rho}^2\right) = 2\pi \int \int dq \, dp \, [W(q,p)]^2 , \qquad (52)$$

and therefore $\int \int dq \, dp \, [W(q,p)]^2 < 1/2\pi$ for a statistical mixture. It is also clear that

$$\int dq \int dp \, W_\psi W_\phi = 0 \quad \text{if } \langle \psi | \phi \rangle = 0 , \qquad (53)$$

which implies that W cannot be always positive. This may be thought as a consequence of the Heisenberg inequalities: since it is not possible to measure simultaneously q and p, one cannot have in quantum mechanics *bona fide* probability distributions in phase space. In fact, one can show that the only pure states leading to positive-definite Wigner functions correspond to Gaussian wave functions [22]. This is the case for coherent states, and also for squeezed states.

One should note that the Husimi distribution, often found in the literature, and defined by $Q(\alpha, \alpha^*) = \langle \alpha | \hat{\rho} | \alpha \rangle / \pi$, is always positive, but does not lead to the correct marginal distributions.

III.F Averages of operators

As shown by Moyal in 1949 [23], the Wigner distribution can be used to calculate averages of symmetric operator functions of q and p, as classical-like integrals in phase space. Thus, for instance,

$$\mathrm{Tr}\left(\hat{\rho}\left\{\hat{q}^2 \hat{p}\right\}_{\mathrm{sim}}\right) = \mathrm{Tr}\left[\hat{\rho}\left(\hat{q}^2 \hat{p} + \hat{q}\hat{p}\hat{q} + \hat{p}\hat{q}^2\right)/3\right] = \int dq \, dp \, W(qp) q^2 p . \qquad (54)$$

The association of a symmetrized quantum operator to a classical function is called *Weyl correspondence*.

This property of the Wigner function can be shown by considering the two equivalent expressions for the characteristic function $\tilde{W}(u,v)$,

$$\tilde{W}(u,v) = \text{Tr}\left[\hat{\rho}e^{-iu\hat{q}-iv\hat{p}}\right],$$

$$\tilde{W}(u,v) = \int\int W(q,p)\exp(-iuq-ivp)\,dq\,dp,$$

from which one gets

$$\text{Tr}\left[\hat{\rho}(\mu\hat{q}+\nu\hat{p})^k\right] = i^k\frac{\partial^k}{\partial\xi^k}\tilde{W}(\xi\mu,\xi\nu)\bigg|_{\xi=0} = \int_{-\infty}^{+\infty} W(q,p)(\mu q+\nu p)^k dq\,dp. \quad (55)$$

Comparing powers of μ and ν, one gets

$$\text{Tr}\left(\hat{\rho}\{\hat{q}^m\hat{p}^n\}_{\text{sim}}\right) = \int_{-\infty}^{+\infty} W(q,p)q^m p^n\,dq\,dp.$$

Of course, the same property holds for the Wigner function expressed in terms of $\hat{a}$ and $\hat{a}^\dagger$. Thus, for instance,

$$\text{Tr}[\hat{\rho}(\hat{a}\hat{a}^\dagger+\hat{a}^\dagger\hat{a})/2] = \int (d^2\alpha/\pi)\,\alpha\alpha^* W(\alpha,\alpha^*). \quad (56)$$

Other distributions in phase space can be introduced, which allow writing as classical-like integrals averages of functions of the operators $\hat{a}$ and $\hat{a}^\dagger$ written in normal and anti-normal order. Thus, for instance, the Husimi distribution can be shown to correspond to operators in anti-normal order. These distributions will not be discussed here [24].

III.G Examples of Wigner functions

Wigner functions corresponding to special states of the electromagnetic field can be obtained from either (46) or (50). We adopt in this section the normalization corresponding to (46).

For the vacuum state, one has the Gaussian

$$W_0(q,p) = \frac{1}{\pi}e^{-q^2-p^2}. \quad (57)$$

The Wigner function for a coherent state can be easily obtained by applying a displacement to the above Gaussian:

$$W_c(q,p) = \frac{1}{\pi}e^{-(q-q_0)^2-(p-p_0)^2}. \quad (58)$$

Application of a scaling transformation to (57) yields the Wigner function for a squeezed vacuum, plotted in Fig. 4(a):

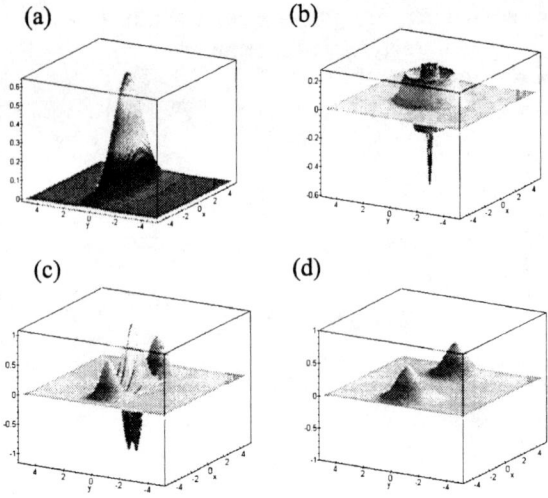

FIGURE 4. Examples of Wigner distributions. (a) Squeezed state; (b) Fock state with $n = 3$; (c) Superposition of two coherent states, $|\psi\rangle \propto |\alpha_0\rangle + |-\alpha_0\rangle$, with $\alpha_0 = 3$; (d) Statistical mixture $(|\alpha_0\rangle\langle\alpha_0| + |-\alpha_0\rangle\langle-\alpha_0|)/2$, also with $\alpha_0 = 3$.

$$W_\xi(q,p) = \frac{1}{\pi}\exp\left(-e^{2\xi}q^2 - e^{-2\xi}p^2\right). \tag{59}$$

For a one-photon Fock state $|1\rangle$, one gets:

$$W_1(q,p) = \frac{1}{\pi}e^{-q^2-p^2}\left(2q^2 + 2p^2 - 1\right). \tag{60}$$

This function vanishes for $\sqrt{q^2 + p^2} = 1/\sqrt{2}$, and is negative at the origin of phase space. This negative value reminds us of the highly non-classical nature of a Fock state. For higher photon numbers, the Wigner function displays more oscillations, the number of zeros coinciding with n. Fig. 4(b) displays the Wigner function corresponding to a Fock state with $n = 3$.

Of special interest for our discussion on the classical limit of quantum mechanics is the state formed by superimposing two coherent states $|\alpha_0\rangle$ and $|-\alpha_0\rangle$ (setting $\alpha_0 = q_0$ real for simplicity):

$$|\psi\rangle = \mathcal{N}[|\alpha_0\rangle + |-\alpha_0\rangle], \tag{61}$$

where $\mathcal{N}$ is a normalization constant, given by

$$\mathcal{N} = \left[2\left(1 + \exp(-2|\alpha_0|^2)\right)\right]^{-1/2}. \tag{62}$$

The corresponding wave function in configuration space is given by

$$\psi(q) = \mathcal{N}\left[e^{-(q-q_0)^2/2} + e^{-(q+q_0)^2/2}\right],$$

while the Wigner function is

$$W(q,p) = (\mathcal{N}/\pi)\left[e^{-(q-q_0)^2-p^2} + e^{-(q+q_0)^2-p^2} + 2e^{-q^2-p^2}\cos(2pq_0)\right]. \quad (63)$$

This function is displayed in Fig. 4(c).

It is interesting to compare this Wigner function with the one corresponding to a statistical mixture of the same coherent states, with equal weights:

$$\hat{\rho} = \tfrac{1}{2}\left(|\alpha_0\rangle\langle\alpha_0| + |-\alpha_0\rangle\langle-\alpha_0|\right), \quad (64)$$

for which

$$W(q,p) = (1/2\pi)\left[e^{-(q-q_0)^2-p^2} + e^{-(q+q_0)^2-p^2}\right]. \quad (65)$$

This function is displayed in Fig. 4(d).

While (65) is just the sum of two Gaussians, corresponding to the coherent states $|\alpha_0\rangle$ and $|-\alpha_0\rangle$ respectively, (63) displays interference fringes around the origin of phase space, which is a clear signature of the coherence between the two states $|\alpha_0\rangle$ and $|-\alpha_0\rangle$ in (61). Therefore, the measurement of the Wigner function of the electromagnetic field would be a clear-cut way of distinguishing between a coherent superposition and a mixture of the two coherent states. It will be seen shortly that these two coherent states may be interpreted, within the framework of recent experiments in cavity QED, as pointers of a measuring apparatus. The mechanism by which a state like (61) loses its coherence, approaching state (64), is thus very relevant for the quantum theory of measurement.

III.H Measurement of the Wigner function

The inverse Radon transform suggests that the Wigner function of an electromagnetic field can be reconstructed by determining $P(q_\theta)$ through homodyne detection [25]. This has actually been achieved in 1993 by Smithey et al. [26]. In view of the low detection efficiency in those experiments, the detected distribution was actually a smoothed version of the Wigner function, closely related to the Husimi distribution. A much better result was achieved by Mlynek's group in 1995 [27], clearly displaying a highly compressed Gaussian, corresponding to the experimentally obtained Wigner function of a squeezed state of light emerging from an optical parametric oscillator. A procedure closely related to the homodyne detection method was used to reconstruct the vibrational state of a molecule by T. J. Dunn et al. [28]

The Wigner function can also be obtained by measuring the populations of displaced states. Indeed, from (50), one has:

$$W(\alpha,\alpha^*) = \mathrm{Tr}\left[\hat{\rho}\hat{D}(\alpha,\alpha^*)e^{i\pi\hat{a}^\dagger\hat{a}}\hat{D}^{-1}(\alpha,\alpha^*)\right] = \sum_n \langle n|\hat{D}^{-1}(\alpha,\alpha^*)\hat{\rho}\hat{D}(\alpha,\alpha^*)e^{i\pi\hat{a}^\dagger\hat{a}}|n\rangle$$
$$= \sum_n (-1)^n \langle n|\hat{D}^{-1}(\alpha,\alpha^*)\hat{\rho}\hat{D}(\alpha,\alpha^*)|n\rangle. \tag{66}$$

If $\hat{\rho}$ corresponds to a mode of the electromagnetic field in a cavity, then $\hat{D}^{-1}(\alpha,\alpha^*)\hat{\rho}\hat{D}(\alpha,\alpha^*)$ can be obtained from $\hat{\rho}$ by injecting a coherent state into the cavity, through for instance the coupling of the cavity with a microwave generator (if the frequency of the mode in the cavity is in the microwave range). One must measure then the population of the displaced states, which can be done for instance by applying the procedure described in Ref. [30].

This method was used by Wineland's group at NIST to measure the Wigner function of vibrational states of a trapped ion [29]. The relevant level scheme is shown in Fig. 5. States $|\downarrow\rangle$ and $|\uparrow\rangle$ correspond to two metastable ground-state hyperfine sublevels ($^2S_{1/2}$, with $F = 2$, $m_F = -2$ and $F = 1$, $m_F = -1$, respectively), separated by $\hbar\omega_{HF}$. The ion is trapped in a harmonic potential, and the vibrational levels associated with each electronic state $|\downarrow\rangle$ and $|\uparrow\rangle$ are also sketched in Fig. 5. Initially, the ion is in the internal state $|\downarrow\rangle$. The displacement of the vibrational state in phase space is obtained by inducing a Raman transition between

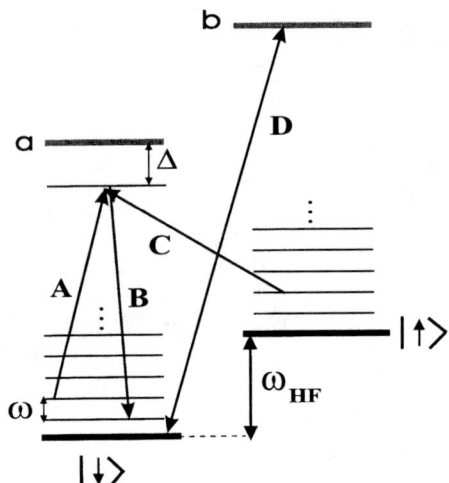

FIGURE 5. Measurement of the Wigner function for a trapped ion, by Leibfried et al. [29] Displacement of the vibrational state associated with the internal state $|\downarrow\rangle$ (a hyperfine structure sublevel) is achieved by applying fields A and B, which induce transitions between neighboring vibrational states corresponding to the electronic level $|\downarrow\rangle$, without changing the ion's internal state. Population of displaced states is measured by inducing a resonant exchange during a time t between states $|\downarrow,n\rangle$ and $|\uparrow,n+1\rangle$ with fields B and C (turning off field A). Probability of finding the atom in $|\downarrow\rangle$ as a function of t, determined by exciting it to level d (with a high fluorescence yield), leads to information on population of displaced vibrational states.

neighboring vibrational states, when the internal state of the ion is $|\downarrow\rangle$. This is accomplished by applying the two fields A and B illustrated in Fig. 5, with a frequency difference equal to the vibrational frequency ω. Beam B is circularly polarized (σ_-), and does not couple $|\uparrow\rangle$ to any virtual $^2P_{1/2}$ state, so that only the motional state correlated with $|\downarrow\rangle$ is displaced. These fields do not lead to transitions between electronic levels of the trapped ion, since they are detuned with respect to the possible electronic transitions, and therefore they affect only the center-of-mass motion. The action of the two fields can thus be modeled by an effective Hamiltonian of the form $H \propto (\hat{a} + \hat{a}^\dagger)$, where $\hat{a}$ and $\hat{a}^\dagger$ are harmonic oscillator lowering and raising operators. The evolution operator corresponding to this Hamiltonian is precisely the displacement operator, therefore the Raman process induces a displacement of the original state in phase space. A resonant exchange between states $|\downarrow\rangle|n\rangle$ and $|\uparrow\rangle|n+1\rangle$ is then induced for a time t, with fields B and C (turning off field A). For each time t and each displacement α the population $P_\downarrow(t,\alpha)$ of the $|\downarrow\rangle$ state is measured. A fourth level b is used for detecting the electronic state of the ion (and also for Doppler precooling): a pulse D resonant with the $|\downarrow\rangle \leftrightarrow |b\rangle$ transition leads to a fluorescence signal if the ion is in $|\downarrow\rangle$, while the absence of fluorescence implies that the ion is in $|\uparrow\rangle$ (the detection efficiency for this process is close to 100%). The internal state at $t=0$ being always equal to $|\downarrow\rangle$, the signal averaged over many measurements is $P_\downarrow(t,\alpha) = \frac{1}{2}[1 + \sum_{n=0}^\infty Q_n(\alpha)\cos(2\Omega_{n,n+1}t)e^{-\gamma_n t}]$, where $\Omega_{n,n+1}$ are the oscillation frequencies, γ_n their experimentally determined decay constants, and $Q_n(\alpha) = \langle n|\hat{D}^\dagger(\alpha)\hat{\rho}\hat{D}(\alpha)|n\rangle$ is the population distribution of the displaced state. The dependence of $\Omega_{n,n+1}$ on n allows the determination of $Q_n(\alpha)$ from $P_\downarrow(t,\alpha)$ [29], and from $Q_n(\alpha)$ one determines the Wigner function through Eq. (66).

It is clear that both methods are highly indirect. It will be shown in the following however that it is possible to conceive a much more direct method for measuring the Wigner function at any point in phase space, for either an electromagnetic field in a cavity or a trapped ion.

IV THE ATOM-FIELD INTERACTION

IV.A The interaction Hamiltonian

We consider now the interaction of atoms and fields. We will be considering situations in which the atom is resonant or quasi-resonant with one of the modes of the electromagnetic field in a cavity. Under these conditions, it is possible to consider just two of the atomic states, and therefore reduce the atom to a two-level system (we will call e the upper level, and g the lower level). The basic Hamiltonian describing this system is

$$H = H_A + H_F + H_{AF}, \tag{67}$$

where

$$H_A = (\hbar\omega_0/2)\sigma_3 \tag{68}$$

is the free-atom Hamiltonian, with σ_3 a Pauli matrix:

$$\sigma_3 = \begin{pmatrix} 1 & 0 \\ 0 & -1 \end{pmatrix}, \tag{69}$$

$$H_F = \hbar\omega \left(a^\dagger a + \tfrac{1}{2}\right) \tag{70}$$

is the free-field Hamiltonian, and

$$H_{AF} = \hbar g(\sigma_+ a + \sigma_- a^\dagger), \tag{71}$$

with

$$\sigma_+ = (\sigma_-)^\dagger = \begin{pmatrix} 0 & 1 \\ 0 & 0 \end{pmatrix}. \tag{72}$$

The real coupling constant $\hbar g$ depends on the transition dipole $\vec{d}_{ab}$ between the two levels, on the polarization vector $\vec{\epsilon}$ and the frequency ω of the electromagnetic field, as well as the effective volume of the mode V. From (6), it follows that

$$\hbar g = -\vec{d}_{ab} \cdot \vec{\epsilon}\sqrt{\frac{\hbar\omega}{V}} u(\vec{R}), \tag{73}$$

where the mode function $u(\vec{R})$ is evaluated on the center-of-mass position ($\vec{R}$) of the atom interacting with the field. One should note that, since only two atomic states are involved, one may always choose their phases so that g is real. If $u(\vec{R})$ is real, then this choice will not depend on $\vec{R}$.

In Eq. (71), we neglected terms of the form $\sigma_+ a^\dagger$ and $\sigma_- a$, which do not conserve energy in first order, and which lead to small corrections in the results to be obtained, as long as $|\omega - \omega_0| \ll \omega_0$.

The above equations define the *Jaynes-Cummings model* [31], a very useful model in Quantum Optics, and which has described with success many experiments in cavity QED. One should note that this model neglects dissipative processes, which will be considered in a while.

IV.B Heisenberg equations of motion for atom and semiclassical approximation

From the above Hamiltonian, one gets the following equations of motion for the atomic operators:

$$\dot{\hat{\sigma}}_+ = i\omega_0 \hat{\sigma}_+ - ig\hat{a}^\dagger \hat{\sigma}_3, \tag{74}$$

$$\dot{\hat{\sigma}}_3 = -2ig\left(\hat{\sigma}_+ \hat{a} - \hat{\sigma}_- \hat{a}^\dagger\right). \tag{75}$$

The *semiclassical approximation* is obtained by setting $\hat{a} \to \alpha$, $\hat{a}^\dagger \to \alpha^*$, where α and α^* are c-numbers.

Adopting this approximation, one gets for the average values of the atomic operators, after applying the transformation $\alpha \to \alpha \exp(-i\omega t)$, $\hat{\sigma}_+ \to \hat{\sigma}_+ \exp(i\omega t)$:

$$\langle \dot{\hat{\sigma}}_+ \rangle = i\delta \langle \hat{\sigma}_+ \rangle - ig\alpha^* \langle \hat{\sigma}_3 \rangle,$$
$$\langle \dot{\hat{\sigma}}_3 \rangle = -2ig\left(\langle \hat{\sigma}_+ \rangle \alpha - \langle \hat{\sigma}_- \rangle \alpha^*\right), \qquad (76)$$

where $\delta = \omega_0 - \omega$ is the detuning between the atom and the field.

IV.C Bloch equations

If one rewrites Eqs. (76) in terms of $r_1 \equiv \langle \hat{\sigma}_1 \rangle$, $r_2 \equiv \langle \hat{\sigma}_2 \rangle$, and $r_3 \equiv \langle \hat{\sigma}_3 \rangle$, with $g\alpha \equiv V/2 \equiv (V_1 - iV_2)/2$, one gets the *Bloch equations*, which can be written as:

$$\frac{d\vec{r}}{dt} = \vec{\Omega} \times \vec{r}, \qquad (77)$$

where $\vec{r} = (r_1, r_2, r_3)$ and $\vec{\Omega} = (V_1, V_2, \delta)$.

In terms of the atomic density matrix,

$$\hat{\rho}_A = \begin{pmatrix} \rho_{ee} & \rho_{eg} \\ \rho_{ge} & \rho_{gg} \end{pmatrix},$$

one may write:

$$r_1 = \text{Tr}(\hat{\sigma}_1 \hat{\rho}_A) = \rho_{eg} + \rho_{ge} = 2\Re e(\rho_{ge}) \qquad (78)$$
$$r_2 = \text{Tr}(\hat{\sigma}_2 \hat{\rho}_A) = i(\rho_{eg} - \rho_{ge}) = 2\Im m(\rho_{ge}) \qquad (79)$$
$$r_3 = \text{Tr}(\hat{\sigma}_3 \hat{\rho}_A) = \rho_{ee} - \rho_{gg} \qquad (80)$$

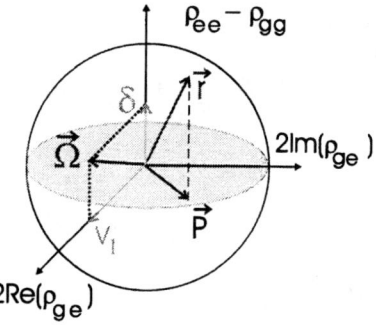

FIGURE 6. Precession of the Bloch vector $\vec{r}$ around the pseudo-magnetic field $\vec{\Omega}$, with V real.

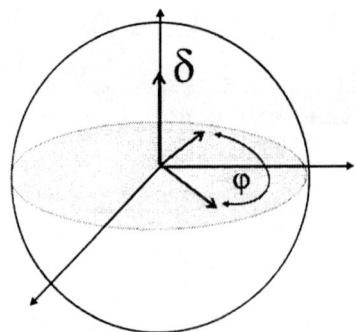

FIGURE 7. Dispersive limit: the Bloch vector precesses around the vertical axis.

These equations represent the atomic state by a pseudo-spin $\vec{r}$ (Bloch vector), which precesses around the pseudo-magnetic field $\vec{\Omega}$, as shown in Fig. 6. The precession frequency, given by

$$\Omega = \sqrt{|V|^2 + \delta^2}, \qquad (81)$$

is the *Rabi frequency*. The vertical component of the Bloch vector represents the atomic population, while the equatorial projection is associated with the atomic polarization ($\vec{P}$ in Fig. 6). This picture of the evolution of a two-level atom is due to Feynman, Vernon, and Hellwarth [32].

For a pure state, $|\psi\rangle = c_e|e\rangle + c_g|g\rangle$, one has

$$\hat{\rho}_A = \begin{pmatrix} |c_e|^2 & c_e^* c_g \\ c_e c_g^* & |c_g|^2 \end{pmatrix}, \quad r_1^2 + r_2^2 + r_3^2 = 1.$$

Therefore, in this case the tip of the Bloch vector is situated on a sphere of unit radius.

Two limiting cases of this expression correspond to the resonant and to the dispersive interaction. When the interaction is resonant, $\delta = 0$, $\vec{\Omega} = (V_1, V_2, 0)$, and the Bloch vector precesses around a vector in the equatorial plane. One gets in this case maximum population transfer. The precession frequency is then $|V|$. In the dispersive limit, $|\delta| \gg |V|$, and $\vec{\Omega} \to (0, 0, \delta)$. The Bloch vector precesses then around a vector parallel to the axis 3 (population axis), with a frequency equal to δ, as shown in Fig. 7.

IV.D Quantum theory: the dressed atom

We go back now to the Hamiltonian (67), and consider the effects resulting from the quantization of the electromagnetic field.

The Hamiltonian (67) defines the *dressed atom* [33]. While H_A has two energy levels, H_F has an infinite number of discrete levels, given by $\hbar\omega(n + 1/2)$, $n =$

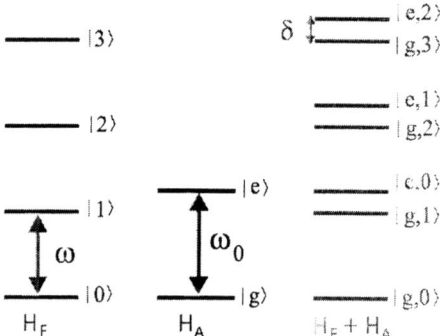

FIGURE 8. Energy level diagram for the uncoupled atom-field system

$0, 1, 2, \ldots$. The interaction H_{AF} couples these levels, leading to a discrete structure of levels of the composed system, which one could call the "atom-field molecule". We will study first the structure of the uncoupled system, and then we will analyze the energy levels of the coupled system.

IV.D.1 Uncoupled states

The state in which the atom is in state e and the field has n photons will be denoted by $|e,n\rangle$. Analogously for $|g,n\rangle$. Let again $\delta = \omega_0 - \omega$ be the detuning between the atom and the field. If $|\delta| \ll \omega_0$, the states $|e,n\rangle$ and $|g,n+1\rangle$ will be very close to each other. If $\delta > 0$, the energy of the state $|g,n+1\rangle$ will be smaller than the energy of the state $|e,n\rangle$ (see Fig. 8). In fact, we can write

$$E_{e,n} = \hbar\omega(n+1) + \frac{\hbar\delta}{2}, \tag{82}$$

$$E_{g,n+1} = \hbar\omega(n+1) - \frac{\hbar\delta}{2}. \tag{83}$$

We have therefore a sequence of subspaces $\mathcal{E}(n) \equiv \{|g, n+1\rangle; |e, n\rangle\}$. Note that $E_{e,0} = \hbar(\omega + \omega_0)/2$, and $E_{g,0} = -\hbar\delta/2 = \hbar(\omega - \omega_0)/2$, consistently with the fact that the zero-point energy is $\hbar\omega/2$ and the energies of the two atomic states are $\pm\hbar\omega_0/2$.

IV.D.2 Coupled states

The Hamiltonian (67), with H_{AF} given by (71), couples only states within the same subspace (this is a consequence of the rotating-wave approximation; the counter-rotating terms, neglected in H_{AF}, connect states belonging to different subspaces, which leads to small corrections to the results considered here, due to

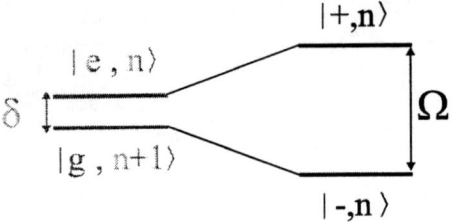

FIGURE 9. Energy displacement of levels, produced by the coupling.

the large energy differences involved). Therefore, in order to calculate the eigenvalues of the complete Hamiltonian, one has to diagonalize a 2 × 2 matrix, given in the subspace $\mathcal{E}(n)$ by:

$$[H]_n = \hbar \begin{pmatrix} \omega(n+1) - \frac{\delta}{2} & g\sqrt{n+1} \\ g\sqrt{n+1} & \omega(n+1) + \frac{\delta}{2} \end{pmatrix}. \tag{84}$$

The eigenvalues of this matrix define the energy levels of the dressed atom:

$$E_{\pm,n} = (n+1)\hbar\omega \pm (\hbar\Omega/2), \tag{85}$$
$$E_{g,0} = \hbar\delta/2, \tag{86}$$

where

$$\Omega = \left[\Omega_0^2(n+1) + \delta^2\right]^{1/2} \tag{87}$$

is the quantum *Rabi frequency* of the system, while $\Omega_0 = 2g$ is the *vacuum Rabi frequency*, which coincides with Ω when $n = 0$ and $\delta = 0$. The quantum Rabi frequency (87) coincides precisely with the classical expression (81) if one identifies $\Omega_0\sqrt{n+1}$ with $|V|$. One should note however that, contrary to the classical expression, the quantum Rabi frequency remains different from zero even when the number of photons in the mode is equal to zero. The remaining contribution is associated with spontaneous emission into the mode, which couples the state $|e, 0\rangle$ to the state $|g, 1\rangle$.

The expression (85) shows that the two states are separated by the coupling, the energy difference between them going from $\hbar\delta$ to $\hbar\Omega$. This effect is displayed in Fig. 9.

The corresponding eigenstates are given by:

$$|+, n\rangle = \sin\theta|g, n+1\rangle + \cos\theta|e, n\rangle, \tag{88}$$
$$|-, n\rangle = \cos\theta|g, n+1\rangle - \sin\theta|e, n\rangle, \tag{89}$$

with

$$\cot 2\theta = \frac{\delta}{\Omega_0\sqrt{n+1}}, \quad 0 \leq 2\theta < \pi. \tag{90}$$

IV.D.3 Resonant interaction

At resonance, $\delta = 0$, and therefore $\theta = \pi/4$, so that

$$|+, n\rangle = \frac{1}{\sqrt{2}} \left(|g, n+1\rangle + |e, n\rangle \right), \tag{91}$$

$$|-, n\rangle = \frac{1}{\sqrt{2}} \left(|g, n+1\rangle - |e, n\rangle \right). \tag{92}$$

In this case, each subspace $\mathcal{E}(n)$ becomes two-fold degenerate, and the dressed states are expressed in terms of the sum and the difference of the corresponding uncoupled states, with equal weights. The corresponding energies are, for $n \neq 0$,

$$E_{\pm, n} = (n+1)\hbar\omega \pm (\hbar\Omega_0\sqrt{n+1}/2). \tag{93}$$

IV.D.4 Dispersive interaction

For large detuning ($|\delta| \gg \Omega_0\sqrt{n+1}$), one gets $\theta \to \pi/2$ if $\delta < 0$, and $\theta \to 0$ if $\delta > 0$, and therefore

$$|+, n\rangle \to |g, n+1\rangle, \quad |-, n\rangle \to |e, n\rangle, \quad (\delta < 0), \tag{94}$$

$$|+-n\rangle \to |e, n\rangle, \quad |-, n\rangle \to |g, n+1\rangle, \quad (\delta > 0). \tag{95}$$

These equations show that, for a dispersive interaction, the coupled states approach the uncoupled states, with an energy shift obtained from (85):

$$E_{\pm, n} \approx (n+1)\hbar\omega \pm (\hbar\omega/2)|\delta| \pm \hbar\Omega_0^2(n+1)/4|\delta|. \tag{96}$$

In any case, we have in this limit:

$$\Delta E_{e,n} \approx \hbar(\Omega_0^2/4\delta)(n+1), \tag{97}$$

$$\Delta E_{g,n} \approx -\hbar(\Omega_0^2/4\delta)n. \tag{98}$$

The two energy levels of each subspace get displaced in opposite directions. This displacements coincide precisely with those which would be obtained using second-order perturbation theory, and constitute the *Stark effect*.

IV.E Dynamics of the interaction

Once the Hamiltonian is diagonalized, one can easily describe the dynamical behavior of the system. From (88), one has:

$$|e, n\rangle = \cos\theta |+, n\rangle - \sin\theta |-, n\rangle, \tag{99}$$

$$|g, n+1\rangle = \sin\theta |+, n\rangle + \cos\theta |-, n\rangle, \tag{100}$$

and therefore, if the initial state of the system is $|\psi(0)\rangle = |e, n\rangle$, we will have at time t:

$$|\psi(t)\rangle = \cos\theta e^{-iE_{+,n}t/\hbar}|+, n\rangle - \sin\theta e^{-iE_{-,n}t/\hbar}|-, n\rangle, \quad (101)$$

or yet, reexpressing in terms of the uncoupled states:

$$|\psi(t)\rangle = e^{-i\omega(n+1)t}\{-i\sin(2\theta)\sin(\Omega t/2)|g, n+1\rangle \\ + [\cos(\Omega t/2) - i\cos(2\theta)\sin(\Omega t/2)]|e, n\rangle\}. \quad (102)$$

The probabilities of finding the system in the states $|e, n\rangle$ e $|g, n+1\rangle$ are given therefore by

$$P_{e,n} = \cos^2(\Omega t/2) + \cos^2(2\theta)\sin^2(\Omega t/2), \quad (103)$$
$$P_{g,n+1} = \sin^2(2\theta)\sin^2(\Omega t/2), \quad (104)$$

oscillating therefore with the Rabi frequency Ω. At resonance, when $\theta = \pi/4$, one gets $P_{e,n} = \cos^2(\Omega t/2)$, $P_{g,n+1} = \sin^2(\Omega t/2)$, so the oscillation has maximum amplitude.

These considerations extend to the quantum case the description of the atomic evolution in terms of the Bloch vector, previously discussed within the semiclassical approximation. One should note that, if one starts from the state $|e, n\rangle$, the quantum system evolves even when the number of photons in the mode is equal to zero, contrary to what would happen if the field is not quantized. This extra quantum feature is again due to spontaneous emission by the atomic excited state into the cavity mode.

One should also note that, in the dispersive limit, neither the number of photons nor the populations of states e and g change, exactly as in the semiclassical treatment. In this case, the dynamics of the atom is well represented by the precession of the Bloch vector around the vertical axis (population axis): if there are n photons in the field, the angle of precession is given by $(\Delta E_{e,n} - \Delta E_{g,n})t/\hbar$, with the Stark energy displacements given by (97).

V COHERENT SUPERPOSITIONS OF MESOSCOPIC STATES IN CAVITY QED

V.A Building the coherent superposition

We show now how, by carefully tailoring the interactions between two-level atoms and one mode of the electromagnetic field in a cavity, one can produce quantum superpositions of distinguishable coherent states of the field, thus mimicking the superposition of two classically distinct states of a pointer.

The method, proposed in Ref. [30], and sketched in Fig. 10, involves a beam of circular Rydberg atoms [34] crossing a high-Q cavity in which a coherent state is

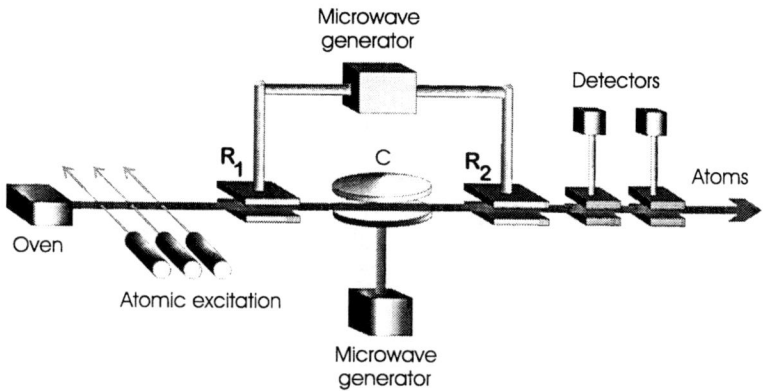

FIGURE 10. Experimental arrangement for producing and measuring a coherent superposition of two coherent states of the field in cavity **C**.

previously injected (this is accomplished by coupling the cavity to a classical source – a microwave generator – through a wave guide). The utilization of circular levels is due to their strong coupling to microwaves and their very long radiative decay times, which makes them ideally suited for preparing and detecting long-lived correlations between atom and field states [35]. On either side of the high-Q cavity there are two low-Q cavities (R_1 and R_2), which remain coupled to a microwave generator. The fields in these two cavities can be considered as classical. This set of two low-Q cavities constitutes the usual experimental arrangement in the Ramsey method of interferometry [35,36]. Two of the atomic levels, which we denote by $|e\rangle$ and $|g\rangle$, are resonant with the microwave fields in cavities R_1 and R_2, the intensity of these fields being such that, for the selected atomic velocity, effectively a $\pi/2$ pulse is applied to the atom as it crosses those cavities. For a properly chosen phase of the microwave field, this pulse transforms the state $|e\rangle$ into the linear combination $(|e\rangle + |g\rangle)/\sqrt{2}$, and the state $|g\rangle$ into $(-|e\rangle + |g\rangle)/\sqrt{2}$.

Therefore, if each atom is prepared in the state $|e\rangle$ just prior to crossing the system, after leaving R_1 the atom is in a superposition of two circular Rydberg states $|e\rangle$ and $|g\rangle$:

$$|\psi_{\text{atom}}\rangle = \frac{1}{\sqrt{2}}(|e\rangle + |g\rangle). \tag{105}$$

On the other hand, the superconducting cavity is assumed not to be in resonance with any of the transitions originating from those two atomic states. This means that the atom does not suffer a transition, and does not emit or absorb photons from the field. This property is further enhanced by the fact that the cavity mode is such that the field slowly rises and decreases along the atomic trajectory, so that, for sufficiently slow atoms, the atom-field coupling is adiabatic. However, the cavity is tuned in such a way that it is much closer to resonance with respect to

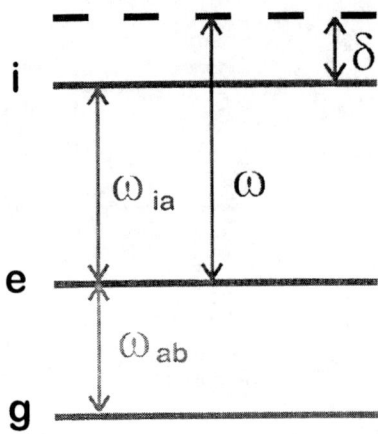

FIGURE 11. Atomic level scheme. The transition $i \leftrightarrow e$ is detuned by δ from the frequency ω of a mode of cavity **C**, while the transition $e \leftrightarrow g$ is resonant with the fields in $\mathbf{R_1}$ and $\mathbf{R_2}$. State $|g\rangle$ is not affected by the field in **C**.

one of those transitions, say the one connecting $|e\rangle$ to some intermediate state $|i\rangle$. The relevant level scheme is illustrated in Fig. 11. This implies that, if the atom crosses the cavity in state $|e\rangle$, dispersive effects can induce an appreciable phase shift on the field in the cavity. The phase shift is negligible, however, if the atom is in state $|g\rangle$. For a principal quantum number equal to 50 in the state e, and the cavity tuned close to the 50 → 51 circular to circular transition (around 50 GHz), a phase shift of the order of π is produced by an atom crossing the centimeter size cavity with a velocity of about 100 m/s [30].

After the atom has crossed the cavity, in a time short compared to the field relaxation time and also to the atomic radiative damping time, the state of the combined atom-field system can be written as

$$|\psi_{\text{atom+field}}\rangle = \frac{1}{\sqrt{2}}(|e; -\alpha\rangle + |g; \alpha\rangle), \quad (106)$$

assuming that the phase shift is π if the atom is in the excited state. The entanglement between the field and atomic states is analogous to the correlated two-particle states in the Einstein-Podolski-Rosen (EPR) paradox [37–39], and is a realization of the Von Neumann entangled state (1). The two possible atomic states e and g are here correlated to the two field states $|-\alpha\rangle$ and $|\alpha\rangle$, respectively, which may be considered as macroscopic pointers (if $|\alpha| \gg 1$). After the atoms leave the superconducting cavity, one can detect them in the e or g states, by sending them through two ionization chambers, the first one having a field smaller than the second, so that it ionizes the atom in the e state, but not in the g state, while the second ionizes the atoms which remain in state g (Fig. 10). This measurement projects the field in the cavity either in the $|\alpha\rangle$ (if the atom is detected in state g), or in the $|-\alpha\rangle$ state (if the atom is detected in state e). However, as in an EPR experiment [39], one may choose to make another kind of measurement, letting the atom cross, after it leaves the superconducting cavity, a second classical microwave field ($\mathbf{R_2}$ in Fig. 10), which amounts to applying to the atom another $\pi/2$ pulse. The state (106) gets transformed then into

$$|\psi'_{\text{atom+field}}\rangle = \tfrac{1}{2}\left(|e;-\alpha\rangle - |e;\alpha\rangle + |g;\alpha\rangle + |g;-\alpha\rangle\right). \tag{107}$$

If one detects now the atom in the state $|g\rangle$ or $|e\rangle$, the field is projected onto the state

$$|\psi_{\text{cat}}\rangle = \frac{1}{N_1}\left(|\alpha\rangle + e^{i\psi_1}|-\alpha\rangle\right), \tag{108}$$

where $N_1 = \sqrt{2\left[1 + \cos\psi_1 \exp(-2|\alpha|^2)\right]}$ and $\psi_1 = 0$ or π, according to whether the detected state is g or e, respectively. One produces therefore a coherent superposition of two coherent states, with phases differing by π. For $|\alpha|^2 \gg 1$, this is a "Schrödinger cat" state.

Superpositions of coherent states of the field were produced in the experiment reported in Ref. [40], and were detected by a procedure which can be considered, as shown in the next Section, as a special case of a method for measuring the Wigner function of the field in the cavity.

V.B Effect of dissipation

Before considering how these states can be detected, we discuss the effect of dissipation, due to imperfections in the mirrors and diffraction losses. A simple model for dissipation is obtained by coupling the field oscillator (of frequency ω) to a bath of harmonic oscillators, which represent the modes of the reservoir. We consider here for simplicity a rotating-wave Hamiltonian, and a zero-temperature bath. The method here exposed, and which results from joint work of the author with V. M. Kenkre, can be easily generalized to account for a finite-temperature bath [41]. The corresponding Hamiltonian may be written as

$$H = \hbar\omega a^\dagger a + \sum_q \hbar\omega_q b_q^\dagger b_q + \hbar \sum_q \left(G_q a^\dagger b_q + G_q^* a b_q^\dagger\right), \tag{109}$$

where G_q are the coupling constants, and the bath oscillators have frequency ω_q.

It is straightforward to write down the evolution equation for the Heisenberg operators $a(t)$, $b_q(t)$, and their adjoints. Since the resulting equations are linear, they may be solved by the Laplace transform method.

The explicit solution for $a(t)$ is given by

$$a(t) = a(0)\xi(t) + \sum_q \eta_q(t) b_q(0). \tag{110}$$

The Laplace transforms of the c-number functions $\xi(t)$ and $\eta_q(t)$ appearing in (110) are given respectively by

$$\tilde{\xi}(s) = \left[s + i\omega + \tilde{\phi}(s)\right]^{-1}, \tag{111}$$

$$\tilde{\eta}_q(s) = -iG_q(s + i\omega_q)^{-1}\left[s + i\omega + \tilde{\phi}(s)\right]^{-1}. \tag{112}$$

The function ϕ which appears in both (111) and (112) reflects the nature of the coupling and is given in the time domain by

$$\phi(t) = \sum_q |G_q|^2 e^{-i\omega_q t} = \int_0^\infty d\nu\, \mathcal{G}(\nu) e^{-i\nu t}. \tag{113}$$

In the second equality in (113) we have introduced the quantity $\mathcal{G}(\nu)$ which equals the product of the coupling $|G_q|^2$ assumed to be a function of the frequency alone, and the density of states $\sum_q \delta(\nu - \omega_q)$ of the b-oscillators.

Since the total number of oscillator excitations in both oscillators, $a^\dagger a + \sum_q b_q^\dagger b_q$, commutes with the Hamiltonian and is therefore an invariant, it follows that

$$|\xi(t)|^2 + \sum_q |\eta_q(t)|^2 = 1. \tag{114}$$

In order to get now the time-dependent Wigner function, it suffices to replace the solution (110) in the characteristic function (48), and then calculate the Wigner function through (49). Assuming that the initial state of the field in the cavity is given by (61), one gets then:

$$W(\alpha, \alpha^*, t) = \frac{2}{\mathcal{N}^2}\left\{e^{-2|\alpha - \xi(t)\alpha_0|^2} + e^{-2|\alpha + \xi(t)\alpha_0|^2} + 2F(t)\right\}, \tag{115}$$

where the normalization factor $\mathcal{N}$ is given by (62). Here, the fringe function $F(t)$ is given by

$$F(t) = \exp\left\{-2\left[|\alpha_0|^2\left(1 - |\xi(t)|^2\right) + |\alpha|^2\right]\right\} \cos\left\{4\alpha_0 \text{Im}\left[\alpha\xi^*(t)\right]\right\}. \tag{116}$$

The Markoffian limit corresponds to setting $\phi(t) = \Gamma\delta(t)$, which implies that

$$\xi(t) = e^{-i\Omega t} e^{-\Gamma t}. \tag{117}$$

More generally, one could have a frequency shift as well, in the Markoffian limit, arising from the fact that the integration in (113) is from zero to infinity. It is clear then from (116) and (117) that, for $\Gamma t \ll 1$, the fringe function decays very fast, in a time scale of the order of $1/2|\alpha_0|^2\Gamma$. This shows explicitly that the term associated with coherence decays at a much faster rate than the energy of the system, if $|\alpha_0|^2 \gg 1$. As the distance between the two coherent states increases, this effect becomes more and more pronounced, and the coherent superposition of the two "pointers" becomes practically indistinguishable from a statistical mixture: this is the reason why such coherent superpositions are extremely difficult to observe in the macroscopic world! The physical origin of the decoherence process is actually very simple: as the field in the cavity leaks into the external reservoir, the states of the field get correlated with states of the reservoir which become approximately orthogonal after the time it takes for one photon to leave the cavity, thus implying the disappearance of interference effects between the two internal states. This can be clearly seen, if one assumes that the decoherence time depends only on the distance between the two states in phase space, upon displacing the initial state in the cavity by α_0, so that the state $|\psi'\rangle = (1/\mathcal{N})[|2\alpha_0\rangle + |0\rangle]$ is obtained (which has the same distance between the two states of the superposition as the original state). For the state $|2\alpha_0\rangle$, a photon leaves the cavity in a time of the order of $(1/4|\alpha_0|^2\Gamma)$, while for the state $|0\rangle$ no photon leaves the cavity. Since the probability for finding the system in each of these states is $1/2$ for $|\alpha_0| \gg 1$, it follows that the effective lifetime of a photon is $(1/2|\alpha_0|^2\Gamma)$, which is precisely the decoherence time: after this time, the state $|2\alpha_0\rangle$ becomes correlated with a state of the reservoir containing approximately one photon, while the state $|0\rangle$ remains correlated with the vacuum. Decoherence of the system under observation is therefore closely connected with entanglement between this system and the reservoir.

VI DIRECT MEASUREMENT OF THE WIGNER FUNCTION

Once the proper state of the field is produced in the cavity, how would one be able to measure it? As shown in [42], it is actually possible to measure the Wigner function of the field by a relatively simple scheme, which provides directly the value of the Wigner function at any point of phase space. This is in contrast with the tomographic procedure, or the method based on the measurement of populations adopted at NIST, which yield the Wigner function only after some integration or summation. Furthermore, and also in contrast with those methods, the present scheme is not sensitive to detection efficiency, as long as one atom is detected within a time shorter than the decoherence time. A similar procedure can be applied to the reconstruction of the vibrational state of a trapped ion [42], and also in some cases to molecules [43]. We will discuss here only the application to the electromagnetic field.

The basic experimental scheme for measuring the Wigner function [42] coincides with the one used to produce the "Schrödinger cat"-like state, illustrated in Fig. 10. A high-Q superconducting cavity **C** is placed between two low-Q cavities (**R$_1$** and **R$_2$** in Fig. 10). The cavities **R$_1$** and **R$_2$** are connected to the same microwave generator, the field in **R$_2$** being dephased by η with respect to the field in **R$_1$**. Another microwave source is connected to **C**, allowing the injection of a coherent state in this cavity, so that the density operator $\hat{\rho}$ of the field to be measured is transformed into $\hat{\rho}' = \hat{D}(z, z^*)\hat{\rho}\hat{D}^{-1}(z, z^*)$. This system is crossed by a velocity-selected atomic beam, such that an atomic transition $e \leftrightarrow g$ is resonant with the fields in $\mathbf{R_1}$ and $\mathbf{R_2}$, while another transition $e \leftrightarrow i$ is quasi-resonant (detuning δ) with the field in $\mathbf{C}$, so that the atom interacts dispersively with this field if it is in state e, while no interaction takes place in $\mathbf{C}$ if the atom is in state g. The relevant level scheme is shown in Fig. 11. Just before $\mathbf{R_1}$, the atoms are promoted to the highly excited circular Rydberg state $|e\rangle$ (typical principal quantum numbers of the order of 50, corresponding to lifetimes of the order of some milliseconds). As each atom crosses the low-Q cavities, it sees a $\pi/2$ pulse, so that $|e\rangle \to [|e\rangle + \exp(i\eta)|g\rangle]/\sqrt{2}$, and $|g\rangle \to [-\exp(-i\eta)|e\rangle + |g\rangle]/\sqrt{2}$, with $\eta = 0$ in $\mathbf{R_1}$. If the atom is in state e when crossing $\mathbf{C}$, there is an energy shift of the atom-field system (Stark shift), which dephases the field, after an effective interaction time t_{int} between the atom and the cavity mode. The one-photon phase shift is given by $\phi = (\Omega^2/\delta)t_{\text{int}}$, where the Rabi frequency Ω measures the coupling between the atom and the cavity mode. The atom is detected and the experiment is repeated many times, for each amplitude and phase of the injected field z, starting from the same initial state of the field $\hat{\rho}$. In this way, the probabilities P_e and P_g of detecting the probe atom in states e or g are determined. It is easy to show that

$$P_g - P_e = \Re e \left\{ e^{i\eta} \text{Tr} \left[\hat{D}(z, z^*) \hat{\rho} \hat{D}^{-1}(z, z^*) e^{i\phi \hat{a}^\dagger \hat{a}} \right] \right\}. \tag{118}$$

Setting $\eta = 0$ and $\phi = \pi$, we can see from (50) that

$$P_g - P_e = W(-z, -z^*)/2. \tag{119}$$

Therefore, the difference between the two probabilities yields a direct measurement of the Wigner function (one should note that, due to the fact that here $|g\rangle$ does not interact with the field in $\mathbf{C}$, this expression differs from the one given in Ref. [42]).

An important feature of this scheme is the insensitivity to the detection efficiency of the atomic counters (of the order of $40 \pm 15\%$ in recent experiments [30]). Indeed, if an atom is not detected after interacting with the cavity mode, the next atom will find a field described by the reduced density operator obtained from the entangled atom-field density matrix by tracing out the atomic states: $\hat{\rho}' \to \hat{\rho}'' = \frac{1}{2}(\hat{\rho}' + \hat{\mathcal{T}}_e \hat{\rho}' \hat{\mathcal{T}}_e^\dagger)$, where $\hat{\mathcal{T}}_e(\phi) = \exp(i\phi \hat{a}^\dagger \hat{a})$ is the phase shift operator associated with level e. The value of $P_g - P_e$ for this second atom is then easily shown to reduce to (119).

The measurement accuracy does depend however on the detector's selectivity, that is, the ability to distinguish between the two atomic states. Another possible source of error is the velocity spread of the atomic beam, which would produce an uncertainty in the angle ϕ and in the angles of rotation in $\mathbf{R_1}$ and $\mathbf{R_2}$. For a 1% velocity spread and for average photon numbers of the order of 10, one can show that the distortion is at most equal to 0.04, in the relevant region of phase space, so that the measured distribution is practically undistinguishable from the true one. In fact, the insensitivity of the proposed scheme to the detection efficiency allows a passive selection of atomic velocity (only the atom which goes through the detectors at the right time after excitation is detected), which can be made with high precision.

One should note that this method allows the measurement of the Wigner function at each time t, allowing therefore the monitoring of the decoherence process "in real time". It is interesting, in this respect, to compare the procedure described above with the one suggested by Davidovich et al [44], with the objective of observing the decoherence of a Schrödinger cat-like state. In that reference, it was proposed that the decoherence of the state $|\pm\rangle = (|\alpha\rangle \pm |-\alpha\rangle)/N_\pm$ could be observed by measuring the joint probability of detecting in states $|e\rangle$ or $|g\rangle$ a pair of atoms, both prepared in the state $|e\rangle$ initially, and sent through the system depicted in Fig. 10. The atomic configuration considered in that reference coincides with the one adopted here. Detection of the first atom prepares the coherent superposition of coherent states, as described above. Detection of the second atom probes the state produced in C. Since no field was injected into the cavity between the two atoms, it is clear now that the experiment proposed in Ref. [44] amounts to a measurement of the Wigner function at the origin, which is non zero for the pure state $|\pm\rangle$, as shown in Fig. 4 (c), vanishes after the decoherence time (shorter than the intensity decay time by the factor $|\alpha|^2$), as shown in Fig. 4 (d), and increases again as dissipation takes place, bringing the field to the vacuum state. Following this proposal, the first observation of decoherence was realized by Brune et al [30]. In the experiment, both $|e\rangle$ and $|g\rangle$ lead to dephasings (in opposite directions) of the field in C. In this case, it is easy to show that the Wigner function is again recovered, as long as the one-photon phase shift is $\phi = \pi/2$ (with opposite signs for e and g), and a dephasing $\eta = \pi/2$ is applied to the second Ramsey zone [42]. This condition was not satisfied however in the experiment reported in Ref. [30]: due to experimental limitations, the angle ϕ was actually smaller than $\pi/2$.

One is led therefore to a natural question: can the Wigner function ($\phi = \pi$) be inferred from that measurement? This question can be answered by using the fact that the Wigner function belongs to a general class of phase-space distributions, parameterized by a complex parameter s, and which can be written as [20]:

$$W(z, z^*, s) = \int e^{z\xi^* - z^*\xi} e^{s|\xi|^2/2} \mathrm{Tr}\left[\hat{\rho}\hat{D}(\xi, \xi^*)\right] \pi^{-1} d^2\xi. \tag{120}$$

Note that $W(z, z^*, s)$ is real when s is real. For $s = 0$, one obtains the Wigner distribution, while $s = -1$ and $s = 1$ correspond respectively to the Q and the Glauber-Sudarshan P representations [20]. Setting $s = -i \cot \phi/2$, (120) becomes [42]:

$$W(z, z^*, \phi) = -2i e^{i\phi/2} \sin\frac{\phi}{2} \mathrm{Tr}\left[\hat{D}(-z, -z^*)\hat{\rho}\hat{D}(z, z^*) e^{i\phi\hat{a}^\dagger\hat{a}}\right]. \tag{121}$$

For $\phi = \pi$ ($s = 0$), we recover Eq. (50).

If the phase shift is different from π, one can see from (118) that by changing η one may detect the real and the imaginary part of $W(z, z^*, \phi)$, given by (121). Therefore, one can measure phase space representations corresponding to imaginary values of s. The connection between $W(z, z^*, \phi)$ and $W(z, z^*) \equiv W(z, z^*, \pi)$ can be obtained in the following way. It is easy to show from (120) that, setting $\tau = is$ and $z = x + iy$,

$$i\frac{\partial W(x,y,\tau)}{\partial \tau} = -\frac{1}{8}\left(\frac{\partial^2}{\partial x^2} + \frac{\partial^2}{\partial y^2}\right)W(x,y,\tau), \qquad (122)$$

so that $W(x, y, \tau)$ obeys a free-particle Schrödinger equation, the parameter τ playing the role of a time. As ϕ changes from 2π to 0, τ changes correspondingly from $\tau = -\infty$ to $\tau = \infty$. The behavior of the real part of $W(x, y, \phi)$ as ϕ changes, for the state $|-\rangle$, is illustrated by Fig. 12. This behavior is easily understandable in terms of the development in time of a free wavepacket. In particular, the vanishing of $W(z, \phi)$ when $\phi = 0$ (and therefore $\tau = \infty$) may be seen as a direct consequence of the wavepacket spreading. The interference fringes at the origin, displayed when $\phi = \pi$ ($\tau = 0$), may be thought as resulting from the collision of the two wavepackets counter-propagating along the x axis and meeting at the origin of the phase

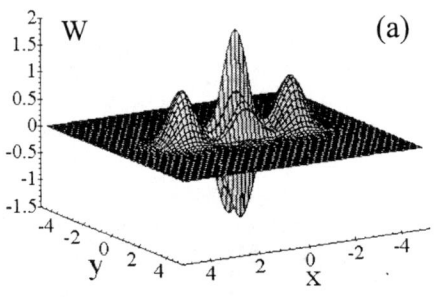

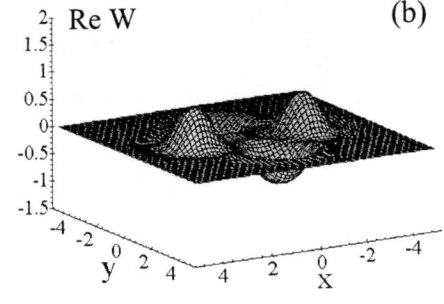

FIGURE 12. Generalized phase-space distribution for the state $(|\alpha_0\rangle + |-\alpha_0\rangle)/\sqrt{2}$, with $|\alpha_0|^2 = 5$. The change of the one-photon phase shift ϕ from π to 0 is equivalent to the time evolution of a wavepacket in phase space from $\tau = 0$ to $\tau = \infty$. (a) $\phi = \pi$, corresponding to the (real) Wigner distribution (initial wavepacket); (b) Real part of $W(\alpha, \phi = \pi/2)$.

space. Equation (122) also implies that $W(x, y, \tau)$ is connected to $W(x, y)$ through the free-particle propagator. Therefore, the reconstruction of the Wigner function from $W(z, z^*, \phi \neq \pi)$ is possible through an integral transform involving that propagator, but would require the knowledge of this generalized distribution for every value of the complex number z.

VII CONCLUSION

In these lectures, I intended to show that techniques used in the field of quantum optics are helpful to discuss the quantum-classical transition, and in particular allow the monitoring of the decoherence process, which is at the heart of the quantum theory of measurement. This does not mean that the problem of the classical limit has been solved. Indeed, one has looked only at the dynamics of linear systems (the field mode coupled to the reservoir oscillators), and therefore we have not discussed the difficult problems related to the classical limit of non-linear systems, where chaotic behavior may play an important role [45]. Furthermore, coherence does not really disappear, and it is still present in entangled states of the cavity field and the rest of the Universe. This fact immediately leads us to the question about the meaning of the wave function of the Universe, and to the seemingly paradoxical application of probability concepts to a Universe which is unique. According to Murray Gell-Man and Jim Hartle, "quantum mechanics is better and more fundamentally understood within the context of quantum cosmology" [46]. Even though fundamental problems related to the classical limit of quantum mechanics and the quantum theory of measurement remain to be solved, I think it is fair to say that quantum optics has helped us to understand and observe an important piece of this puzzle.

ACKNOWLEDGMENTS

This work was partially supported by PRONEX (Programa de Apoio a Núcleos de Excelência), CNPq (Conselho Nacional de Desenvolvimento Científico e Tecnológico), and FUJB (Fundação Universitária José Bonifácio). It is a pleasure to acknowledge the collaboration on the subjects covered by these lecture notes with M. Brune, S. Haroche, L.G. Lutterbach, J.M. Raimond, and N. Zagury.

REFERENCES

1. Letter from Albert Einstein to Max Born in 1954, cited by Joos, E., in *New Techniques and Ideas in Quantum Measurement Theory*, edited by Greenberger, D. M. (New York Academy of Science, New York, 1986); see also Zurek, W. M., Habib, S., and Paz, J. P., *Phys. Rev. Lett.* **70**, 1187 (1993).

2. Schrödinger, E., *Naturwissenschaften* **23**, 807 (1935); **23**, 823 (1935); **23**, 844 (1935). English translation by Trimmer, J. D.,*Proc. Am. Phys. Soc.* **124**, 3235 (1980).
3. Von Neumann, J., *Die Mathematische Grundlagen der Quantenmechanik* (Springer-Verlag, Berlin, 1932); english translation by Beyer, R. T.: *Mathematical Foundations of Quantum Mechanics*, (Princeton University Press, Princeton, NJ, 1955)
4. Wigner, E., in *The Scientist Speculates*, edited by Good, I. J. (William Heinemann, London, 1962), p. 284, and also in *Symmetries and Reflections* (Indiana University Press, Bloomington, 1967), p. 171; Ludwig G. in *Werner Heisenberg und die Physik unserer Zeit* (Braunschweig, Friedrich Vieweg und Sohn, 1961). See also Wigner, E.,*Am. J. of Phys.* **31**, No. 1 (1963).
5. *Quantum Theory and Measurement*, edited by Wheeler, J. A., and Zurek, W. H. (Princeton Univ. Press, Princeton, 1983); Zurek, W. H., *Phys. Today* **44**, No. 10, 36 (1991); Omnès, R., *The Interpretation of Quantum Mechanics* (Princeton University Press, Princeton, NJ, 1994); Giulini, D., Joos, E., Kiefer, C., Kupsch, J., Stamatescu, I.-O., and Zeh, H. D., *Decoherence and the Appearance of a Classical World in Quantum Theory* (Springer, Berlin, 1996).
6. Hepp, K., *Helv. Phys. Acta* **45**, 237 (1972); Bell, J. S., *Helv. Phys. Acta* **48**, 93 (1975).
7. Gottfried, K., *Quantum Mechanics* (Benjamim, Reading, MA ,1966), Sec. IV.
8. Zeh, H. D., *Found. Phys.* **1**, 69 (1970); Zurek, W. H., *Phys. Rev. D* **24**, 1516 (1981); **26**, 1862 (1982); Unruh, W. G., and Zurek, W. H., *Phys. Rev. D* **40**, 1071 (1989); Zurek, W. H., *Phys. Today* **44**, No. 10, 36 (1991); Hu, B. L., Paz, J. P., and Zhang, Y., *Phys. Rev. D* **45**, 2843 (1992).
9. Dekker, H., *Phys. Rev. A* **16**, 2116 (1977); Caldeira, A. O., and Leggett, A. J.,*Physica* (Amsterdam) **121A**, 587 (1983); *Phys. Rev. A* **31**, 1059 (1985).
10. Joos, E., and Zeh, H. D.,*Z. Phys. B* **59**, 223 (1985); Milburn, G. J., and Holmes, C. A.,*Phys. Rev. Lett.* **56**, 2237 (1986); Haake, F., and Walls, D.,*Phys. Rev. A* **36**, 730 (1987).
11. Haroche, S., "*Cavity Quantum Electrodynamics*", in *Fundamental Systems in Quantum Optics*, Dalibard, J., Raimond,J.M., and Zinn-Justin, J., eds., *Proc. Les Houches Summer School, Session LIII* (Elsevier, Amsterdam, 1992); see also *Cavity Quantum Electrodynamics*, edited by Berman, P., (Academic Press, New York, 1994).
12. Glauber, R. J., *Phys. Rev.* **131**, 2766-2788, 1963.
13. Pauli, W. ,*Die allgemeinen Prinzipien des Wellenmechanik*, in *Handbuch der Physik*, ed. Geiger, H., and Scheel, K. (Springer, Berlin, 1933). English translation: Pauli, W., *General Principles of Quantum Mechanics* (Springer, Berlin, 1980).
14. Leonhardt, U., *Measuring the Quantum State of Light* (Cambridge University Press, 1997).
15. Loudon, R., and Knight, P., *J. Mod. Opt.* **34**, 709 (1987).
16. Yuen, H. P., and Shapiro, J. H.,*IEEE Trans. IT* **26**, 78 (1980); Caves, C. M., *Phys. Rev. D* **23**, 1693 (1981); Schumaker, B. L., *Opt. Lett.* **9**, 189 (1984).
17. Bertrand, J., and Bertrand, P., *Found. Phys.* **17**, 397 (1987).
18. Radon, J., *Berichte über die Verhandlungen der Königlich-Sächsischen Gesellschaft der Wissenschaften zu Leipzig, Mathematisch-Physiche Klass* **69**, 262 (1917).
19. Wigner, E., *Phys. Rev.* **40**, 749 (1932).

20. Cahill, K. E., and Glauber, R. J., *Phys. Rev.* **177**, 1857 (1969); *ibid* **177**, 1882 (1969).
21. Hillery, M., O'Connell, R.F., Scully, M.O., and Wigner, E. P., *Phys. Rep.* **106**, 121 (1984).
22. Hudson,R. L., *Rep. Math. Phys.* **6**, 249 (1974).
23. Moyal, J. E., *Proc. Cambridge Phil. Soc.* **45**, 99 (1949).
24. See, for instance, Nussenzveig, H. M., *Introduction to Quantum Optics* (Gordon and Breach, N.Y., 1973); Gardiner, C. W., *Quantum Noise* (Springer, Berlin, 1991).
25. Vogel, K., and Risken, H., *Phys. Rev. A* **40**, 2847 (1989).
26. Smithey, D. T., Beck, M., Raymer, M. G., and Faridani, A., *Phys. Rev. Lett.* **70**, 1244 (1993).
27. Breitenbach, G., Müller, T., Pereira, S. F., Poizat, J.-Ph., Schiller, S., and Mlynek, J., *J. Opt. Soc. Am. B* **12**, 2304 (1995).
28. Dunn, T. J., Walmsley, I. A., and Mukamel, S. *Phys. Rev. Lett.* **74**, 884 (1995).
29. Leibfried, D., Meekhof, D. M., King, B. E., Monroe, C., Itano, W. M., and Wineland, D. J., *Phys. Rev. Lett.* **77**, 4281 (1996); see also *Physics Today* **51**, No. 4, 22 (1998).
30. Brune, M., Haroche, S., Raimond, J.M., Davidovich, L., and Zagury, N., *Phys. Rev. A* **45**, 5193 (1992).
31. Jaynes, E. T., and Cummings, F. W., *Proc. IEEE* **51**, 89 (1963).
32. Feynman, R. P., Vernon Jr., F. L., and Hellwarth,R. W., *J. Appl. Phys.* **28**, 49 (1957).
33. Grynberg, G., Dupont-Roc,J., and Cohen-Tannoudji, C., *Atom-Photon Interactions: Basic Processes and Applications* (Wiley, N.Y., 1992).
34. Hulet, R. J., and Kleppner, D., *Phys. Rev. Lett.* **51**, 1430 (1983); Nussenzveig, A., Hare, J., Steinberg, A. M., Moi, L., Gross, M., and Haroche, S., *Euro. Phys. Lett.* **14**, 755 (1991).
35. Brune, M., Nussenzveig, P., Schmidt-Kaler, F., Bernardot, F., Maali, A., Raimond, J. M., and Haroche, S., *Phys. Rev. Lett.* **76**, 1800 (1996).
36. Ramsey, N. F., *Molecular Beams* (Oxford Univ. Press, N. Y., 1985).
37. Einstein, A., Podolski, B., and Rosen, N., *Phys. Rev.* **47**, 777 (1935).
38. Bell, J. S., *Physics* (Long Island City, N.Y.) **1**, 195 (1964).
39. Freedman, S. J., and Clauser, J. S., *Phys. Rev. Lett.* **28**, 938 (1972); Aspect, A., Dalibard, J., and Roger, G., *Phys. Rev. Lett.* **49**, 1804 (1982).
40. Brune, M., Hagley, E., Dreyer, J., Maître, X., Maali, A., Wunderlich, C., Raimond, J. M., and Haroche, S., *Phys. Rev. Lett.* **77**, 4887 (1996).
41. Davidovich, L., and Kenkre,V. M., to be published.
42. Lutterbach, L. G., and Davidovich, L., *Phys. Rev. Lett.* **78**, 2547 (1997); *ibid Optics Express* **3**, 147 (1998).
43. Davidovich, L., Orszag, M., and Zagury, N., *Phys. Rev. A* **57**, 2544 (1998).
44. Davidovich, L., Maali, A., Brune,M., Raimond, J. M., and Haroche, S., *Phys. Rev. Lett.* **71**, 2360 (1993); Davidovich, L., Brune, M., Raimond, J. M., and Haroche, S., *Phys. Rev. A* **53**, 1295 (1996).
45. Habib, S., Shizume, K., and Zurek, W. H.,*Phys. Rev. Lett.* **80**, 4361 (1998).
46. Gell-Mann, M., and Hartle, J. B., "Quantum mechanics in the light of quantum cosmology", in *Complexity, Entropy, and the Physics of Information*, edited by Zurek, W. H., p. 425 (Addison-Wesley, Reading, 1990).

Cavity Quantum Electrodynamics: a review of Rydberg atom-microwave experiments on entanglement and decoherence

Serge Haroche

Laboratoire Kastler Brossel,
Département de Physique de l'Ecole Normale Supérieure,
24 rue Lhomond, 75231, Paris Cedex 05, France

I INTRODUCTION

Cavity Quantum electrodynamics (CQED) studies the properties of single atoms interacting with photons in cavities [1–3]. Two domains of CQED can be distinguished. In the "low-Q cavity regime", the photons emitted by the atoms are very rapidly and irreversibly dissipated in the cavity walls. As it has been remarked early on by E.Purcell [4], and later on by D.Kleppner [5], the presence of these walls changes the field distribution around the atom and alters the atom's spontaneous emission rates. Atomic frequencies are also modified by the coupling to the cavity walls [2]. Many experiments demonstrating spontaneous emission enhancement and inhibition have been realized in the 1970's and 1980's on various atomic and molecular systems [6]. These low-Q regime experiments have been extended more recently to the study of the radiative properties of excitonic systems in microcavity solid state structures [7].

In the "high Q cavity regime ", to which this course is restricted, the photons are stored in the cavity long enough to interact coherently with the atom, leading to a reversible atom–field evolution. The photon and atomic decay times are in this regime much longer than the atom–cavity coupling time. Under these conditions, the coherent atom–field coupling produces interesting new quantum effects [1]. We will describe various experiments in which atoms interact one by one with a cavity either empty (vacuum field) or containing a few photons. The atoms couple coherently to this quantum field before escaping from the cavity and being eventually detected.

A single atom crossing an initially empty cavity resonant with the atomic transition undergoes a coherent Rabi evolution, emitting and reabsorbing periodically a single photon. If the field is initially in a superposition of photon number states, the Rabi oscillation presents several components of different frequencies, one for each photon number in the cavity. The frequencies of these components scale as the square roots of successive integers. The Fourier analyzis of the Rabi oscillation signal makes it possible to reconstruct the photon number distribution in the cavity and provides a direct and striking illustration of field quantization in a box [8].

The atom–field coupling generally leads to a quantum correlation between the atom's energy and the field photon number and creates quantum states of the atom–field system quite similar to those described by Einstein, Podolsky and Rosen (EPR) in their famous 1935 paper [9]. The correlation between the atom and the field survives after the atom and the cavity have separated, leading to a situation known as entanglement. The wave function of the atom–cavity system is then a non–local object, extending over macroscopic distances. Any detection performed on the atom outside the cavity has an immediate effect on the state of the field inside the cavity. Moreover, this entanglement can be swapped between the field and the state of successive atoms crossing the cavity one after the other. Interatomic entanglement can then be created and studied [10].

If the atom and the cavity field have slightly different frequencies, their coupling becomes dispersive and can be described in terms of a refractive index [1]. Both the atomic transition and the cavity mode frequencies are shifted by the non–resonant interaction. The shifts are dependent upon the atom's state and the field photon number. If the atom or the field are prepared in a superposition of states, the dispersive effect leads again to entanglement between the atom and field variables.

The notion of entanglement is a basic one in quantum physics. It is at the heart of the theory of measurement [11]. The first stage of a measurement process always implies the entanglement of the microscopic system to be measured to another, presumably larger system, which, in turn is coupled to a still larger system, leading finally to the truly macroscopic scale of the "meter" in direct contact with the observer. In a CQED experiment, one can see the entanglement between the atom and the photon field as the first step in this measuring chain. One can say, for example, that the atomic states are microscopic meters allowing us to read out the photon number in the cavity; or, conversely, that the field in the cavity is a small pointer allowing us to measure the energy of the atom.

Analyzing CQED experiments in this way, one gains a very interesting insight into the basic process of a quantum measurement. One can for instance design ideal quantum non demolition (QND) measuring devices which detect the field photon number without changing it [12]. One can also prepare fields containing several photons in linear superpositions of states corresponding to different phases, realizing what are called Schrödinger cat states [13]. These states are "mesoscopic" in the sense that they are intermediate between the truly microscopic single photon states and the macroscopic classical fields with very large photon numbers.

The Schrödinger cat situation results from the dispersive entanglement between the coherent states of the field and the energy levels of an atom crossing the cavity. This situation models the first step in a measurement of the atom's energy in which the field plays the role of a mesoscopic meter. Our experiment has allowed us to observe how the meter state superposition evolves into a mere statistical mixture under the irreversible effect of the system's coupling to its environment [13,14]. This is the "decoherence" phenomenon [15] which quite generally explains why, in a measurement process, one never observes the meter in a linear superposition of position states.

The experiments performed at Ecole Normale Supérieure involve circular Rydberg atoms [16] and microwave superconducting cavities. These ingredients are ideal to achieve the high atom–field coupling required, and to minimize the causes of relaxation which tend to destroy the coherence in the system. Their coupling realizes with a very good approximation the ideal situation of a two–level atom coupled to a single field mode, the quintessential model of the quantum theory of radiation [17]. We start (section II) by describing the atom and the cavity systems, and we recall (section III) the main properties of the two–level atom–single field mode system. We then present a brief description of our very versatile Cavity QED set–up (section IV), before analyzing the basic quantum Rabi oscillation experiment which illustrates the main features of the atom–field coupling in the resonant case (section V). We then present the resonant (section VI), dispersive (section VII) and combined resonant–dispersive (section VIII) methods to realize entanglement with this system. The last sections (section IX and X) are devoted to the description of the Schrödinger cat and decoherence experiments. A short review of some perspectives for future work concludes the chapter (section XI).

II THE ATOM–CAVITY SYSTEM

Circular Rydberg atoms interacting with very high Q superconducting millimeter wave cavities offer ideal conditions to study matter–field coupling. The circular Rydberg atoms [16] combine a large principal quantum number ($N = 51$ and 50 in the experiments described here), with maximum orbital and magnetic quantum numbers ($\ell = |m| = N - 1$). The electron is confined along the classical circular orbit of radius $a_0 N^2$, where a_0 is the Bohr radius. The lifetime of these levels is extremely long (30 ms for $N = 50$). The atom coupling to millimeter wave radiation on a transition between neighboring circular states is very large. The dipole matrix element is $d = a_0 N^2 / 2 = 1250$ atomic units for the transition at $\omega_0/2\pi = 51.099$ Ghz between the $N = 51$ and $N = 50$ circular Rydberg states of Rubidium. We will call e and g respectively these two states in the following.

Circular atomic levels are efficiently produced from laser prepared low angular momentum Rydberg states by multiphoton radiofrequency transitions. The process is a time resolved adiabatic rapid passage involving Stark sublevels of the Rydberg state manifold [16,18]. The circular levels are efficiently and selectively detected by field ionization. Note also that the circular states are stable only in a weak applied electric or magnetic field which provides a quantization axis and maintains the electron orbit in a plane.

Our cavity is a Fabry–Perot resonator made of two spherical mirrors facing each other. The cavity sustains a low order longitudinal mode with a transverse Gaussian profile. The waist of the mode is 6 mm and there are nine antinode of the field between the mirrors separated by 2.7 cm. Careful preparation and polishing of the mirrors yields very high Q values, in the range 10^8 to 10^9. Most experiments described in this chapter correspond to Q in the 10^8 range, and to photon damping times $T_{cav} = Q/\omega$ of the order of 100 to 200 microseconds. Recently, about ten time larger T_{cav} values, of the order of a few millisecond, have been obtained, opening promising perspective for new generations of experiments.

Both the atom and field subsystems have a lifetime much longer than their characteristic interaction time t (about 20 microseconds for an atom crossing the cavity mode waist at a thermal velocity of the order of 300 m/s). This condition is essential for the preparation of non–local entanglement between atom and field variables. The cavity walls are cooled to 0.6 K to suppress thermal field effects (there is less than 0.05 blackbody photon in the mode) and to optimize the intrinsic superconducting mirror reflectivity.

The effective cavity mode volume V is small (about 0.7 cm^3), corresponding to a high photon confinement. The field of a single photon in the mode is $E_0 = \sqrt{\hbar\omega/2\varepsilon_0 V}$. It reaches the value of 1.6 10^{-3} V/m. This results in a very high atom–field coupling, measured by the single photon Rabi nutation period at cavity center $\Omega_0 = 2dE_0/\hbar = 2\pi \times 50$ kHz. The dephasing produced at resonance by the atom–field interaction, $\Omega_0 t$, can reach very large values of the order of several π.

III THE TWO–LEVEL ATOM–SINGLE FIELD MODE MODEL

The model of a two–level atom coupled to a single mode of the electromagnetic field has been introduced in the early days of quantum optics. It describes the atom as a spin 1/2 like particle, interacting with a harmonic oscillator representing the field mode. The evolution of this simple model system is ruled by the so–called Jaynes Cummings hamiltonian [17]:

$$H = \frac{\hbar\omega_0}{2}(|e\rangle\langle e| - |g\rangle\langle g|) + \hbar\omega(a^\dagger a + \frac{1}{2}) - \frac{\hbar\Omega}{2}(a|e\rangle\langle g| + a^\dagger|g\rangle\langle e|) \qquad (1)$$

in which the three terms represent respectively the atomic hamiltonian, the field one (a and $a^\dagger$ being the photon annihilation and creation operators) and the atom–field coupling. The two–level atom and field angular frequencies, denoted by ω_0 and ω are equal at exact resonance and differ by a quantity $\delta = \omega - \omega_0$ in off–resonant situations. The atom–field coupling describes the resonant or nearly resonant elementary processes of absorption or emission of photons by the atom undergoing transitions between levels e and g. The hamiltonian described by Equation (1) is obtained in the rotating wave approximation, which amounts to neglecting far off–resonant processes in which the atom and the field simultaneously loose or gain energy. The atom field coupling, whose magnitude is measured by the Rabi frequency Ω, is in fact a quantity depending upon the position of the atom in the field mode. The Rabi frequency takes the maximum value Ω_0 at cavity center and vanishes when the atom is outside the cavity. We describe here classically the motion of the atom across the cavity (the atom's position being a time dependent parameter). This is justified because the de Broglie wavelength of the atom moving at thermal velocities is very small compared to all the relevant lengths in the problem (microwave wavelength notably).

The hamiltonian (1) has been studied in details in very many articles and text books. Its eigenstates are the "dressed states" of the atom–field mode system [19,20]. We only recall here the main properties of these states. The non–degenerate ground state of the atom–cavity system is the state $|g,0\rangle$ representing the atom in the lower circular state g in an empty cavity (vacuum field). Obviously, this state is not sensitive, within the rotating wave approximation, to the atom–field coupling. All the other dressed states are organized in doublets, separated in energy by one field quantum. The $(n+1)$-th doublet consists of the levels $|+,n\rangle$ and $|-,n\rangle$, linear combinations of the uncoupled states $|e,n\rangle$ and $|g,n+1\rangle$, representing e with a cavity containing n photons and g with a cavity containing $n+1$ photons.

Note that all the other atomic states (non-circular states of the $N=50$ and 51 manifolds as well as states belonging to other manifolds) are, to a very good approximation, not coupled to the states e and g by the resonant or slightly off- resonant atom–cavity coupling. Similarly, all the other modes sustained by the cavity can be disregarded. Our experimental system thus constitutes a very good realization of the Jaynes–Cummings model.

At exact resonance ($\delta=0$), the dressed states $|+,n\rangle$ and $|-,n\rangle$ are symmetric and antisymmetric superpositions of the uncoupled states. The energy splitting of the doublet, for an atom at cavity center, is $\hbar\Omega_0\sqrt{n+1}$. The corresponding Bohr frequency, $\Omega_0\sqrt{n+1}$, is the Rabi nutation frequency of the two–level atom at cavity center in an n photon field. The $\sqrt{n+1}$ factor appearing in this Rabi frequency is proportional to the root mean square amplitude of the field.

For "large" detunings ($\omega_0 \gg \delta \gg \Omega$), the dressed levels almost coincide with the uncoupled states, with small level contaminations given by second order perturbation theory. The non–resonant interaction then shifts slightly the levels position [1]. At cavity center, the state $|e,n\rangle$ is displaced by the amount:

$$\Delta_{e,n} = -\hbar\Omega_0^2 \frac{n+1}{4\delta} \tag{2}$$

and the level $|g,n\rangle$ by:

$$\Delta_{g,n} = \hbar\Omega_0^2 \frac{n}{4\delta}. \tag{3}$$

These shifts are the sum of light shifts terms (proportional to n) and Lamb shift contributions (term remaining for $n=0$). They are also related, as we show in section IX below, to a frequency shift of the mode which can be interpreted as a refractive index effect produced on the cavity field by a single atom.

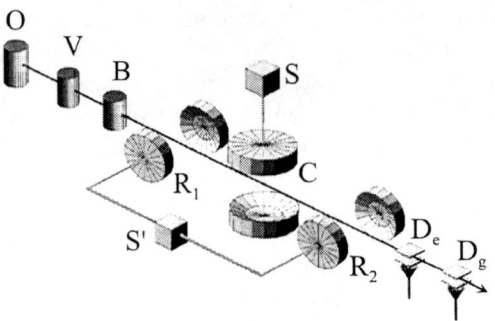

FIGURE 1. Scheme of the experimental set–up.

We have described so far the case of a motionless atom. In an actual experiment, the atoms fly at a few hundred m/s across the cavity mode. The atom–field coupling then varies with time, reflecting the variation of the field mode amplitude along the atom's trajectory. Taking into account the Gaussian mode structure, we have $\Omega(z) = \Omega_0 \exp(-z^2/w^2)$ where z is the atom's position ($z = 0$ at cavity center) and w the mode's waist. The dressed states and their energies are thus also position dependent. A force acting on the atom results from these position dependent energy shifts. Although it might have very interesting manifestations for very slow atoms [21,22], this force is very small and thermal atoms cross the cavity along a line at an almost constant velocity. Cavity relaxation introduces a coupling between different dressed states manifolds. Energy decreasing transitions, corresponding to photon absorption in the mirrors and energy increasing ones, corresponding to the creation of thermal photons, are easily described by a master equation formalism [2].

IV THE EXPERIMENTAL SET–UP

We now turn to a brief description of the experimental set–up schematized in Fig. 1. The atoms, effusing from an oven O, are velocity selected by optical pumping in zone V (typical velocity of the order of 400 m/s). The velocity selection makes use of the Doppler effect. The atoms are first removed from one of the hyperfine levels of the Rubidium ground state by a laser exciting the atomic beam at right angle, which pumps all the atoms in the other hyperfine state. Then, a second laser beam, making a 58^0 angle with the atom propagation direction, is used to selectively repopulate the previously emptied hyperfine level with atoms having only one velocity. By tuning the frequency of this second laser, the atom's velocity can be adjusted at will. The velocity selected atoms are then prepared in box B by a pulsed combined laser and radiofrequency excitation into either one of the two circular Rydberg states e or g. The atoms then cross the cavity C midway between the two mirrors.

The cavity field is excited either by the atoms themselves, or by a classical source of radiation S, coupled into C by a waveguide. This field is either resonant or slightly off–resonant with the transition between the states e and g. The resonance condition can be modified at well defined times, by applying a pulse of electric field on the mirrors, making use of the Stark effect which tunes the atomic transition in or out of resonance with the cavity mode.

The atom–cavity interaction lasts for an adjustable time t between 10 and 100 μs. The circular atoms then drift out of the cavity and are counted by two high efficiency field ionization detectors D_e and D_g sensitive to atoms in levels e and g respectively. It is possible to coherently mix levels e and g with the help of classical microwave pulses of radiation before they interact with C (radiation applied in zone R_1) and after this interaction, just before the detection (zone R_2). These auxiliary microwave fields (frequency ν) are generated by the source S'. The whole set–up is placed at the bottom of a He^3-He^4 cryostat at 0.6 K. The atomic excitation is reduced to a level such that at most one atom is prepared in a single pulse of Rydberg state excitation. Sequences of atoms with variable and adjustable delay can be sent through the set–up.

Since the velocity of each atom is known as well as its time of preparation, its position at each moment is perfectly controlled. It is thus possible to manipulate the atoms individually, by applying convenient sequences of pulses in R_1 and R_2 and by Stark switching the cavity C in and out of resonance when each atom crosses it. An experiment consists in repeating a sequence of events and accumulating statistics of atom counts in both detectors. In this way, one atom transition probabilities and two–atom correlation signals have been studied. Extensions of the experiments to higher order atom correlations are being planned.

V RESONANT QUANTUM RABI OSCILLATION

When the cavity mode is tuned in exact resonance with the $e \rightarrow g$ atomic transition, photons can be emitted or absorbed by each atom while it crosses C. The exchange of energy between the atom and the field exhibits an oscillatory behavior, the Quantum Rabi oscillation [8]. This phenomenon reveals directly the quantization of the field in the cavity and can also be used to generate entanglement between successive atoms crossing the cavity.

Let us send an atom in level e in the cavity and measure, by repeating the experiment many times, the probability P_g that the atom is transferred to level g after crossing C. The auxiliary microwave zones R_1 and R_2 are not used in this experiment. The sequence of measurement is performed for various atom cavity interaction times, which are obtained either by changing the velocity of the detected atoms, or by Stark tuning the atomic transition in resonance with the cavity mode during a fraction of the atom–cavity crossing time (in this experiment, the atom's velocity was not actively selected by optical pumping, as described above, but only passively determined by a time of flight measurement).

Figure 2(a) shows the Rabi oscillation signal obtained when the cavity is empty (save for the very small residual thermal field with an average number of 0.05

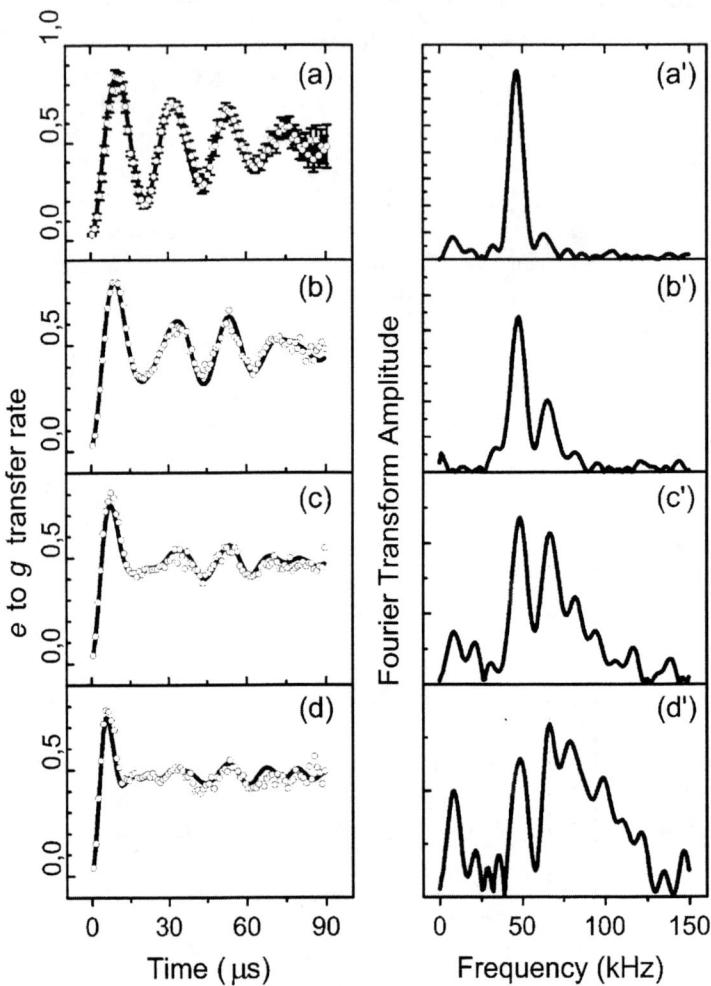

FIGURE 2. The Quantum Rabi oscillation signal. (a), (b), (c) and (d): Probability P_g to find the atom in level g as a function of the effective time of interaction with the cavity field, recorded with no injected field (a) and with a coherent field corresponding to 0.4, 0.85 and 1.77 photons on average (b to d). The points are experimental and the curves are theoretical fits. (a'), (b'), (c'), (d'): corresponding Fourier transforms. Frequencies ranging as the square roots of successive integers are indicated by vertical lines (from Ref. [8]).

photon). Four complete Rabi oscillations are observed, at the expected frequency $\Omega_0/2\pi$ close to 50 kHz. The oscillation corresponds to the coupling of the $|e, 0\rangle$ and $|g, 1\rangle$ states of the "atom + field" system and describes the reversible evolution of the atom between e and g, correlated to the emission and absorption of one photon in C. It can be interpreted in the coupled oscillator model of the Thomson theory of atom–matter interaction, as resulting from the resonant coupling of the atom and field "oscillators" which reversibly exchange their energy.

Note that the variable represented on the horizontal axis of Fig. 2 is an effective atom–cavity interaction time. In fact, the Rabi precession angle is obtained by integrating over the atom's trajectory the spatial variation of the atom–field coupling. It is equivalent to express this integral by the product of Ω_0, the Rabi frequency at cavity center, with an effective atom cavity interaction time t. The damping of the oscillation is due to the inhomogeneity of the Rabi frequency across the atomic beam diameter as well as to various decoherence processes, including atom and cavity damping and Rydberg atom collisions with background gas.

Figures 2(b) to 2(d) show the Rabi oscillation signal when the cavity contains a coherent field with a small average photon number increasing from top to bottom. The signal becomes then a non–sinusoidal superposition of several frequency components beating with each other, which correspond to the various photon numbers present in the coherent field. The beating between these components gives rise to collapse and revival of the oscillations, which have been predicted in [23]. The Fourier transforms of the signals, shown in Fig 2(a') to 2(d'), exhibit peaks at the frequencies $\Omega_0\sqrt{n+1}$ corresponding to the Rabi frequency in the field of n photons ($n = 0$ to 3). From the relative heights of these peaks one can reconstruct the photon number distribution in the coherent field. (Note that the peak around zero frequency is an artefact due to our non perfect data acquisition procedure). The Fourier transform signals demonstrate clearly that the Rabi frequency, classically proportional to the field amplitude, is in fact a discrete quantity providing a visceral and direct evidence of field quantization in a cavity.

Let us note finally that we restrict in this section our attention to a situation where atoms interact one by one with a pre–existing cavity field. Interesting effects do also occur when atoms interact successively with the field they contribute to build up in the cavity. One thus realizes micromasers [24,25] in which the quantum Rabi oscillation plays a very important role. Reviews of micromaser experiments can be found elsewhere [26].

VI ENGINEERING ENTANGLEMENT BY RESONANT ATOM–FIELD COUPLING

Let us assume again that C is exactly resonant with the $e \to g$ transition and that there is now no photon in C at the beginning of each run. An atom is prepared in level e before entering C, the auxiliary resonators R_1 and R_2 being left inactive for the time being. The quantum Rabi oscillation produces atom–field entanglement by the simple process which coherently mixes the atom–field states $|e,0\rangle$ and $|g,1\rangle$. The state of the atom–field combined system evolves into:

$$|\Psi\rangle = \cos(\Omega_0 t/2)|e,0\rangle + i\sin(\Omega_0 t/2)|g,1\rangle \tag{4}$$

where t is again the effective atom–cavity interaction time. In general, the system's wave function, when the atom emerges form C, appears to be an entangled combination of atom and field states. The weights of the superposition can be adjusted by modifying t, i.e. the atom's velocity.

Let us assume now that a second atom, prepared in level g, is sent across C after the first one has left the cavity. The interaction time t' of this second atom with C is set so that $\Omega_0 t' = \pi$. This atom will in turn undergo a Rabi oscillation in the field left by the first atom, which remains in the process a mere spectator. The π-pulse condition means that the second atom has a unit probability to absorb the photon if there is one in C, while nothing happens if C is empty. As a result, the quantum state of the system made of the two atoms and the cavity field evolves into:

$$|\Psi'\rangle = \cos(\Omega_0 t/2)|e,g,0\rangle - \sin(\Omega_0 t/2)|g,e,0\rangle \tag{5}$$

(the first and second symbol in the kets correspond to first and second atom respectively). The system is then prepared in an entangled state of the two atoms, the field factorizing out in the vacuum state. In the process, the cavity has played the role of a catalyst, the field starting and ending in the same state. Note that the entanglement, first occurring between the atom and the field, has been "swapped" from the cavity field into the second atom. The two subsystems which get finally correlated have never interacted together directly. A different kind of entanglement swapping has been recently demonstrated [27]. The weights of the components in the entangled superposition of Equation (5) can be adjusted. If t is chosen so that $\Omega_0 t = \pi/2$, $|\Psi'\rangle$ is the tensor product of the field vacuum by the maximally entangled atom pair:

$$|\Psi_{12}\rangle = \frac{1}{\sqrt{2}}(|e,g\rangle - |g,e\rangle) \tag{6}$$

analogous to the entangled two–spin system described in the Einstein–Podolsky–Rosen problem [9].

In most experimental realizations of the EPR situation so far, the pair of particles are made of photons produced in a spontaneous cascade or down–conversion process [28]. At variance, we are dealing here with pairs of massive particles prepared in a deterministic fashion in an entangled state.

The two–atom system is an exact analog of a pair of entangled spins. One can assimilate each atom to a spin one–half particle, the e and g states corresponding to the $+1/2$ and $-1/2$ states quantized along an arbitrary direction Oz. The entangled state described by Equation (6) is the rotationally invariant "spin zero" state of the combined system. Such a state can equivalently be expressed in any basis corresponding to another quantization direction. If the new axis is taken in the xOy plane instead of Oz, in a direction making an angle ϕ with Ox, the new spin eigenvectors are of the form $|e\rangle \pm e^{i\phi}|g\rangle$ and the same entangled pair can be written (within an irrelevant overall phase factor) as:

$$|\Psi_{EPR}\rangle = \frac{1}{\sqrt{2}}\left[(|e_1\rangle + e^{i\phi}|g_1\rangle)(|e_2\rangle - e^{i\phi}|g_2\rangle) - (|e_1\rangle - e^{i\phi}|g_1\rangle)(|e_2\rangle + e^{i\phi}|g_2\rangle)\right] \quad (7)$$

where the indices 1 and 2 label the two atoms. Equations (6) and (7) describe a perfect anticorrelation between the states of the two atoms, whichever basis is chosen for the detectors. This anticorrelation cannot be explained by classical arguments, which is the essence of the EPR paradox.

We have performed two-atom counting experiments to demonstrate these non-classical correlations [10]. To analyze the correlations in the energy basis, we directly detect the state of the atoms after they leave C, without using the auxiliary cavities R_1 and R_2. Ideally, we should find that the joint probabilities to detect the atoms in any combination of e and g levels are $P_{eg} = P_{ge} = 1/2$, $P_{ee} = P_{gg} = 0$. We find instead experimentally $P_{eg} = 0.44$, $P_{ge} = 0.27$, $P_{ee} = 0.06$, $P_{gg} = 0.23$. The difference with the ideal case is due to the photon decay in the cavity between the two atoms as well as to imperfections in the Rabi pulses. A quantitative analyzis of the experiment shows that we prepare in this experiment EPR pairs with a 63% purity [10].

In order to detect the "transverse" entanglement described by Equation (7), we subject both atoms, after they have interacted with C and before detecting their energy, to a $\pi/2$ pulse in R_2. Note that R_1 is not used in this experiment. The succession of R_2, followed by D_e and D_g amounts to a detector of the atomic coherent superpositions $|e\rangle \pm e^{i\phi}|g\rangle$, where ϕ is related to the phase of the pulses applied in R_2. If both atoms were crossing R_2 simultaneously, one would again expect, according to Equation (7), a perfect anticorrelation between the e and g detectors. In fact, the detection of the first atom in e immediately "projects" the second atom into the anticorrelated state corresponding to an instantaneous detection in g. The second atom coherence precesses however during the time interval between the two detection events. Depending upon the relative phase accumulation between the atomic coherence and the microwave pulses in R_2, the probability of detecting the second atom in e varies between 0 and 1. This modulation reveals the non-local correlations between the two atoms.

This oscillation, shown in Fig. 3, is observed in the conditional probability P_{ee} of detecting the second atom in e, knowing that the first was found in e (solid line), and also in the conditional probability P_{ge} of finding the second atom in e knowing that the first was detected in g (dashed lines). The two modulations are in phase opposition. Detecting the first atom in g sets indeed the second atom coherence with an initial phase opposite to the one obtained if the first atom was detected in e. Note that the fringe contrast is only 25%, far from the ideally expected value

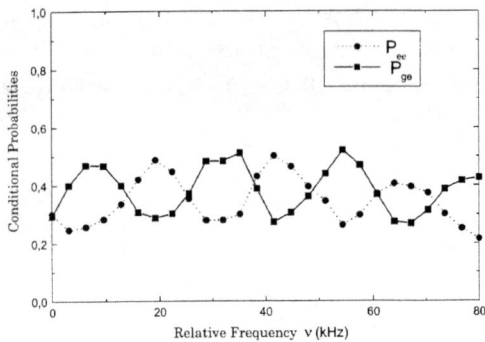

FIGURE 3. Observation of EPR atomic entanglement: conditional probabilities P_{ee} (circles) and P_{ge} (squares) of measuring the second atom in level e when the first has been found in e or g respectively, plotted versus the frequency ν of the pulses in R_2. The lines connecting the experimental points are for visual aid (from Ref. [10]).

of 100to photon decay in C between the two atoms, as well to imperfections in the various Rabi pulses.

This experiment is to our knowledge the first one in which atoms have been entangled at a distance (the maximum separation in this experiment was 1.5 cm). Note that proposals to perform entanglement between mercury atoms and to investigate Bell's inequalities with these atom pairs have been made [29]. By improving on our CQED entanglement experiment, it should also be possible to perform a Bell's inequality test with Rydberg atoms.

VII ENGINEERING ENTANGLEMENT BY DISPERSIVE ATOM–CAVITY COUPLING

When the atomic transition and the cavity field are slightly off–resonant (detuning δ larger than Ω), photon emission and absorption are forbidden by energy conservation. The atom–field quantum states undergo however the small dispersive energy shifts expressed by Equations (2) and (3). The transition frequency between e and g in the presence of n photons in C is thus displaced, at cavity center, by $-(\Omega^2/4\delta)(2n+1)$. This is the well known light shift, essentially proportional to the field intensity, which we consider here at the limit of very small fields, down to the vacuum. For $n = 0$, the remaining shift, equal to $-\Omega^2/4\delta$, can be considered as the Lambshift produced on the atomic transition by the vacuum field in the cavity mode [30].

In order to observe these shifts, we have measured the $e \to g$ transition frequency using Ramsey interferometry [31]. We have subjected each atom crossing the apparatus and initially prepared in level e to a sequence of two $\pi/2$ pulses of classical microwave applied in R_1 and R_2. We have then measured the probability $P_g(\nu)$ of detecting it in g, as a function of the frequency ν of these microwave pulses.

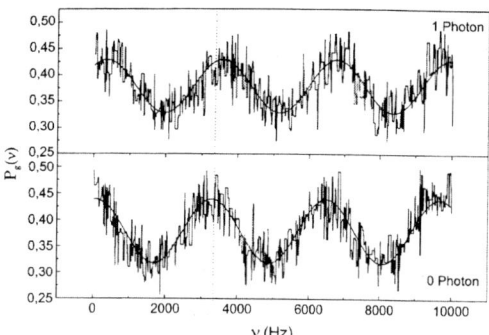

FIGURE 4. Ramsey fringe signal versus the frequency ν applied in R_1 and R_2, exhibiting the light shift produced on the atoms by a small coherent field. Lower Trace: signal with no photon injected in the cavity. Upper trace: signal observed when a field with an average number of 1 photon is injected in C (from Ref. [30]).

The probability versus ν exhibits fringes typical of a quantum interference effect. Each atom can indeed undergo a transition from e to g either in R_1 or in R_2, with equal probabilities. As long as one cannot tell which way the atom has followed, the two corresponding probability amplitude interfere. The relative phase between the two amplitudes is proportional to the frequency difference between the applied pulse and the atomic frequency, thus explaining the modulation of $P_g(\nu)$.

When the atomic transition frequency is shifted by a cavity–induced effect, the fringes are translated. The vacuum shift effect has been actually observed with our atom–cavity set-up, as well as the tiny light shifts produced by coherent fields in C containing on average one photon [30] (see Fig. 4). This experiment was performed on an earlier version of our set-up, with a cavity having a lower Q factor. The magnitude of the atom–cavity dispersive effect was quite large, but the cavity relaxation time too short to use this effect in order to entangle atom and field states.

Consider now theoretically the situation where $(\Omega_0^2/2\delta)t = \pi$, t being the effective duration of the dispersive interaction of the atom with the cavity field. In this case, a change of one photon in C shifts the phase of the fringes by π, turning a maximum of $P_g(\nu)$ equal to 1 into a minimum equal to 0. If ν is set so that $P_g(\nu) = 1$ for $n = 0$, then $P_g(\nu) = 0$ for $n = 1$. The final state of the atom is correlated unambiguously to the photon number in C and the Ramsey interferometer appears as a measuring device for the field. This is a Quantum Non Demolition (QND) device, since the dispersive atom–field coupling does not change the photon number in C [12].

If the field in C is prepared in a superposition $C_0|0\rangle + C_1|1\rangle$ of 0 and 1 photon states, the final state of the atom-field system, after the atom exits the second microwave zone R_2, is:

$$|\Psi\rangle = C_0|g,0\rangle + C_1|e,1\rangle \ . \tag{8}$$

Note the analogy with Equation (4). We have again an entanglement between the atom and the field, but now this entanglement is obtained by a dispersive interaction. In order to realize this situation in the laboratory, we need of course, as in the resonant experiment, a cavity with a damping time much longer than the atom–cavity interaction time t. We are confident that we will soon be able to resume this experiment with a very high Q cavity and to prepare the entangled atom–field state described by Equation (8).

VIII MANY ATOM ENTANGLEMENT BY COMBINED RESONANT AND DISPERSIVE INTERACTIONS

A combination of resonant and dispersive atom–field interaction can be used to entangle more than two atoms. Assume that the first atom resonantly interacts with an initially empty cavity, according to the process described above, preparing an atom–field state of the form given by Equation (4). Before this field has time to decay, the cavity is detuned from the atomic transition and the microwave zones R_1 and R_2 are activated, the Ramsey interferometer being set to provide a π-shift per photon and a maximum in the fringe signal for $n = 0$. A second atom, initially prepared in e, is then sent through the system. If there is 0 photon, this second atom will end up, with unit probability in g and if there is 1 photon in e. Thus, after this second atom exits from R_2, the combined system is in the state:

$$|\Psi''\rangle = \cos(\Omega_0 t/2)|e, g, 0\rangle + i\sin(\Omega_0 t/2)|g, e, 1\rangle \tag{9}$$

which exhibits an entanglement between three subsystems (the two atoms and the cavity field). One can then deactivate the zones R_1 and R_2 and tune the cavity back into resonance. A third atom, initially in g, is then sent across the set–up, with a velocity corresponding to a π pulse condition for the Rabi pulse in C. This atom will then absorb with unit probability the photon, if there is one and end up in level e, whereas it will stay in g if there is no photon. Finally, the state of the three atoms and field will be:

$$|\Psi'''\rangle = \cos(\Omega_0 t/2)|e, g, g, 0\rangle - \sin(\Omega_0 t/2)|g, e, e, 0\rangle \tag{10}$$

We have now three atoms entangled in a state of the kind first described by Greenberger, Horne and Zeilinger (GHZ) [32], with the cavity field factoring out in the vacuum state. We have just described the blue print of an experiment we are preparing at ENS. Slightly different schemes leading to similar GHZ entangled triplet states have been described in earlier publications [33]. Although triplets of entangled photons have been very recently prepared [34], this would be the first time that an ensemble of three massive particles are entangled in a deterministic way. The generalization of this experiment to more than three atoms is straightforward.

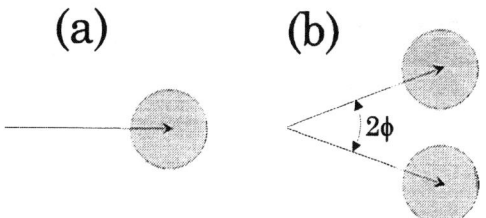

FIGURE 5. (a) Pictorial representation in phase space of a coherent field state. (b) The two components of the field in Equation (12) which are correlated to the two atomic states e and g.

IX DISPERSIVE ENTANGLEMENT BETWEEN AN ATOM AND A MESOSCOPIC FIELD: SCHRÖDINGER CATS

Let us now assume that the atom in level e is injected in a cavity containing a coherent field of complex amplitude α made of a superposition of n-states [35]. The initial state of the system is then $|\Psi_i\rangle = \exp(-|\alpha|^2/2) \sum_n \alpha^n/\sqrt{n!}|e,n\rangle$. The cavity is off–resonant so that each $|e,n\rangle$ state in the superposition is affected by a phase factor $\exp(i(\Omega^2/4\delta)(n+1)t)$ and the system state turns into:

$$|\Psi_f\rangle = \exp(-|\alpha|^2/2) \sum_n \alpha^n \exp(i(\Omega^2/4\delta)(n+1)t)/\sqrt{n!}|e,n\rangle . \qquad (11)$$

We recognize in Equation(11) the expression of a coherent field with an amplitude $\alpha \exp(i\Phi)$ whose phase has been shifted by the angle $\Phi = (\Omega^2/4\delta)t$ from its initial value. (There is also a global phase factor $\exp(i(\Omega^2/4\delta)t)$ multiplying the system's state which can be interpreted as due to the cavity induced Lamb shift. We will reintroduce this phase in the definition of $|e\rangle$ and omit it in the following). The non–resonant atom crossing C plays the role of a microscopic refractive index medium which dephases the cavity field (provided field relaxation is negligible during time t). In the same way, an atom in level g dephases the same field by the opposite angle $-\Phi$.

The quantum phase shift effect can be used to generate quantum superposition of field states with different phases [12]. A single atom is prepared in a linear superposition of states e and g by a $\pi/2$ pulse in R_1, while a coherent field corresponding to a complex amplitude α is injected in C. When the atom crosses C, it imparts to the field two opposite phase kicks, $\pm \Phi$, depending upon whether it is in e or g. As a result, the combined atom field system becomes:

$$|\Psi\rangle = \frac{1}{\sqrt{2}}(|e,\alpha e^{i\Phi}\rangle + |g,\alpha e^{-i\Phi}\rangle) , \qquad (12)$$

which describes an entangled atom–cavity state in which the energy of the atom is correlated to the phase of the field.

A coherent field can be represented as an arrow in phase space whose length and direction are associated to the amplitude and phase of the field (see Fig. 5). The tip of the arrow lies in a circle of unit radius describing the conjugated uncertainties in field amplitude and phase. The length of the arrow is equal to the square root of the average photon number. Equation (12) indicates that this arrow is in fact a meter which assumes two different directions when the atom is in e or g. One can say that the dispersive interaction realizes an essential step in a "measurement" process [11] in which the "field arrow" is used to determine the atom's energy. One can also adopt Schrödinger's metaphor [36] and say that the $+\Phi$ and $-\Phi$ field components are laboratory versions of the "live" and "dead" states of the famous cat trapped in a box with an atom in a linear superposition of its excited and ground states. Since the field in the cavity may contain several photons on average, these superpositions can be considered as "mesoscopic".

After C, the atom undergoes another $\pi/2$ pulse in R_2, phase coherent with the pulse in R_1 and is detected by D_e or D_g. Repeating the experiment many times, we reconstruct the probability of detecting the atom in g, versus the frequency ν applied in R_1 and R_2. The experiment [13] is performed either with an empty cavity detuned by $\delta/2\pi = 712$ kHz (Fig 6(a)), or with a cavity containing initially a coherent field with an average of 9.5 photons, with decreasing values of the detuning δ (from Fig 6(b) to 6(d)). The fringes observed when the cavity is empty are a typical Ramsey signal [31], which can, as we have already discussed, be interpreted as an atomic interference effect. The atom can be transferred from e to g either in R_1 (in which case it crosses C in level g) or in R_2 (it then crosses C in level e). Since the two "paths" cannot be distinguished, the corresponding amplitudes interfere, leading to fringes in the final probability (Fig. 6(a)).

When a coherent field is initially present in C, it gets a phase kick which could allow us to determine in principle the state of the atom when it was in C. Such a measurement, even if it remains virtual, must according to the notion of complementarity destroy the interference effect and wash out the Ramsey fringes (similar complementarity experiments in atomic interferometry have been described by several authors [37]). If δ is relatively large, and Φ accordingly small (Fig 6(b)), the field components overlap so that the "measurement" of the atom's energy remains ambiguous. The potential knowledge of the atomic path is only partial and the fringes remain visible, albeit with a reduced contrast. This contrast decreases further when δ becomes smaller (Fig. 6(c)) and vanishes altogether when δ is so small that the overlap between the field components is negligible (Fig. 6(d)). The vanishing of the fringe contrast demonstrates that a field with non overlapping components has been prepared in C. A quantitative analyzis shows that the fringe signal is fully described by the overlap integral between the two field components, its modulus yielding the fringe contrast and its phase fixing the phase of the Ramsey fringes. From this phase shift (clearly observable in Fig. 6 when δ is changed) we can deduce the average number of photons in C ($\bar{n} = 9.5$ in this experiment).

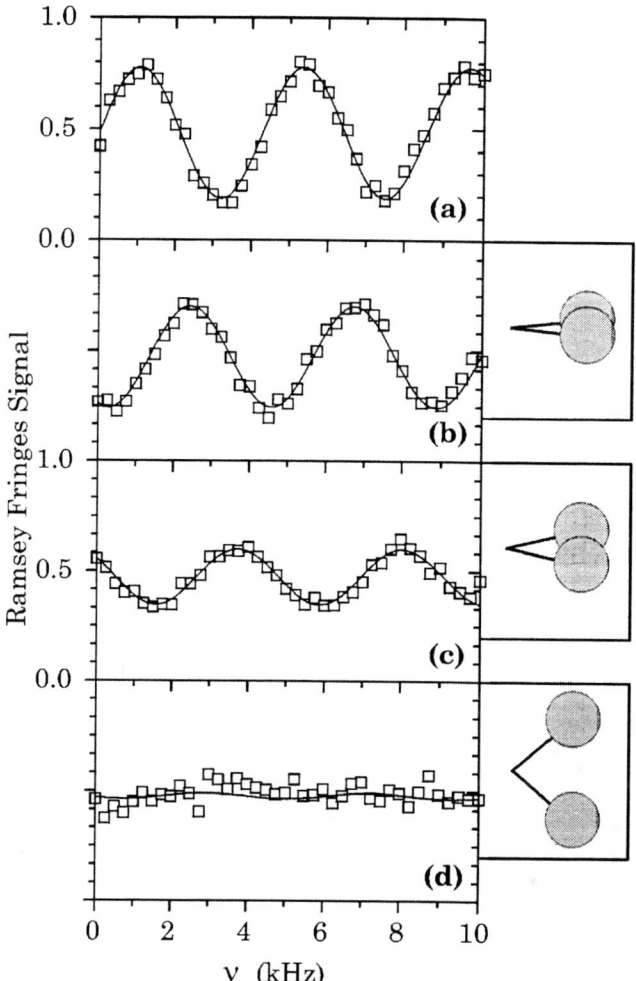

FIGURE 6. The probability of detecting the atom in level g exhibits Ramsey fringes versus ν. (a): C is empty, $\delta/2\pi = 712$ kHz; (b) to (d): C stores a coherent field with $\alpha = \sqrt{9.5} = 3.1$, $\delta/2\pi = 712, 347$ and 104 kHz respectively. Points are experimental and curves are sinusoidal fits. Inserts show for each signal the phase space representation of the field components left in C. The decrease, from top to bottom, of the fringe contrast is a demonstration of complementarity in this atomic interferometry experiment: as the information left in the cavity allows us to determine in principle better and better the "path" of the atom in the interferometer, the fringes progressively disappear (from Ref. [13]).

X DECOHERENCE OF SCHRÖDINGER CAT STATES

Theory predicts that coherent field states superpositions of the kind described by Equation (12) are very fragile and subject to decoherence, when the number of photons, or the angle Φ between the field components become large [15]. In order to check the coherence of the superposition and to study how it gets transformed with time into a mere statistical mixture, we have probed the "cat state" with a second atom, crossing the cavity after a delay [13], according to a scheme first proposed in [38]. The probe has the same velocity as the first atom and produces identical phase shifts. Since it is also prepared into a superposition of e and g, it again splits into two parts each of the field components produced by the first atom.

The final field state exhibits then four components, two of which coincide in phase. Whether the two atoms have crossed C in the e,g combination, or in the g,e one, the net result is indeed to bring back in both cases the phase of the field to itsinitial value. After the atomic states have been mixed again in R_2, there is no way to tell in which state the atoms have crossed C (e,g or g,e combination), since the second atom has partially erased [39] the information left by the first one in the field. As a result, two "paths" associated with the atom pair are undistinguishable. The contributions corresponding to the e,g and g,e paths lead, in the joint probabilities P_{ee}, P_{eg}, P_{ge} and P_{gg}, to the presence of interfering terms. It is convenient to define an atomic correlation signal η by the following combination of joint probabilities:

$$\eta = \frac{P_{ee}}{P_{ee} + P_{eg}} - \frac{P_{ge}}{P_{ge} + P_{gg}} \qquad (13)$$

This correlation signal is directly linked to the quantum interference resulting from the overlap of the components in the final field in C. If the state superposition survives during the time interval between the atoms, η ideally takes the value $1/2$, whereas it vanishes when the state superposition is turned into a statistical mixture. The result of the η measurement versus the time interval T between the two atoms is shown in Fig. 7 for two different "cat" states produced by the first atom in C (these states are depicted in the inserts). The points are experimental and the curves theoretical [14]. The maximum correlation signal is 0.18, and not 0.5 because of the limited fringe contrast of our Ramsey interferometer. We see that decoherence occurs within a time much shorter than the cavity damping time and is more efficient when the separation between the cat components is increased. The agreement between experiment and theory is quite good.

The decoherence process is due to the loss of photons escaping from the cavity via scattering on mirror imperfections. Each escaping photon can be described as a small "Schrödinger kitten" copying in the environment the phase information contained in C [40]. The mere fact that this "leaking" information could be read out to determine the phase of the field is enough to wash out the interference effects related to the quantum coherence of the "cat" state. In this respect, we understand that decoherence is also a complementarity phenomenon. The short decoherence

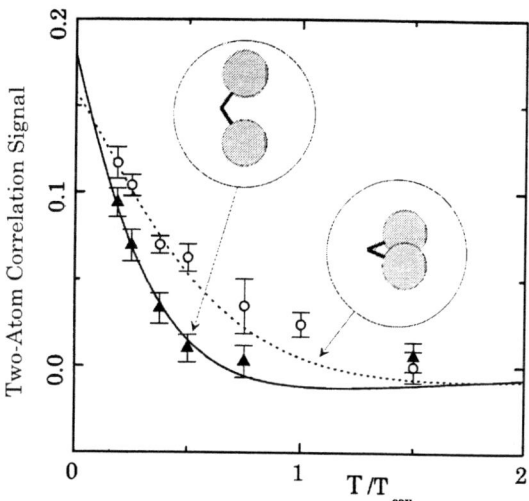

FIGURE 7. Decoherence of a "Schrödinger cat": two-atom correlation signals η versus T/T_{cav} for two different "cat" states corresponding to two values of Φ (Inserts: pictorial representations of the corresponding field components separated by 2Φ). Dashed and solid lines are theoretical. The vanishing of the correlation demonstrates decoherence, which occurs faster when the two components are more separated (from Ref. [13]).

time of our Schrödinger cat, of the order of $T_{cav}/\bar{n}$ where $\bar{n}$ is the average photon number in C, is also explained by this approach. The larger the photon number, the shorter is the time required to leak a single "photon-copy" in the environment. This experiment verifies the basic features of decoherence and clearly exhibits the fragility of quantum coherences in large systems.

XI PERSPECTIVES

We are presently trying to extend the experiments described here to more complex situations, involving more atoms, more photons or more cavities. Particularly interesting and intriguing will be the study of entangled states involving mesoscopic fields with many photons entangled between two spatially separated cavities.

Let us give here a brief description of a possible experiment based on the concept of a quantum switch [41]. A classical microwave source S is coupled to the cavity C. The frequency of S is tuned slightly off resonance, so that no field is injected in C. When an atom (which is resonant neither with C nor with the source) crosses the cavity, its index shifts the mode frequency. The detunings can be arranged so that an atom in e tunes the cavity in resonance with the source while an atom in g leaves the cavity off resonance. If the atom is prepared by R_1 in a superposition of e and g, the cavity interacts with a medium in a superposition of two indices, one of which corresponds to an open switch (photons flow in the cavity tuned into resonance) while the other corresponds to a closed switch (the cavity is off resonance and cannot accept photons). After the interaction, the atom–cavity system is in the combined state

$$|\Psi\rangle = \frac{1}{\sqrt{2}}(|e,\alpha\rangle + |g,0\rangle) \,, \tag{14}$$

describing an atom in e or g, correlated to a field either in a state α containing several photons, or in the vacuum. Consider now that the atom is crossing in turn two identical cavities C_1 and C_2, both coupled by a waveguide to the same source S, which is again slightly off resonance with the cavities. The "switch atom" flies first through C_1, the "atom + C_1" system evolving into the state described by Equation (14). Then, before entering C_2, the atom undergoes in an auxiliary cavity a π pulse which exchanges e and g. If the atom crossed C_1 in e, it will cross C_2 in g and leave this cavity off resonance with the source. If the atom had crossed C_1 in g, it will cross C_2 in e and act as an open switch for the field to flow into C_2. As a result, the combined "atom + C_1+ C_2" system, after the atom has crossed C_2, has become

$$|\Psi\rangle = \frac{1}{\sqrt{2}}(|g,\alpha,0\rangle + |e,0,\alpha\rangle) \,, \tag{15}$$

where the second and third symbols in each ket refers to the state of the fields in C_1 and C_2 respectively. We could in this way prepare a non–local mesoscopic field suspended in a coherent way between two spatially separated cavities.

Our atom–cavity experiments can also be related to the active field of quantum information processing. Two–level atoms can indeed be considered as "qubits" carrying quantum information, on which elementary logical operations can be performed. By submitting trains of atoms to well defined resonant or non–resonant interactions, either in the cavity C or in the auxiliary cavities R_1 and R_2, it is possible to map information from one qubit to another (thus realizing a quantum memory [42]) or to produce elementary quantum gates (blue prints for such gates are described in [43]). Quantum teleportation [44] of an atomic state between two cavities could also be performed as has been proposed in [45]. Teleportation has already been demonstrated with photons [46–48], but, here again, CQED should make it possible to realize similar operations on massive particles. Note that there are similarities between the microwave cavity QED experiments discussed here and optical cavity QED experiments performed at Caltech [49]. Strong similarities also exist between our experiments and those being performed with ions oscillating in a trap [50]. In the latter case, the internal states of the ions are entangled to the vibrational degrees of freedom of the ions, which replace the field excitation of the Cavity QED experiments.

In all these experiments the goal is to push as far as possible towards larger and larger systems the quantum–classical boundary. It will certainly be possible, in not too a distant future, to entangle more than two or three particles and fields containing many photons together and to perform entanglement experiments with these systems. These studies will certainly teach us more about decoherence, the theory of quantum measurement and the yet somewhat mysterious quantum-classical boundary.

REFERENCES

1. Haroche, S., and Raimond, J. M., in *Cavity Quantum Electrodynamics*, Berman, P., editor, Academic Press, New York (1994).
2. Haroche, S., in *Fundamental Systems in Quantum Optics*, Les Houches Session LIII, Dalibard, J., Raimond, J. M., and Zin-Justin, J., editors, Elsevier (1992).
3. Haroche, S., and Kleppner, D., *Physics Today*, p. 24, January (1989); Haroche, S., and J.M.Raimond, *Scientific American*, p. 54, April (1993).
4. Purcell, E. M., *Phys.Rev.* **69**, 681 (1946).
5. Kleppner, D., *Phys. Rev. Lett.* **47**, 233 (1981).
6. Drexhage, K. H., in Progress in Optics XII, Wolf, E., editor (North Holland, Amsterdam, 1974); Goy, P., et al., *Phys. Rev. Lett.* **50**, 1903 (1983); Gabrielse, G., and Dehmelt, H., *Phys. Rev. Lett.* **55**, 67 (1985); Hulet, R. G., Hilfer, E. S., and Kleppner, D., *Phys. Rev. Lett.* **55**, 2137 (1985); Jhe, W., et al., *Phys. Rev. Let.* 58, 666 (1987); De Martini, F., et al., *Phys. Rev. Lett.* **59**, 2955 (1987); Heinzen, D. J., et al *Phys. Rev. Lett.* **58**, 1320 (1987).
7. See for example *Confined Electrons and Photons: New Physics and Applications*, edited by Burstein, E., and Weisbuch, C., Plenum Press (New York, 1995).
8. Brune, M., et al., *Phys. Rev. Lett.* **76**, 1800 (1996).
9. Einstein, A., Podolsky, B., and Rosen, N., *Phys. Rev.* **47**, 777 (1935).
10. Hagley, E., et al., *Phys. Rev. Lett.* **79**, 1 (1997).
11. Wheeler, J. A. and Zurek, W. H., *Quantum Theory of measurement* , Princeton University Press, Princeton, New Jersey (1983).
12. Brune, M., et al., *Phys. Rev A* **45**, 5193 (1992).
13. Brune, M., et al., *Phys. Rev. Lett.* **77**, 4887 (1996).
14. Raimond, J. M., Brune, M., and Haroche, S., *Phys. Rev. Lett.* **79**, 1964 (1997).
15. Zurek, W. H., *Physics Today* **44**, 36 (1991); Zurek, W. H., *Phys. Rev. D* **24**, 1516 (1981); **26**, 1862 (1982); Caldeira, A. O., and Legget, A. J., *Physica A* **121**, 587 (1983); Joos, E., and Zeh, H. D., *Z. Phys. B* **59**, 223 (1985); Omnès, R., *The Interpretation of Quantum Mechanics* , Princeton University Press, Princeton, N.J. (1994).
16. Hulet, R. G., and Kleppner, D., *Phys. Rev. Lett* **51**, 1430 (1983).
17. Jaynes, E. T., and Cummings, F. W., *Proc. IEEE* **51**, 89 (1963).
18. Nussenzveig, P., et al , *Phys. Rev. A* **48**, 3991 (1993).
19. Cohen-Tannoudji, C., Dupont-Roc, J., and Grynberg, G., *Atom-Photon Interactions*, (Wiley, New York, 1992).
20. Haroche, S., *Annales de Physique, Paris* **6**, 189 and 327 (1971).
21. Haroche, S., Brune, M., and Raimond, J. M., *Europhysics Letters* **14**, 19 (1991).
22. Englert, B. G., et al., *Europhysics Letters* **14**, 25 (1991).

23. Sanchez–Mondragon, J. J., Narozhny, N. B., and Eberly, J. H., *Phys. Rev. Lett.* **51**, 550 (1983).
24. Meschede, D., Walther, H., and Müller, G., *Phys. Rev. Lett.* **54**, 551 (1985).
25. Brune, M., et al., *Phys. Rev. Lett.* **59**, 1899 (1987).
26. Raithel, G., et al., in *Cavity Quantum Electrodynamics*, Berman, P., editor, Academic Press, New York (1994).
27. Jan-Wei Pan et al., *Phys. Rev. Lett* **80**, 3891 (1998).
28. Freedman, S. J., and Clauser, J. F., *Phys. Rev. Lett.* **28**, 938 (1972); Clauser, J. F., *Phys. Rev. Lett.* **36**, 1223 (1976); Fry, E. S., and Thompson, R. C., *Phys. Rev. Lett.* **37**, 465 (1976); Aspect, A., et al., *Phys. Rev. Lett.* **47**, 460 (1981); Aspect, A., et al., *Phys. Rev. Lett.* **49**, 1804 (1982); Ou, Z. Y., and Mandel, L., *Phys. Rev. Lett.* **61**, 50 (1988); Kwiat, P. G., et al., *Phys. Rev. Lett.* **75**, 4337 (1995).
29. Fry, E. S., Walther, T., and Li, S., *Phys. Rev. A* **52**, 4381 (1995).
30. Brune, M., et al., *Phys. Rev. Lett.* **72**, 3339 (1994).
31. Ramsey, N., *Molecular Beams*, Oxford University Press, New York (1985).
32. Greenberger, D. M., Horne, M. A., and Zeilinger, A., *Am. J. Phys.* **58**, 1131 (1990).
33. Haroche, S., in *Fundamental Problems in Quantum Theory*, Greenberger, D., editor, Annals New York Academy of Sciences (1995).
34. Zeilinger, A., private communication.
35. Glauber, R. J., *Phys. Rev.* **130**, 2529 (1963).
36. Schrödinger, E., *Naturwissenschaften* **23**, 807, 823, and 844 (1935); reprinted in English in (11).
37. Scully, M. O., et al., *Nature* (London) **351**, 111 (1991); Haroche, S., et al., *Appl.,Phys. B* **54**, 355 (1992); Pfau, T., et al., *Phys. Rev. Lett.* **73**, 1223 (1994); Chapman, M. S., et al., *Phys. Rev. Lett.* **75**, 3783 (1995).
38. Davidovich, L., et al., *Phys. Rev. A* **53**, 1295 (1996).
39. Scully, M. O., and Druhl, K., *Phys. Rev. A* **25**, 2208 (1982); Herzog, T. J., et al., *Phys. Rev. Lett.* **75**, 3034 (1995); Seager, W., *Philos. Sci.* **63**, 81 (1996); Ou, Z. Y., *Phys. Lett. A* **226**, 323 (1997).
40. Zurek, W. H., *Physics World*, p.25, Jan (1997).
41. Davidovich, L., et al., *Phys. Rev. Lett.* **71**, 2360 (1993).
42. Matre, X., et al., *Phys. Rev. Lett.* **79**, 769 (1997).
43. Barenco, A., et al., *Phys. Rev. Lett.* **74**, 4083 (1995); Sleator, T., and Weinfurter, H., *Phys. Rev. Lett.* **74**, 4087 (1995); Domokos, P., et al., *Phys. Rev. A* **52**, 3554 (1995).
44. Bennet, C. H., et al., *Phys. Rev. Lett.* **70**, 1895 (1993).
45. Davidovich, L., et al., *Phys. Rev. A* **50**, 895 (1994).
46. Boschi, D., et al., *Phys. Rev. Lett.* **80**, 1121 (1998).
47. Bouwmeester, D., et al., *Nature* **390**, 575 (1997).
48. Furusawa, A., et al., *Science* **282**, 706 (1998).
49. Turchette, Q. A., et al., *Phys. Rev. Lett.* **75**, 4710 (1995).
50. Monroe, C., et al., *Phys. Rev. Lett.* **75**, 4714 (1995); Monroe, C., et al., *Science* **272**, 1131 (1996); Turchette, Q. A., et al., *Phys. Rev. Lett.* **81**, 3631 (1998).

Laser Cooling and Trapping of Neutral Atoms

Luis A. Orozco

*Department of Physics and Astronomy,
State University of New York at Stony Brook,
Stony Brook, NY 1794-3800,
United States*

Abstract. The forces felt by atoms when illuminated with resonant radiation can reduce their velocity dispersion and confine them in a region of space for further probing and experimentation. The forces can be dissipative or conservative and allow manipulations of the external degrees of freedom of atoms and small neutral particles. Laser cooling and trapping is now an important tool for many spectroscopic studies. It enhances the density of atoms in phase space by many orders of magnitude reducing the need of large samples. These lecture notes review the fundamental principles of the field and show some of the applications to the study of the spectroscopy of radioactive atoms.

I INTRODUCTION

These notes are based on the lectures I gave at the Escuela Latinoamericana de Física in México City during the summer of 1998. The purpose of the course was to familiarize the participants with the exciting new developments in atomic physics during the last decade. We have gained unprecedented abilities to control the positions and velocities of neutral atoms, that have opened new possibilities in the investigation of their spectroscopy and collective behavior.

There are excellent reviews and summer school proceedings in the literature [1-4]. In these notes I treat only very general aspects of laser cooling and trapping without the careful detail given in the above reviews. The covered material follows the presentation of Ref. [5]. The aim is to develop an intuitive understanding of the principles and the basic mechanisms for laser cooling and trapping of neutral atoms.

Last century the electromagnetic theory of Maxwell gave a quantitative explanation to the pressure associated with light. This idea was not new, it had been proposed at least in the XVII century, to explain why comet tails point away from the sun. At the beginning of this century Einstein studied the thermodynamics of emission and absorption of radiation in his paper on blackbody radiation [6]. He remarked on the transfer of momentum in spontaneous emission that 'the smallness of the impulses transmitted by the radiation field implies that these can almost always be neglected in practice'. At that time, given the available light sources, any mechanical effects were extremely difficult to detect. Frisch observed the deflection of an atomic beam of Na by resonant light from a Na lamp in 1933 [7]. Ideas about using light to manipulate atoms and particles continued to appear in the literature and the invention of the laser helped trigger some of them. Hänsch and Schawlow [8] and Wineland and Dehmelt [9] realized that high brightness sources can exert a substantial force on atoms or ions, potentially cooling their velocity distributions. The advent of tunable lasers during the 1970s with very narrow linewidths made pioneering experiments possible. Since then a long list of people have contributed to advances in the development of laser cooling and trapping. Among the spectacular achievements facilitated by the new techniques is the Bose-Einstein condensation of a dilute gas of alkali atoms, (see the lectures of S. Rolston). Finally, the field of laser cooling and trapping received the 1997 Physics Nobel Prize in the persons of Steve Chu, Claude Cohen-Tannoudji and William Phillips [10]. It is possible to say that laser trapping and cooling is now part of the cannon of physics.

In the course of this lectures we will try to understand how to cool and trap neutral atoms using forces derived from the interaction of light with atoms. Section II introduces the light forces. Section III shows the velocity dependent force and the associated cooling mechanisms. The position dependent force is discussed in section IV. Section V shows how the forces combine to form an optical trap. Finally, in section VI I have included some examples drawn from the work with radioactive neutral atoms where I have been involved.

II THE LIGHT FORCES

The origin of the light force is the momentum transferred when an atom absorbs a photon from a laser beam. The momentum of the atom changes by $\hbar \mathbf{k}$, where $\mathbf{k}$ is the wave vector of the incoming photon. After the emission of a photon by an atom the atom recoils. The associated recoil velocity v_{rec} and recoil energy E_{rec} for an atom of mass M are:

$$v_{\text{rec}} = \frac{\hbar \mathbf{k}}{M}, \tag{1}$$

$$E_{\text{rec}} = \frac{\hbar^2 |\mathbf{k}|^2}{2M}. \tag{2}$$

II. A Spontaneous emission force

If the excitation is followed by spontaneous emission, the emission can be in any direction, but because the electromagnetic interaction preserves parity, the emission will be in a symmetric pattern with respect to the incoming photon. In this case the recoil momentum summed over many absorption and emission cycles will average of the two-level transition $(3\lambda^2/2\pi)$. The rate of fluorescence (see Eq. (6)) depends on the detuning Δ between the atom and the laser.

At low intensities the scattering rate is proportional to the saturation parameter, but as the intensity grows it shows power broadening and the rate saturates at $1/2\Gamma$. The FWHM of the Lorentzian goes from Γ at low intensities $S_0 << 1$ to $\Gamma\sqrt{1+S_0}$ for $S_0 > 1$. Power broadening can be thought as arising from the absorption-stimulated emission cycles that do not contribute to the force because the emission is into the same laser beam. The on-resonance atoms are already saturated and it is only those off resonance that can contribute and broaden the width.

The force is small but a two-level atom it returns to the ground state after emission of a photon and can be re-excited by the same laser beam. When such a transition exists in real atoms it is called a cycling transition. Alkali atoms with nuclear spin I and total angular momentum F have the transition from the $S_{1/2}$ ground state with $F = I + 1/2$ to the $P_{3/2}$ state with $F = I + 3/2$ that satisfies the cycling condition. The excited state can not decay to the other hyperfine level ($F = I - 1/2$) of the ground state because of the $\Delta F = 0, \pm 1$ selection rule. This transition in the D_2 line is commonly used for trapping alkali atoms. The saturation intensities of their cycling transitions are in the range of 1 mW/cm$^2 < I_{\text{sat}} < 10$ mW/cm^2.

II. B Stimulated emission force

If the absorption of a photon is followed by stimulated emission into the same laser beam, the outgoing photon will again carry away $\hbar\mathbf{k}$, so there is no momentum transferred. However, if the emission is into another laser beam, there is a redistribution of laser photons causing a force proportional to the difference between the two $\mathbf{k}$ vectors $\Delta\mathbf{k} = \mathbf{k_1} - \mathbf{k_2}$. The absorption an emission are correlated events and they are coherent scattering of photons. This redistributon of momentum is what happens in an optical lens and a positive lens will be drawn towards regions of high intensity as a consequence of the third law of Newton.

To calculate the index of refraction of an atom it is necessary to add the amplitude of the incident light field with the dipole field generated by the driven atomic electrons. An optical field $\mathbf{E}$ of the light induces a dipole moment $\mathbf{d}$ on the atom. Considering the electron as a harmonic oscillator, the induced dipole moment can be in phase or out of phase depending on the detuning of the driving frequency with respect to resonance. When it is in phase, the interaction energy between the dipole and the field $U = -\mathbf{d} \cdot \mathbf{E}$ is lower in high field regions. When it is out of phase U increases with $\mathbf{E}$ and a force will eject the atom out of the field. On resonance the oscillator is orthogonal to $\mathbf{E}$ and there is no force.

If the atom is illuminated only with a plane wave the stimulated force will be zero as all the $\mathbf{k}$ vectors are the same. A force from stimulated emission needs a gradient in the intensity of the light such that the $\mathbf{k}$ vectors point in different ways. This force is sometimes called the dipole or stimulated force. A force will act on an to zero. The atom then gains momentum in the direction of the wave vector of the incoming laser beam. The resulting force is sometimes called Doppler, radiation pressure, scattering, or the spontaneous force. The variance of the momentum transferred does not vanish, and the atom performs a random walk in momentum space as it emits spontaneously. These fluctuations limit the lowest temperature achievable when the laser beam is present.

$$\mathbf{F} = \mathbf{F}_{abs} + \mathbf{F}_{em}, \tag{3}$$

$$\mathbf{F} = <\mathbf{F}_{abs}> + <\mathbf{F}_{em}> + \delta\mathbf{F}_{em} \tag{4}$$

$$<\mathbf{F}> = R_{sp}\hbar\mathbf{k}, \tag{5}$$

where $\mathbf{F}$ is the force on the atom, R_{sp} is the rate of fluorescence scattering in cycles per second, and $\delta\mathbf{F}$ represents the random fluctuations from the recoiling atoms. The repeated transfer of momentum from a light beam to the atom by absorption and spontaneous emission provides the spontaneous light force.

The mean number of fluorescence cycles per second from a two level atom illuminated by a laser beam near or at resonance with the transition is equal to the population in the excited state times the Einstein A coefficient Γ. To precisely calculate the population it is necessary to include off diagonal elements in the density matrix and solve in steady state the optical Bloch equations (see for example [11]). Here we present the result without deriving it. The fluorescence depends on the amount of power available for the excitation (governed by the saturation parameter S_0) and the full width at half maximum (FWHM) Γ of the Lorentzian lineshape. The radiative lifetime of the transition $\tau = 1/\Gamma$ is the inverse of the Einstein A-coefficient. The fluorescent rate is:

$$R_{sp}(\Delta) = \frac{\Gamma}{2}\frac{S_0}{1+S_0+(2\Delta/\Gamma)^2}, \tag{6}$$

where Δ is the laser detuning from resonance,

$$\Delta = \omega_{laser} - \omega_{atom}, \tag{7}$$

and the on-resonance saturation parameter $S_0 = I_{exp}/I_{sat}$ is the ratio between the available intensity I_{exp} and the saturation intensity I_{sat}. At $S_0 = 1$ and on resonance the atom scatters at half of the maximum possible rate. There are different definitions of S_0 in the literature depending on particular definitions of I_{sat} and the reader has to pay attention to the particular one used. Here we follow the work of Citron et al. [12].

$$I_{sat} = \frac{\hbar\pi c\Gamma}{3\lambda^3}. \tag{8}$$

With this definition, an intensity of I_{sat} corresponds to providing the energy of one photon ($\hbar\omega$) every two lifetimes ($2/\Gamma$) over the area of the radiative cross section induced dipole if there is a gradient in the intensity, it can be attractive or repulsive depending on the drive detuning with respect to resonance. Any material with an index of refraction feels a force in the presence of a gradient of the intensity. The dipole force acts cells, organelles and even DNA, providing 'optical tweezers' for their manipulation . (See for example the Nobel lecture of S. Chu [10]).

III VELOCITY DEPENDENT FORCE

The spontaneous force $\mathbf{F}_{\text{spont}}$ is a velocity dependent force because the resonance condition of an atom depends on its velocity $\mathbf{v}$ through the Doppler shift $\mathbf{k} \cdot \mathbf{v}$.

$$\mathbf{F}_{\text{spont}} = \hbar \mathbf{k} \frac{\Gamma}{2} \frac{S_0}{1 + S_0 + (2(\Delta - \mathbf{k} \cdot \mathbf{v})/\Gamma)^2}. \tag{9}$$

This force saturates at $\hbar k \Gamma/2$ and is limited by the spontaneous decay time of the atomic level. The force felt by an atom when the intensities are large ($S_0 \approx 1$) are more complicated since stimulated emission is significant. We limit the discussion to the case where those processes are negligible. The velocity range of the force is significant for atoms with velocity such that their Doppler detuning keeps them within one linewidth of the Lorentzian of Eq. (9). See Fig. 1. This condition states that:

$$|\Delta - \mathbf{k} \cdot \mathbf{v}| \leq \frac{\Gamma}{2}\sqrt{1 + S_0}. \tag{10}$$

III. A Deceleration of an atomic beam

The maximum acceleration of a sodium atom interacting with resonant laser light in the D_2 cycling transition shows that light can decelerate an atom in a very short time.

$$\begin{aligned} a_{\text{max}} &= \frac{\hbar \mathbf{k}}{M} \frac{1}{2\tau}, \\ &= \frac{v_{\text{rec}}}{2\tau}, \\ &\approx \frac{3 \times 10^{-2} \text{m/s}^2}{2 \times 16 \times 10^{-9} \text{s}}, \\ &\approx 10^6 m/s^2, \\ &\approx 10^5 g. \end{aligned} \tag{11}$$

The thermal velocities of atomic beams are in the order of a thousand meters per second, so the stopping time is about one millisecond at a_{rmmax}, stopping in about one meter. However, these estimates do not consider that the force will be different for atoms with changing velocities through the Doppler effect. The spontaneous

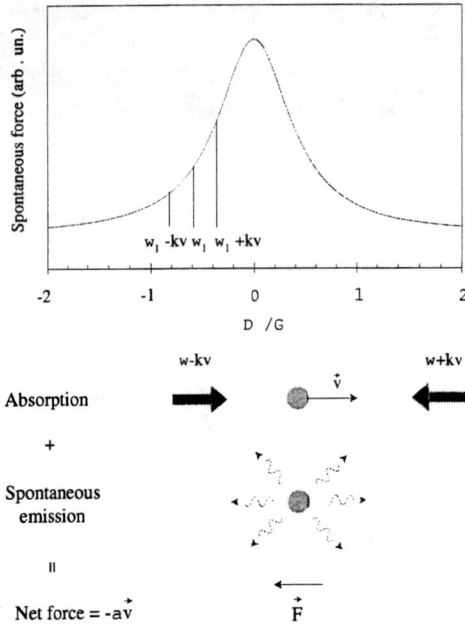

FIGURE 1. Velocity dependent force

force can act on atoms that have a velocity range where the force is significant: A Doppler shift of the order of the linewidth of the transition.

$$v_{\text{Dop}} = \frac{\Gamma}{k}, \tag{12}$$

where $k = |\mathbf{k}|$, and for Na $v_{\text{Dop}} \approx 6$ m/s which is two orders of magnitude smaller than the thermal velocity and three orders of magnitude higher than the recoil velocity.

A laser beam red detuned with respect of the resonant transition and counter-propagating with a beam of atoms at velocity v can decelerate a velocity class of atoms with a width of v_{Dop} and pile them at a lower velocity. To compensate for the resonant changing transition it is necessary to either tune the energy level of the atom in space or to change the frequency of the laser in time to keep it resonant with a group of atoms while they decelerate. Real atoms have more complications, cycling transitions are not perfect. For example, there is hyperfine structure in alkali atoms and some of the off resonant excitation can optically pump the atom into the non-cycling ground state ($F=1$ for Na). Then the atom no longer feels the force. The methods developed for deceleration maintain the atom in a cycling transition. They use the selection rules from the polarization of the light in the presence of a magnetic field, take advantage of the Clebsh-Gordan coefficients between the levels, and sometimes require excitation at other frequencies.

III. B Zeeman slowing

One approach to slowing atoms uses the Zeeman effect in a spatially varying magnetic field to tune the atomic energy levels with the changing velocity. The magnetic field is shaped to optimize the match between velocity and Zeeman detuning and keep a strong scattering of photons along the solenoid [13]. The method works if the g-factors of the levels that scale the Zeeman shifts of the ground and excited states are different so that their resonant frequency shifts. The largest ground state m sub-level in the D_2 line of Na shifts 1.4 MHz/G while the excited state shifts by 2.8 MHz/G. As a result of their difference the magnetic field can shift the transition energy and can compensate for the Doppler shift along the path of a moving atom.

Assuming that the atoms decelerate with a constant acceleration a from an initial velocity v_0, the position dependent velocity $v(z)$ is:

$$v(z) = \sqrt{v_0^2 - 2az}. \tag{13}$$

We take the changing Doppler shift $\mathbf{k} \cdot \mathbf{v}(z)$ equal it to the Zeeman shift $\vec{\mu} \cdot \mathbf{B}(z)$, where $\vec{\mu}$ is the magnetic moment of the transition, to find the shape of the compensating magnetic field.

$$B(z) = B_0 \sqrt{1 - \frac{z}{z_0}}, \tag{14}$$

$$B_0 = \frac{kv_0}{\mu}, \tag{15}$$

$$z_0 = \frac{v_0^2}{2a}, \tag{16}$$

the field B_0 induces a Zeeman shift equal to the Doppler shift of an atom having velocity v_0. A tapered solenoid produces a field of such spatial dependence. In certain applications it may be necessary to add a uniform bias field B_b to keep the field high enough to avoid optical pumping [13].

The atomic beam comes from a thermal source with a dispersion of velocities comparable to its mean velocity. It enters a tapered solenoid where the field is higher at the oven side. The laser is resonant with atoms of a given velocity v_0, usually around the mean of the thermal distribution, but this transition is modified by the Zeeman shift at the entrance and by the Doppler shift. These atoms at v_0 decelerate. As their velocity changes, their Doppler shift changes but it is compensated by a different Zeeman shift. The initially fast atoms continue to be on resonance. As they decelerate and move downstream in the magnet more atoms come on resonance and start feeling the light force of the opposing laser beam. At the end of the tapered solenoid all the atoms with velocities smaller than v_0 are decelerated to a final velocity that depends on the details of the solenoid and the laser detuning. The result is a significant enhancement of density in phase space;

TABLE 1. Trapping and cooling parameters for alkali atoms from a source at 1000 K.

Atom	A	λ_{D_2} [nm]	τ_{D_2} [nsec]	$T_{Dop.}$ [μK]	l_{Zeeman} [cm]
Na	23	589	16.2	235	40
K	39	766	26.3	145	84
Rb	87	780	26.2	145	85
Cs	133	852	30.4	125	108
Fr	210	718	21.0	181	63

despite the fact that the diffusion process associated with the cooling increases significantly the divergence in the transverse direction. Table 1 gives lengths for Zeeman slowers, required to bring different alkali atom with velocities $v_{\text{thermal}} = \sqrt{2k_BT/M}$ to a halt by driving it on a fully saturated transition.

III. C Frequency chirping

Another method to slow atoms in a beam is to chirp the frequency of the laser maintaining the resonant interaction with a group of atoms and leaving the others without deceleration [14]. The instantaneous acceleration is negative and the varying laser detuning compensates for the changing Doppler shift.

$$\Delta'(t) = -kv(t) + \Delta, \qquad (17)$$

where $\Delta'(t)$ is the time varying laser detuning of the laser frequency. In the deceleration frame the force on an atom at velocity v is:

$$F(v) = \frac{\hbar k \Gamma}{2} \left[\frac{-S_0}{1 + S_0 + \frac{2(\Delta+kv)^2}{\Gamma}} + \frac{S_0}{1 + S_0 + (\frac{2\Delta}{\Gamma})^2} \right], \qquad (18)$$

expanding near $v = 0$

$$F(v) = \left[2\hbar k^2 S_0 \frac{2\Delta/\Gamma}{[1 + S_0 + (2\Delta/\Gamma)^2]^2} \right] v. \qquad (19)$$

The force is proportional to the velocity and the proportionality constant is a friction coefficient. The method is self correcting and works in batches of atoms. All velocities near $v(0)$ damp towards $v(t)$. Any velocities not initially near $v(0)$ become close to $v(t)$ at a later time. Changes in the saturation parameter from the attenuation of the laser beam as it propagates through the beam can be compensated. The chirp rate of the laser frequency to obtain deceleration ($\Delta < 0$) is

$$\frac{d\Delta'}{dt} = ka, \tag{20}$$

$$a = \frac{F(v)}{M}. \tag{21}$$

A chirp rate of 780 MHz/ms can stop an initially thermal Na atom.

III. D Optical molasses

Velocity dependent forces are necessary to cool an atom and reduce its velocity. They do not confine the atom, but they provide what has been termed 'optical molasses'. The damping felt by the atoms is substantial and the study of the cooling mechanisms has been discussed in the literature. (See for example the review paper of Metcalf and van der Straten [1]).

An atom subject to two laser beams in opposite directions will feel a force $\mathbf{F}(\mathbf{v})$ coming from its interaction with both beams. If the counterpropagating laser beams are detuned to the red of the zero velocity atomic resonance, a moving atom will see the light of the opposing beam blue shifted in its rest frame (See Fig. 1). The beam in the same direction as the atom will be further red shifted in its rest frame. Considering only one dimension and $S_0 \ll 1$, the force opposing the motion will always be larger than the force in the direction of the motion, and this leads to Doppler cooling.

$$\mathbf{F}(\mathbf{v}) = \mathbf{F}_{\text{spont}}(\mathbf{k}) + \mathbf{F}_{\text{spont}}(-\mathbf{k}). \tag{22}$$

The sum of the two forces, with the semiclassical assumption that the recoil shift is negligible $kv_{\text{rec}} \ll \Gamma$ gives in the limit where $v^4 \ll (\Gamma/k)^4$,

$$\mathbf{F}(\mathbf{v}) \approx \frac{8\hbar k^2 S_0 \Delta}{\Gamma(1 + S_0 + (2\Delta/\Gamma)^2)^2}\mathbf{v}, \tag{23}$$

$$\mathbf{F}(\mathbf{v}) \approx \alpha \mathbf{v}. \tag{24}$$

The force is proportional to the velocity of the atom through the friction coefficient α and depends on the sign of the laser detuning Δ. Figure 2 shows the Doppler cooling force in one dimension as a function of velocity and detuning for the D_2 line of francium. This force is limited by the spontaneous decay time of the atomic level. An estimate for the maximum velocity an atom can have and still feel the light force is when the Doppler shift is equal to the laser detuning from the transition: $v_{max} \approx \Delta/k$. Only a very small fraction of the thermal distribution of atoms at room temperature can be cooled in optical molasses.

III. E Cooling Limits

Optical molasses provides a velocity dependent or viscous force. In the three-dimensional configuration atoms get slowed wherever they are in the region defined

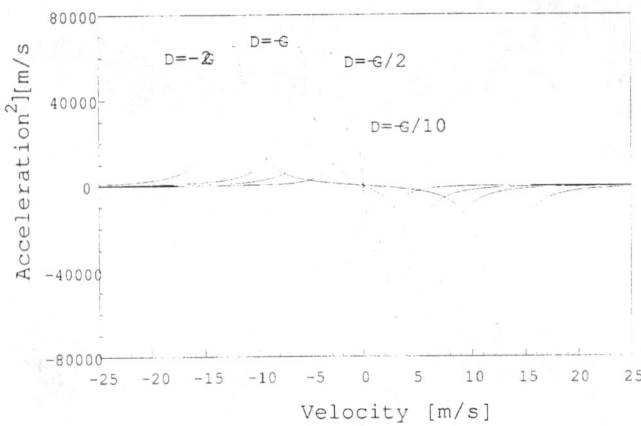

FIGURE 2. Doppler cooling in one-dimensional optical molasses. The numerical values are for the francium D_2 line at $S_0 = 1$. (From Ref. [21]).

by the overlap of the six orthogonal beams. Large laser beams will increase the total number of cooled atoms, but the atomic density remains constant. Because of the variance of the momentum coming from the repeated random spontaneous emission, atoms can diffuse out of the molasses region because this is not a trap. The competition between the cooling process and the diffusion of the momentum reaches an equilibrium that determines the lowest temperature of the atoms [1].

III. E. 1 The Doppler Cooling Limit

The atomic momentum and energy change by $\hbar k$ and E_{rec} after each interaction with the laser beam. Following the one dimensional treatment of the force above, the change of the energy has an associated change in the frequency of the transition such that $E_{\text{rec}} = \hbar \omega_{\text{rec}}$. Then the average frequencies of absorption and emission are:

$$\omega_{\text{abs}} = \omega_{\text{atom}} - \omega_{\text{rec}}, \tag{25}$$

$$\omega_{\text{em}} = \omega_{\text{atom}} + \omega_{\text{rec}}. \tag{26}$$

The light field losses every cycle an average energy of:

$$\hbar(\omega_{\text{abs}} - \omega_{\text{em}}) = 2\hbar\omega_{\text{rec}}, \tag{27}$$

and the power lost by the laser field becomes atomic kinetic energy. The rate of heating should equal the rate of cooling in thermal equilibrium and

$$\mathbf{F} \cdot \mathbf{v} = \frac{\hbar \omega_{\text{rec}}}{1/R_{sc}}, \tag{28}$$

$$\alpha v^2 = \frac{\hbar \omega_{\text{rec}}}{1/R_{sc}}. \tag{29}$$

The cooling force in the optical molasses is proportional to the velocity through the friction coefficient α. The temperature associated with the kinetic energy is:

$$k_B T = \frac{\hbar \Gamma}{4} \left[\frac{1 + 2S_0 + (2\Delta/\Gamma)^2}{2\Delta/\Gamma} \right]. \qquad (30)$$

This expression becomes independent of S_0 in the limit of low intensity and has a minimum for $\Delta = -\Gamma/2$. This temperature is called the Doppler cooling limit T_{Doppler}.

$$T_{\text{Doppler}} = \frac{\hbar \Gamma}{2 k_B}. \qquad (31)$$

The lowest temperature in optical molasses is independent of the optical wavelength, atomic mass, and, in the limit of low intensity, also of laser intensity. The only atomic parameter that enters is the rate of spontaneous emission Γ. The value for Na is 240 μK which corresponds to an average velocity of 30 cm/s four orders or magnitude smaller than the typical thermal velocities produced out of an effusive oven (See Table 1).

III. E. 2 Beyond Doppler cooling

In 1988 the NIST group [15] discovered that the temperature of sodium atoms in optical molasses was a factor of six lower than the Doppler cooling limit. The quantitative understanding of this result requires the inclusion of all the energy levels that are present in an atom, the effects of the polarization of the different laser beams, and the non-adiabatic response of a moving atom to the light field [3,16,17].

The atom has a finite response time τ_{int} to adjust its internal state σ to a new environment. σ depends on the position z and velocity v of the atom and in general lags behind the steady state of an atom which would be at rest in z

$$\sigma(z, v) \approx \sigma_{st}(z) - v \tau_{int} \frac{d}{dz} \sigma_{st}(z). \qquad (32)$$

The non-adiabaticity parameter in the problem is:

$$\epsilon = \frac{v \tau_{int}}{\lambda/2\pi}, \qquad (33)$$
$$= k v \tau_{int}. \qquad (34)$$

The frictional force is going to be linear in v as long as $\epsilon < 1$. The equilibrium temperature of the system is:

$$k_B T \approx \frac{\hbar}{\tau_{int}}. \qquad (35)$$

For a two level atom there is a single internal time $\tau_{int} = 1/\Gamma$, the radiative lifetime of the excited state. The non-adiabaticity parameter is the ratio of the Doppler shift divided by the natural width of the transition. The temperature reachable is of the order of $\hbar\Gamma$. This result is in agreement with the T_{Doppler} calculated based on the change in the energy of the laser field from Eq. (31). The Doppler limit is independent of the intensity.

However; a multilevel atom, for example an alkali, has hyperfine splitting and Zeeman sublevels. There is a new internal time: The optical pumping time between ground state sublevels. Let Γ' be the absorption rate from $|g>$, this number depends on the intensity and will give a different value for the lowest temperature than Doppler cooling. At low intensities $S_0 \ll 1$ and $\Gamma' \ll \Gamma$. The associated ϵ', which is the ratio of the Doppler shift to the optical pumping rate, will be very large.

III. E. 3 Sisyphus cooling

When the intensity and detuning of the laser beams are significant, a different mechanism can cool an atom. It requires an AC Stark Shift of the atomic ground state. The dressed atom formalism of the atom + photon interaction shows (see for example the contribution of Cohen-Tannoudji in Ref. [3]) that the light shift for the ground state δ' in the presence of a field with Rabi frequency Ω much smaller than the absolute value of the detuning between the laser and the atomic transition Δ is:

$$\delta' = \frac{\Omega^2}{4\Delta}. \qquad (36)$$

The light shifts are proportional to the intensity (Ω^2), the sign depends on the detuning Δ of the laser with respect to the atomic transition. If an atom is illuminated by two detuned laser beams counterpropagating but one with horizontal polarization and the other with vertical polarization, the atom will feel a very different force from the spontaneous force. The resulting field has polarization gradients. The field has negative circular helicity in one point in space, a distance $\lambda/4$ away has positive helicity, and is elliptically polarized in between with linearly polarized light exactly at $\lambda/8$ of the point with purely circular light. (See the Nobel lecture of Cohen-Tannoudji [10]). For a case where the ground state has two sublevels $J_g = 1/2$ and the excited state four $J_e = 3/2$ the optical pumping rates are the largest from the highest sublevel of the ground state to the lowest sublevel of the ground state. If $v\tau_{int} \approx \lambda/2\pi$ the atom can climb a potential hill and reach the top before being pumped back to a valley. The atom is always climbing in analogy to the Greek Sisyphus. There is a decrease of the kinetic energy and the dissipation of potential energy is by spontaneous anti-Stokes Raman photons. The equilibrium temperature comes when the atoms gets trapped in one of the potential wells formed by the position dependent AC Stark Shift, then the equilibrium temperature is of the order of the well depth:

$$k_B T \approx \frac{\hbar \Omega^2}{|\Delta|}, \tag{37}$$

further cooling in the well is possible using adiabatic expansion by lowering the laser intensity at a rate slow compared to the frequency of oscillation of the trapped atom in the potential well.

Another way to understand Sisyphus cooling is the following (See Ref. [4] and the contribution of S. Chu in [2]). The induced electric dipole **d** of an atom in the presence of an off-resonant field minimizes its energy when it aligns with the optical electric field **E**. If an atom at a point of linearly polarized light moves a distance $\lambda/8$ the polarization is now circular because of the way the opposite polarizations add at each point in space. The atom can only follow a change in field alignment with a finite time delay characteristic of the damping process. This process changes kinetic energy into potential energy which is lost from damping as the dipole relaxes to the new state of polarization.

There are other configurations that produce Sisyphus cooling, for example two counterpropagating beams with σ^+ and σ^- polarizations. The polarization of the field is always linear but it changes directions continuously over one wavelength. The atomic dipole sees a change in the direction it should oscillate.

All the mechanisms described before rely on absorption and spontaneous emission of photons. A natural limit to the lowest achievable temperature is given by the recoil energy $k_b T_{\text{rec}}/2 = E_{\text{rec}}$. Finding a way to 'protect' the atoms from light can bypass this limit. Two laser cooling methods are known to reduce the temperature of the atoms beyond T_{rec}: velocity selective coherent population trapping (VSCPT) and Raman cooling. VSCPT prepares the atom in a 'dark state' that does not absorb any light eliminating the possibility of recoil. This state is stationary and an atom that diffuses into it will be trapped (See [18] and the Nobel lecture of Cohen-Tannoudji [10]).

In Raman cooling, a series of light pulses, with well defined frequency and duration, produces an excitation profile that constitutes a 'trap' in velocity space for the atom. (See S. Chu in [2]).

IV POSITION DEPENDENT FORCE

The position dependent force is necessary to construct a trap but is more subtle than the velocity dependent force. A series of no trapping theorems constrain the distribution of electric and magnetic fields for capturing neutral atoms. (See the contribution by S. Chu in Ref. [2]).

Earnshaw theorem states that is is impossible to arrange any set of static charges to generate a point of stable equilibrium in a charge-free region. The electrostatic potential ϕ satisfies $\nabla^2 \phi = 0$, then $\phi(x, y, z)$ at any point is the average of ϕ on the surface of the sphere centered at (x, y, z). There can not be an extremum of ϕ and since the electrostatic energy is proportional to the potential, there is no minimum of the energy. Similarly: $\nabla \cdot \mathbf{E} = 0$ and all the lines of force that go in are balanced by lines that go out of it. The optical Earnshaw theorem uses the Poynting vector of the field $\mathbf{S}$ and it applies to the scattering force. The light flux can not point inward everywhere, so a light trap is unstable ($\nabla \cdot \mathbf{S} = 0$).

A solution is not to use static light beams, but alternate them in time to generate a trap following the ideas of the Paul trap. Another way to circumvent the optical Earnshaw theorem is to exploit the internal structure of the atoms. The effective atomic polarizability $\mathbf{P}$ can be position dependent through the presence of an external magnetic field $\mathbf{B}$ resulting in a negative divergence of the spontaneous light force, since the force is proportional to $\mathbf{P}$.

J. Dalibard proposed a solution to the neutral atom trapping using the spontaneous light force. His idea became the basis of the Magneto-Optical Trap (MOT). The solution of Dalibard was to add a spatially varying magnetic field, so that the shifts in the energy levels make the light force dependent on the position. Soon afterwards this scheme was generalized to three dimensions and it was successfully demonstrated with Na atoms by Raab *et al.* [19]. Despite many new developments the MOT remains the workhorse of laser trapping due to its robustness, large volume and capture range. The next section discusses this trap in more detail since this type has been used in the successful trapping of radioactive atoms [5].

V OPTICAL TRAPS

V. A The Magneto-Optical Trap

This section presents a simplified one-dimensional model to explain the trapping scheme in a $J=0 \to J=1$ transition.

Figure 3 shows a configuration similar to optical molasses. Two counterpropagating, circularly polarized beams of equal helicity are detuned by Δ to the red of the transition. In addition there is a magnetic field gradient, splitting the $J=1$ excited state into three magnetic sublevels. If an atom is located to the left of the center, defined by the zero of the magnetic field, its $J=0 \to J=1, m=1$ transition is closer to the laser frequency than the transitions to the other m-levels. However, $\Delta m = +1$ transitions are driven by σ^+ light. Atoms on the left are more in resonance with the beam coming from the left, pushing them towards the center. The same argument holds for atoms on the right side. This provides a position dependent force. The Doppler-cooling mechanism is also still valid, providing the velocity dependent force. Writing the Zeeman shift as βx, where x is the coordinate with respect to the center, the total force is:

$$F_{\text{MOT}} = \frac{\hbar k \Gamma}{2} \left[\frac{S_0}{1 + S_0 + (2(\Delta - \xi)/\Gamma)^2} - \frac{S_0}{1 + S_0 + (2(\Delta + \xi)/\Gamma)^2)} \right], \quad (38)$$

where

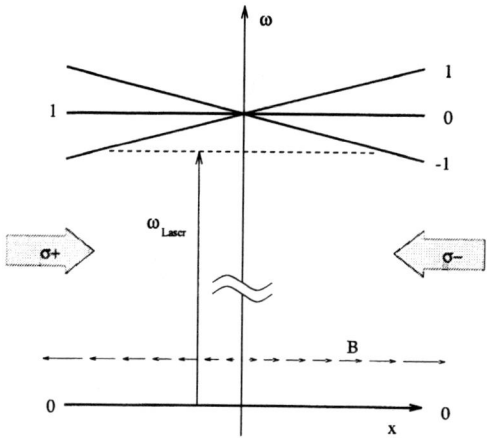

FIGURE 3. Simple 1-D model of the MOT. (From Ref. [21]).

$$\xi = kv + \beta x. \tag{39}$$

For small detunings, expansion of the fractions in the same way as in Eq. (24), shows the force proportional to ξ (see W.D. Phillips in, [2]). In the small-field, low-velocity limit the system behaves as a damped harmonic oscillator subject to the force:

$$F(v,x) = \frac{4\hbar k S_0 (2\Delta/\Gamma)(kv+\beta x)}{[1+(2\Delta/\Gamma)^2]^2}, \tag{40}$$

and

$$\ddot{x} + \gamma \dot{x} + \omega_{trap}^2 x = 0, \tag{41}$$

with

$$\gamma = \frac{4\hbar k^2 S_0 (2\Delta/\Gamma)}{M[1+(2\Delta/\Gamma)^2]^2}, \tag{42}$$

$$\omega_{trap}^2 = \frac{4\hbar k \beta S_0 (2\Delta/\Gamma)}{M[1+(2\Delta/\Gamma)^2]^2}. \tag{43}$$

The motion of the atom in the harmonic region of the trap is overdamped since $\gamma^2/4\omega_{trap}^2 > 1$. This same ratio in terms of the recoil energy and the Zeeman shift over one waveleght is:

$$\frac{\gamma^2}{4\omega_{tap}^2} = \frac{\pi E_{\text{rec}}}{4\lambda \hbar \beta}. \tag{44}$$

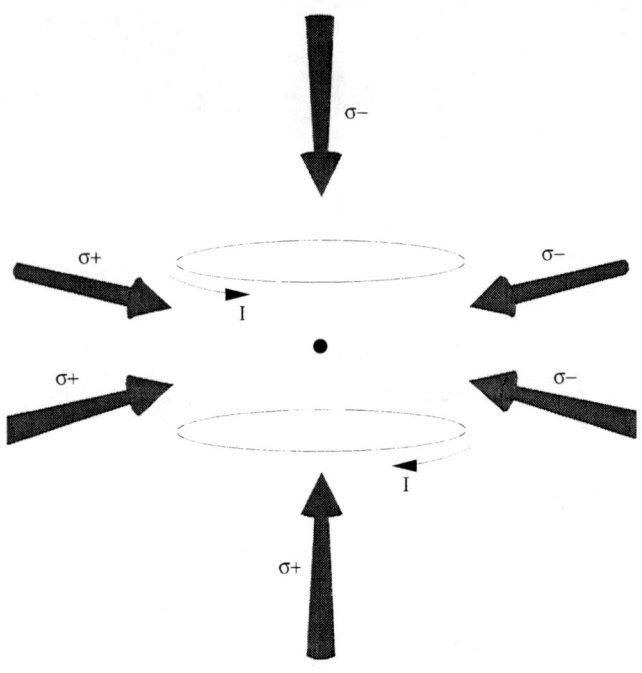

FIGURE 4. Laser beams and coils for a MOT. (From Ref. [22]).

A trap with a magnetic field gradient that produces a Zeeman shift of $\beta = 14$ MHz/cm has a trapping frequency of a few kilohertz and an Eq. (44) of the order of 10.

The real world requires three-dimensional trapping, and in alkalis a $J=0 \to J=1$ transition is hard to find. For an alkali atom with non-vanishing nuclear spin the ground state $(nS_{1/2})$ splits into two levels. The transition to the first $P_{3/2}$ excited state has four levels (for $J<I$), yet the trap works quite well under these conditions. Ideally, the transition from the upper ground state to the highest excited state F-level is cycling, and one can almost ignore the other states. Due to finite linewidths, off resonance excitation, and other energy levels the cycling is not perfect. An atom can get out of the cycling transition and an extra beam, a weak 'repump' laser, can transfer atoms from the 'dark' lower ground state to the upper one.

A magnetic quadrupole field, as produced by circular coils in the anti-Helmholtz configuration, provides a suitable field gradient in all three dimensions. The exact shape of the field is not very critical, and the separation between the two coils does not have to be equal to the radius. Typical gradients are 10 G/cm.

A large variety of optical configurations are available for the MOT. The main condition is to cover a closed volume with areas normal to the **k** vectors of the laser beams with the appropriate polarized light. (See Fig. 4). The realization with three retro-reflected beams in orthogonal directions requires quarter-wave plates before

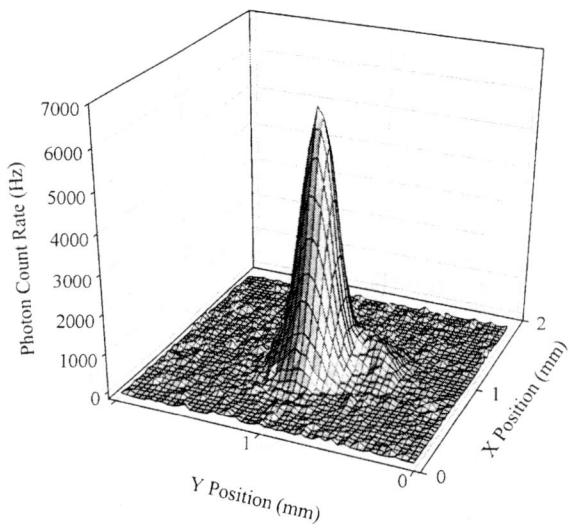

FIGURE 5. Two-dimensional CCD image of the fluorescence from francium atoms trapped in a MOT. (From Ref. [23]).

entering the interaction region. In order to have the appropriate polarization on the retro-reflected beam the phase has to advance half a wavelength. The usual arrangement is to place a quarter wave plate in front of a plane mirror, but two reflections can also provide the same phase shift [20].

The intensity of the laser beams should provide a saturation parameter $S_0 \approx 1$. The MOT can work with significantly less intensity but it becomes more sensitive to alignment. In general the MOT is a very forgiving trap as far as polarization and intensities. The retro-reflecting technique for traps, despite the scatter losses in the windows and the beam divergence as it propagates, works very well.

The well depth of a MOT is set by the maximum capture velocity v_{max}. For alkali atoms and $\Delta \approx 2\Gamma$ it is close to 1 K. The background pressure around the MOT limits its lifetime and consequently the maximum number of atoms in steady state. A pressure of 1×10^{-8} Torr produces a trap lifetime of the order of 1 s. The characteristic size x_0 of the trapping volume is set by the gradient and the detuning of the specific realization of a MOT: $x_0 = \Delta/\beta$. x_0 is about 1 cm and to obtain larger volumes larger laser beams are required. The captured atoms concentrate in a region much smaller than the trapping volume. The size of the fluorescing ball of less than 10^6 captured atoms is smaller than 1 mm in diameter. It depends on the temperature and is related to the laser beams shape, magnetic environment, and polarization. The shape of the fluorescence when integrated in a charge couple device (CCD) camera is usually Gaussian (see Fig. 5).

If the alignment of the laser beams is not good there can be a torque impressed into the trap and satellites can form. Fringes in the beams can also generate satellites. As the number of atoms increases there is a limit to the size of the trap. A similar effect to space charge appears. The optical density is thick enough to create an imbalance in the two counterpropagating beams; also the atoms can absorb spontaneously emitted light that is not red-detuned from neighboring atoms. The trap is no longer optically transparent with an extra internal radiation pressure that may eject the atoms out of the trap.

To increase the density and the number of atoms beyond the point where the repelling force turns on, Ketterle et al. [24] developed the dark MOT. The repumping beams are blocked from the central region of the trap. The trap maintains the atoms in a non-cycling state and only repumps them to the cycling transition when they stray to the edge of the trapping volume. This approach works with alkali atoms since the ground state hyperfine splitting already requires a repumping laser.

The first experiments with a MOT by Raab et al. [19] reported the capture of atoms from the residual background gas in the vacuum chamber without need of deceleration. In 1990 Monroe et al. [25] showed trapping in a glass cell from the residual vapor pressure of a Cs metal reservoir. If the vapor pressure of an element is sufficiently high, a MOT inside a cell filled with a vapor continuously captures atoms from the low-velocity tail of the Maxwell-Boltzmann distribution. The remaining atoms thermalize during wall collisions and form a new Maxwell-Boltzmann distribution. From this the MOT can again capture the low velocity atoms. The trapping efficiency depends on the number of wall contacts that an atom can make before leaving the system. Since alkali atoms tend to chemisorb in the glass walls, special coatings can prevent the loss of an atom [26]. If the wall is coated, the atom physi-sorbs for a short time, thermalizes and then is free to again cross the capture region and fall into the trap.

The capture range of the MOT is enhanced with the help of large and intense laser beams. Gibble et al. [27] reported that for their large trap they captured atoms with initial velocities below about 18 % of the average thermal velocity at room temperature. However, the fraction of the Maxwell-Boltzmann distribution of atom velocities below the capture velocity of the trap is too small to capture a significant fraction of scarce radioactive atoms on a single pass through the cell. Wall collisions are critical to provide multiple opportunities for capture in the vapor cell technique. On the one hand they provide the thermalization process, but they also increase the possibility of losing the atom by chemical adsorption onto the wall.

No significant vapor pressure of stable alkali atoms normally builds up unless the walls of the glass cell are coated by a mono-layer of the atom to be trapped. For most radioactive samples this is impossible, and also not desired since that will create a source of background for the study of the decay products. An alternative is to coat the cell with a special non-stick coating. The coatings are in general silanes and have been extensively studied for optical pumping applications of alkali atoms. Collisions with the bare glass walls destroy the atomic polarization and the coatings can provide a 'soft surface' for reflection. The Stony Brook group uses one commercially identified by the name of Dryfilm (a mixture of dichlorodimethylsilane and methyltrichlorosilane). The coating procedure follows the techniques of Swenson *et al.* [26]. The choice of a particular coating depends on many issues. For example: The difficulties in the application of the coating to the surface, how well the coating withstands high temperatures present nearby in the experimental apparatus. The coating of choice constrains the attainable background pressure in the cell and the geometry of the vacuum container. Nevertheless the vapor cell is appealingly simple. As long as a coating is known to work for a stable alkali it seems to work for the radioactive ones. The Colorado group has studied different coatings extensively [28], and have developed curing procedures to optimize the performance of the coatings.

The glass cell method relies on the non-stick coatings and works well for radioactive alkali atoms, but for other radioactive elements it may not be so easily implemented and the Zeeman slower could prove more effective to load atoms into a MOT.

The group of the University of Colorado has published a resource letter on laser trapping and cooling [29]. They also published a detailed explanation, including electronic diagrams, on how to build a glass cell MOT for Rb or Cs using laser diodes [30].

V. B The dipole force trap

An electric or magnetic dipole in an inhomogeneous electric or magnetic field feels an attractive or repulsive force depending on the specific conditions. A strong laser field can induce an electric dipole in an atom. In 1968 Letokhov [31]proposed laser traps based on the interaction of this induced electric dipole moment with the laser field. Later, Ashkin [32] proposed a trap that combined this dipole force and the scattering force. The first laser trap for neutral atoms was of this type [33]. The trap depth is proportional to the laser intensity divided by the detuning $\hbar\Omega^2/|\Delta|$. In order to minimize heating from spontaneous emission, the frequency of the intense laser is tuned hundreds of thousands of linewidths away from resonance. The heating is greatly reduced since the emission rate is proportional to the laser intensity divided by the square of the laser detuning. The off-resonance nature of the trap requires very intense beams with an extremely tight focus, and is often referred to as a Far Off Resonance Trap (FORT). A single laser red-detuned tightly focused has a gradient large enough to capture atoms from a MOT. The well depth is very small, fractions of a milliKelvin, depending on precooled atoms and very good vacuum for an extended residence in the trap. The atoms reside in a conservative trap and can cool down further by other mechanisms like evaporative cooling [34]. This kind of trap has found applications in the manipulation of extended objects as a form of optical tweezers.

V. C Other traps and further manipulation

Although the MOT is a proven trap for radioactive atoms, it may not be the ideal environment for some of the experiments now planned. The atoms are not polarized because there are all helicities present in the laser field, and the magnetic field is inhomogeneous. There have been a series of traps developed in conjunction with the pursuit of Bose Einstein condensation (BEC) [35–38]. that may have application in the field of radioactive atom trapping. In this quest for even higher phase space densities, new techniques for transport and manipulation of cold atoms have also appeared.

V. C. 1 Cold atom manipulation

To move the accumulated atoms in a MOT to a different environment requires some care. Simply turning the trapping and cooling fields off will cause the atoms to fall ballistically. The trajectories out of the trap will map out the original velocity distribution of the captured atoms, dispersing the atoms significantly as they fall. An auxiliary laser beam can push the atoms in one direction, but it has a limited interaction range since the atoms accelerate until they are shifted out of resonance by their Doppler shift. The acceleration is in only one direction and there is still ballistic expansion of the cold atoms. Gibble et al. [39] created a moving molasses with the six beams of the MOT. By appropriate shifting of the frequencies of the beams, the atoms accelerate in the 111 direction (along the diagonal of the cube formed by the beams), but they are kept cold by the continuous interaction with the six beams.

VI COOLING AND TRAPPING OF FR

Francium is the heaviest of the alkali atoms and has no stable isotopes. It occurs naturally from the α decay of actinium or artificially from fusion or spallation nuclear reactions in an accelerator. Its longest lived isotope has a half-life of 22 minutes. Previously, experiments to study the atomic structure of francium were possible only with the very high fluxes available at a few facilities in the world [40], or by use of natural sources [41].

Because of its large number of constituent particles, electron correlations and relativistic effects are important, but its structure is calculable with many-body perturbation theory (MBPT). Its more than two hundred nucleons and simple atomic structure make it an attractive candidate for a future atomic parity non-conservation (PNC) experiment. (See Ref. [42] for the most recent results in Cs). The PNC effect is predicted to be 18 times larger in Fr than Cs [43].

The present francium spectroscopy serves to test the theoretical calculations in a heavier alkali. This ensures that the Cs structure, calculated with the same techniques, is well understood.

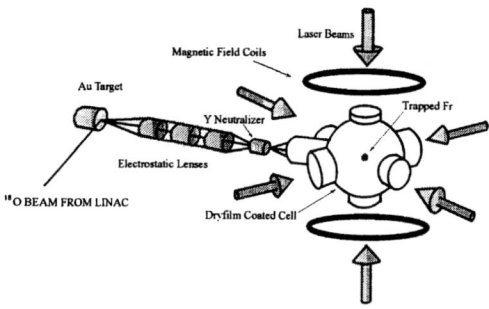

FIGURE 6. Schematic view of target, ion transport system, and magneto optical trap. (From Ref. [23])

Heavy-ion fusion reactions can, by proper choice of projectile, target and beam energy, provide selective production of the neutron deficient francium isotopes. Gold is an ideal target because it is chemically inert, has clean surfaces, and a low vapor pressure. The ^{197}Au(^{18}O,xn) reaction at 100 MeV produces predominantly ^{210}Fr, which has a 3.2 min half-life. Changing the energy and the isotope of the oxygen beam maximizes the production of isotopes 208, 209 or 211. The reaction ^{198}Pt(^{19}F,5n) produces ^{212}Fr.

Fig. 6 shows the apparatus to trap and produce Fr at Stony Brook 10^{12} ^{18}O ions/s on Au produce ^{210}Fr in the target, with less than 10% of other isotopes. The target is heated to $\approx$ 1200 K by the beam power and by an auxiliary resistance heater. The elevated temperature is necessary for the alkali elements to rapidly diffuse to the surface and be surface ionized.

Separation of the production and the trapping regions is critical in order to operate the trap in a UHV environment. Extracted at 800 V, the $\approx 1 \times 10^6$/s ^{210}Fr ions travel about one meter where they are deposited on the inner surface of a cylinder coated with yttrium which is heated to 1000 K and located 0.3 cm away from the entrance of the cell. Neutral Fr atoms evaporate from the Y surface and form an atomic beam directed towards an aperture into the vapor cell MOT.

The physical trap consists of a 10 cm diameter Pyrex bulb with six 5 cm diameter windows and two viewing windows 3 cm in diameter. The MOT is formed by six intersecting laser beams each with $1/e^2$ (power) diameter of 4 cm and power of 150 mW, with a magnetic field gradient of 6 G/cm. The glass cell is coated with a non-stick Dry-film coating [26] to allow the atoms multiple passes through the trapping region after thermalization with the walls [25]. The trapping laser operates in the D_2 line of francium, while the repumper may operate in the D_1 or in the D_2 lines depending on the measurement. The ground state hyperfine splitting of ^{210}Fr is 46.7 GHz.

We have recently captured francium atoms [44] in a magneto optical trap (MOT), opening the possibility for extensive studies of its atomic properties. (See Fig. 4 for an image of the fluorescence of Fr atoms in a MOT).

We have been studying the spectroscopy of francium in a magneto optical trap on-line with an accelerator. The captured atoms are confined for long periods of time moving at low velocity in a small volume, an ideal environment for precision spectroscopy. Our investigations have included the location of the $8S$ and $9S$ energy levels [45,48]. We have also made the first measurements of any radiative lifetime in Fr. The precision of our lifetime measurements of the D_1 and D_2 lines are comparable to those achieved in stable atoms [46,47]. They test atomic theory in a heavy atom where relativistic and correlation effects are large.

ACKNOWLEDGMENTS

During the years that I have been working in cooling and trapping of atoms I have benefited from the interaction with many people. Among them I would like to mention Jesse Simsarian, Jeff Ng, Gerald Gwinner, Jürgen Gripp, Steve Mielke, Greg Foster, Joshua Grossman, Gene Sprouse, Hal Metcalf, Tom Bergeman, Simone Kulin, and Steven Rolston. I would like to thank the organizers of the 1998 Escuela Latinoamericana de Física for their invitation and hospitality.

Support for the experiments with radioactive atoms has come from the National Science Foundation and the National Institute of Standards and Technology.

REFERENCES

1. Metcalf, H., and van der Staten, P., *Phys. Rep.* **244**, 203 (1994).
2. Arimondo, E., Phillips, W. D., and Strumia, F., eds., *Laser Manipulation of Atoms and Ions*, Amsterdam: North Holland, 1992.
3. Dalibard, J., Raimond, J.-M., and Zinn-Justin, J., eds., *Fundamental Systems in Quantum Optics*, Les Houches 1990, Session LIII, Amsterdam: North Holland, 1992.
4. Adams, C. S., and Riis, E.,*Progress in Quantum Electronics* (1996).
5. Sprouse, G. D., and Orozco, L. A., *Annu. Rev. Nucl. Part. Sci.* **47**, 429 (1997).
6. Einstein, A., *Physikalishe Zeit.* **18**, 121 (1917); translation in, *Sources of Quantum Mechanics*, van der Waerden, B., ed. Amsterdam: North-Holland, 1967.
7. Frisch, O., *Zeit. f. Phys.* **86**, 42 (1933).
8. Hänsch, T., and Schawlow, A., *Opt. Commun.* **13**, 68 (1975).
9. Wineland, D., and Dehmelt, H., *Bull. Am. Phys. Soc.* **20**, 637 (1975).
10. Chu, S., *Rev. Mod. Phys.* **70**, 685 (1998); Cohen-Tannoudji, C. N., *Rev. Mod. Phys.* **70**, 707 (1998); Phillips, W. D., *Rev. Mod. Phys.* **70**, 721 (1998).
11. Allen, L., and Eberly, J. H., *Optical resonance and two-level atoms*, New York: Wiley, 1975.
12. Citron, M. L., Gray, H. R., Gabel, C. W., and Stroud Jr., C. R., *Phys. Rev. A* **16**, 1507 (1977).
13. Phillips, W. D., and Metcalf, H., *Phys. Rev. Lett.* **48**, 596 (1982).
14. Ertmer, W., Blatt, R., Hall, J. L., and Zhu, M., *Phys. Rev. Lett.* **54**, 996 (1985).

15. Lett, P., Watts, R., Westbrook, C., Phillips, W., Gould, P., and Metcalf, H., *Phys. Rev. Lett.* **61**, 169 (1988).
16. Dalibard, J., and Cohen-Tannoudji, C., *J. Opt. Soc. Am. B* **6**, 2023 (1989).
17. Ungar, P., Weiss, D., Riis, E., and Chu, S., *J. Opt. Soc. Am. B* **6**, 2058 (1989).
18. Lawall, J., Kulin, S., Saubamea, B., Bigelow, N., Leduc, M., and Cohen-Tannoudji, C., *Phys. Rev. Lett.* **75**, 4194 (1995).
19. Raab, E. L., Prentiss, M., Cable, A., Chu, S., and Pritchard, D. E., *Phys. Rev. Lett.* **59**, 2631 (1987).
20. Gwinner, G., Behr, J. A., Cahn, S. B., Ghosh, A., Orozco, L. A., Sprouse, G. D., and Xu, F., *Phys. Rev. Lett.* **72**, 3795 (1994).
21. Gwinner, G., Ph. D. Thesis, SUNY Stony Brook, (1995), unpublished.
22. Ng, J., M. Sc. Thesis, SUNY Stony Brook (1996), unpublished.
23. Simsarian, J. E., Ph. D. Thesis, SUNY Stony Brook (1998), unpublished.
24. Ketterle, W., Martin, A., Joffe, M.A., and Pritchard, D. E., *Phys. Rev. Lett.* **69**, 2483 (1992).
25. Monroe, C., Swann, W., Robinson, H., and Wieman, C. E., *Phys. Rev. Lett.* **65**, 1571 (1990).
26. Swenson, D. R., and Anderson, L. W., *Nucl. Instr. Meth. B* **29**, 627 (1988).
27. Gibble, K. E., Kasapi, S., and Chu, S., *Opt. Lett.* **17**, 526 (1992).
28. Stephens, M., Rhodes, R., and Wieman, C., *J. Appl. Phys.* **76**, 3479 (1994).
29. Newbury, N. R., Wieman, C. E., *Am. J. Phys.* **63**, 317 (1995).
30. Wieman, C. E., Flowers, G., Gilbert, S., *Am. J. Phys.* **64**, 18 (1996).
31. Letokhov, V. S., Pisma, *Zh. Eksp. Teor. Fiz* **7**, 348 (1968) [*JETP Lett.* **7**, 272 (1968)].
32. Ashkin A. *Phys. Rev. Lett.* **40**, 729 (1978).
33. Chu, S., Bjorkholm, J. E., Ashkin, A., and Cable, A., *Phys. Rev. Lett.* **57**, 314 (1986).
34. Ketterle, W., and Van Druten, N. J., in *Advances in Atomic Molecular and Optical Physics* **37**, 181, Academic Press (1996).
35. Anderson, M. H., Ensher, J. R., Matthews, M. R., Wieman, C. E., and Cornell, E. A., *Science* **269**, 198 (1995).
36. Davis, K. B., Mewes, M. O., Andrews, M. R., van Druten, N. J., Durgee, D. S., Kurn, D. M., and Ketterle, W., *Phys. Rev. Lett.* **75**, 3969 (1995).
37. Bradley, C. C., Sackett, C. A., Tollett, J. J., and Hulet, R. G., *Phys. Rev. Lett.* **75**, 1687 (1995).
38. Bradley, C. C., Sackett, C. A., and Hulet, R. G., *Phys. Rev. Lett.* **78**, 985 (1997).
39. Gibble, K. E., and Chu, S., *Phys. Rev. Lett.* **70**, 1771 (1993).
40. Arnold, E., Borchers, W., Duong, H. T., Juncar, P., Lermé, J., Lievens, P., Neu, W., Neugart, R., Pellarin, M., Pinard, J., Vialle, J. L., Wendt, K., and the ISOLDE Collaboration, *J. Phys. B* **23**, 3511 (1990).
41. Andreev, S. V., Mishin, V. I., and Letokhov, V. S., *J. Opt. Soc. Am. B* **5**, 2190 (1988).
42. Wood, C. S., Bennett, C. C., Cho, D., Masterson, B. P., Robers, J. L., Tanner, C. E., Wieman, C. E., *Science* **275**, 1759 (1997).
43. Dzuba, V. A., Flambaum, V. V., and Sushkov, O. P., *Phys. Rev. A* **51**, 3454 (1995).
44. Simsarian, J.E., Ghosh, A., Gwinner, G., Orozco, L. A., Sprouse, G.D., and Voytas,

P., *Phys. Rev. Lett.* **76**, 3522 (1996).
45. Simsarian, J. E., Shi, W., Orozco, L. A., Sprouse, G. D., and Zhao, W. Z., *Opt. Lett.* **21**, 1939 (1996).
46. Zhao, W. Z., Simsarian, J. E., Orozco, L. A., Shi, W., and Sprouse, G. D., *Phys. Rev. Lett.* **78**, 4169 (1997).
47. Simsarian, J. E., Orozco, L. A., Sprouse, G. D., and Zhao, W. Z., *Phys. Rev. A* **57**, 2448 (1998).
48. Simsarian, J. E., Zhao, W. Z., Orozco, L. A., and Sprouse, G. D., *Phys. Rev. A* in press (1999).

Magnetic Trapping, Evaporative Cooling, and Bose Einstein Condensation

S. L. Rolston

National Institute of Standards and Technology PHY A167, Gaithersburg, MD 20899, United States

Abstract. In these lecture notes I will discuss the techniques of magnetic trapping of neutral atoms and evaporative cooling methods. I will further discuss the observation of Bose Einstein Condensation in weakly interacting gases and summarize some of the experiments that have been performed on the condensates.

I INTRODUCTION

In my lectures in Mexico City I spent one day lecturing on optical lattices, and the next four on the techniques and physics of Bose Einstein Condensation (BEC). In these written lectures I will omit the optical lattices work to maintain some coherence in the presentation. For those interested in optical lattice research, I recommend reading two review articles [1,2], and the references therein.

These written lectures will follow the structure of my oral lectures: first will be a review of the principles and techniques of trapping atoms with magnetic fields. The third section will cover the technique of evaporative cooling, which is critical for the formation of BEC in weakly-interacting gases. The next section will deal with BEC, including the physics of the phase transition, and the experimental methods used to form it. The last section will be a synopsis of a number of the recent experiments that have been performed on BECs in the last three years. I will not attempt to cover all of the multitudinous experiments performed to date, but will select some of what I consider among the most important and interesting. The field of research in BEC in alkali gases is rapidly expanding. At this writing there are at least 15 different Bose condensates in laboratories around the world.

II MAGNETIC TRAPPING

Laser cooling (see the lecture in this volume by L. Orozco, and the recent Nobel lectures [3]) has been very successful in creating samples of cold atoms at relatively high densities. Even with some stunning accomplishments, these methods have not been able to simultaneously produce the high density and low temperature necessary to achieve Bose Einstein Condensation (high phase space density). Although the exact values depend on the atom, the conditions necessary for BEC are typically temperatures of order $1\mu K$, and at the same time densities of order 10^{14}cm^{-3}. Laser cooling has produced temperatures close to this, but in general the densities achieved have been limited to the range of $10^{10-12} \text{cm}^{-3}$ To achieve the lowest laser cooling temperatures normally requires working at the low end of the density range. This limitation is predominantly due to two mechanisms: collisions between laser-cooled atoms, and reabsorption of light. The presence of light tends to greatly increase the inelastic collision processes that lead to loss of atoms, resulting in limitations to the density of the sample. The $1/R^3$ long range dipole-dipole molecular potentials that govern collisions in the presence of light allow many partial waves to contribute to the collisions, with correspondingly large collision rates. The second limitation for laser-cooling to BEC comes from the high densities necessary, which generally end up producing optically thick samples, i.e. $n\sigma l \gg 1$, where n is the density, l is the sample thickness, and σ is the optical absorption cross section (of order $\lambda^2/2\pi$). The effect of photon rescattering is a net repulsive force [4] as atoms in the sample absorb light that was emitted from other atoms. This creates a limit to the density. It has been found that for typical magneto-optical traps (MOTs) increasing the loading rate leads to an operating regime where the density of the sample remains constant while the physical size increases. Recall that for BEC it is density, not number, that is important. In addition, it has been found that the rescattering of photons leads to heating, so that the temperature generally rises as the number of atoms in the sample (and the optical depth) is increased.

The criteria for BEC sets a condition on the phase space density, which is proportional to $nT^{-3/2}$, where T is the temperature. From this expression it is evident that one may hope to achieve BEC by simply getting very cold, even at low density. There have been a number of laser-cooling experiments that have demonstrated temperatures below the photon recoil limit [5]. They have mainly been in one or two dimensions, and were performed with low densities of atoms. These techniques, Raman cooling and velocity-selective coherent population trapping, rely on trapping the atoms in a non-absorbing zero-momentum state, which is very susceptible to rescattered photons. Although there are still experimental efforts to achieve BEC with all-optical techniques, I will concentrate on the successful techniques, employing magnetic trapping and evaporative cooling, that do not use light, thereby avoiding the aforementioned problems.

When we eliminate light as a trapping mechanism, we are left with magnetic or electric fields. The polarizibility of ground state atoms is quite small, making trapping with electric fields quite difficult. On the other hand, the magnetic moments of alkali atoms in particular, are large enough that is is possible to magnetically trap atoms with quite reasonable magnetic fields. Magnetic trapping relies on the interaction of the magnetic dipole of an atom with a spatially-varying magnetic field. The force exerted by the magnetic trapping fields can be written as

$$\vec{F} = \vec{\mu} \cdot \nabla \vec{B} \qquad (1)$$

where μ is the magnetic moment. For alkali atoms the magnetic moment will be of order the Bohr magneton, μ_B, whose size can be expressed in terms of temperature as $\mu_B/k_B = 672 \mu K/mT$. With laser cooling temperatures of $10-100 \mu K$, and easily achievable laboratory magnetic fields of 10-100 mT, it is clear that the magnetic interaction is more than sufficient.

Because we will be considering trapping slow, laser-cooled atoms, the adiabatic approximation is generally valid. The atomic spin (magnetic dipole) will follow the changes in the magnetic field direction as the atom orbits in the trap. This implies that the force is proportional to the angle between the spin and the field, given quantum mechanically by the m quantum number, and is proportional to the magnitude of the field. The force relevant for magnetic trapping can then be expressed as

$$\vec{F} = m g_J \mu_B \nabla |\vec{B}| \qquad (2)$$

where g_J is the Landé g-factor. Maxwell's equations determine the field configuration: in particular they prohibit a field maximum in a region without any magnetic sources (Wing's Theorem) (assuming static fields). This determines which states can be magnetically trapped: those whose energy increases with increasing field (sometimes called weak-field seekers). Fig. 1 shows a Zeeman energy vs. magnetic field plot (Breit-Rabi diagram) for an atom with I=3/2 and J=1/2, such as Na or [87]Rb. The states that are potentially trappable are those whose energy increases with field: the F=2, m=2,1 and F=1, m=-1 states. The F=2, m=0 state is also in principle trappable due to the quadratic Zeeman effect, although this produces a much weaker trap. This state has not been magnetically trapped to date. The particular state that is trapped is usually determined by collisional processes that limit the lifetime of some states due to collisions that change m.

To form a trap therefore requires a magnetic field configuration that has a minimum where the atoms can collect. The simplest geometry, known as a spherical quadrupole, consists of two current carrying coils aligned in the Helmholtz fashion (although not necessarily at the Helmholtz separation), with equal currents flowing in *opposite* directions in the two coils. This produces a field with a zero at the center of the coils, that linearly increases in any direction away from the center. Near the center the field can be written as

$$\vec{B} = B'_q \vec{\rho} + 2 B'_q \vec{z} \qquad (3)$$

where B'_q is the linear radial gradient. The coils are oriented along the $\hat{z}$ axis. Note that the field gradient is twice as large along the symmetry axis as along a radial direction (a consequence of $\nabla \cdot \vec{B} = 0$). If the coils are located at $\pm a$, the axial field can be written to first order as:

$$\vec{B}(\rho = 0) = B \vec{z} = \frac{3 \mu I r^2 a}{(a^2 + r^2)^{5/2}} z \hat{z} \qquad (4)$$

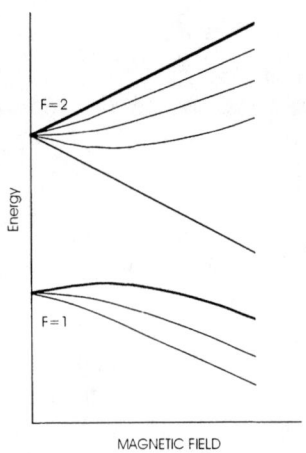

FIGURE 1. Breit-Rabi Diagram of energy vs magnetic field for an atom with $J = \frac{1}{2}$ and $I = \frac{3}{2}$. The two weak-field-seeking states that have been magnetically trapped are shown in bold.

Such a spherical quadrupole trap was the first magnetic trap used to trap atoms, in work done in 1985 at NIST [6], where they trapped sodium atoms loaded from an atomic beam.

The spherical quadrupole trap has one obvious problem. At the center of the trap the field reverses direction, going through zero. An atom with a trajectory that takes it directly through the center will be unable to adiabatically follow the changing direction of the field, since there is no guiding field precisely in the place where the field direction changes most rapidly. These nonadiabatic transitions are known as Marjorana transitions (keep in mind that the spin stays aligned with a space-fixed axis, it just fails to flip when the field does). If the magnetic field seen by an orbiting atom changes direction by an angle θ, the requirement for adiabatic following can be written as :

$$\frac{d\theta}{dt} << \frac{U_i - U_j}{\hbar} = \omega_{\text{Larmor}} \qquad (5)$$

Clearly if an atom's trajectory takes it directly through the center, where $\omega_{\text{Larmor}} = 0$, this condition cannot be satisfied. The probability of such a direct hit is of course quite small. The effect of the hole in the quadrupole trap only became apparent in experiments [7] where the temperatures of the clouds became low enough so that there was a reasonable probability of atoms getting close enough to the center so as to be unable to follow the changing field. This loss channel can be viewed as a Landau-Zener curve crossing problem, where the coupling between the curves corresponding to to the trapped and untrapped states is given by the residual magnetic field, which is proportional to the distance from the origin. From such an argument we can determine that the effective size of the hole under typical conditions is about 1μm.

There have been a number of solutions applied to the problem of the hole in the middle of this magnetic trap. Ketterle's group "plugged" the hole by focusing a blue-detuned far-off resonance laser into the center of the quadrupole trap [8], creating a small repulsive light shift potential that prevented the atoms from ever seeing the zero of the magnetic field. While this method worked, and allowed the MIT group to observe BEC, it was quickly abandoned, in part due to the difficulty of alignment, the problem of laser beam vibration causing heating, and the fact that the resultant potential was not simple.

The most successful approach to solve the problem of the hole was developed at JILA [9]. If we add a uniform field to a quadrupole field, it is easy to see that the zero of the field is just shifted in space, and does not vanish. We have just translated the problem to a different point in space. The JILA group developed a time-varying field approach that circumvents this problem. In a time-orbiting potential (TOP) trap a uniform field is applied, but its direction rotates in space. The zero is displaced from the origin, and can be arranged to describe an orbit that is always outside of the cloud of trapped atoms. If the rotation frequency is high enough, the atoms feel a time-averaged harmonic potential. For the JILA geometry the bias field rotated in a plane orthogonal to the quadrupole symmetry axis, so that the instantaneous field applied was:

$$\vec{B} = [B'_q x + B_b \sin \omega t]\hat{x} + [B'_q y + B_b \cos \omega t]\hat{y} + [2B'_q z]\hat{z} \qquad (6)$$

where B_b is the rotating bias field. If we assume the atoms will be near the origin, we can Taylor expand the magnitude of the magnetic field (remember that it is the magnitude of the field that is important for magnetic trapping) about the origin and time average to find an effective potential:

$$U = \mu_i \frac{B'^2_q}{4B_b}[x^2 + y^2 + 8z^2] \qquad (7)$$

For this expression to be applicable, and for such a trap to work, we must choose the frequency of rotation of the bias field so that $\omega_x \leq \omega_{\text{TOP}} \leq \omega_{\text{Larmor}}$, where ω_x is the orbital frequency in the trap, and ω_{TOP} is the bias field rotational frequency. The first inequality assures that the atoms will see a time averaged potential, while the second inequality assures that they will adiabatically follow the field. Typical values are $\omega_x/2\pi = 100\,\text{Hz}$, $\omega_{\text{TOP}}/2\pi = 10\,\text{kHz}$, and $\omega_{\text{Larmor}}/2\pi = 20\,\text{MHz}$. From Eq. (7) it may appear that we can create an arbitrarily stiff trap simply by *reducing* the bias field, B_b. There is, however, an additional requirement that the orbit of the field zero (colorfully known as the "circle of death") must be larger than the size of the cloud, i.e. $R_{\text{cod}} = \frac{B_b}{B'_q} \gg \langle x^2 \rangle^{1/2}$, which limits how small of a value for B_b we can choose.

In a variation of the TOP trap, our group at NIST uses a trap where the plane of rotation of the bias field includes the quadrupole symmetry axis (see Fig. 2) [10]. The same principles apply, but the ratio of spring constants becomes 1:2:4 instead of 1:1:8. Some typical experimental parameters for TOP traps are quadrupole coils

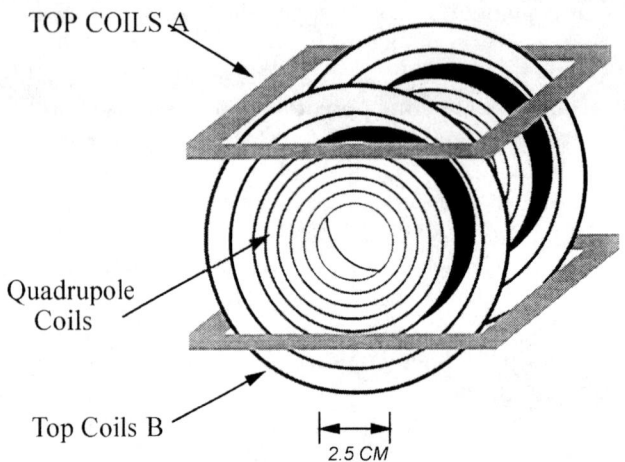

FIGURE 2. A schematic of the TOP trap used at NIST-Gaithersburg. The rotating bias field is created by two pairs of coils, and the plane of rotation includes the quadrupole symmetry axis, unlike the JILA design.

with 20-50 turns, with characteristic dimensions of 1-3 cm, and currents of 200-300 A. This will produce linear field gradients of order 5-10 T/m. The bias coils consist of 2-6 turns, currents of 10-100 A, and produce rotating fields of 0.001 - 0.01 T.

Another solution to the problem of the hole in a quadrupole trap is to dispense with this type of trap altogether, and use a geometry with higher order gradients that has a non-zero field at the center. This approach has been very successful, and is becoming the trap of choice for BEC experiments. All of the traps that have been developed for BEC of this type can be considered a type of Ioffe trap [11]. The Ioffe trap geometry consists of four current bars with current flowing in alternating directions (producing a two-dimensional linear quadrupole field) with a set of coils around the bars with current flowing in the same direction (see Fig. 3). It is important that these so-called "pinch" coils be separated by more than the Helmholtz condition (a = r) so that their field has a minimum at the center. The pinch coils actually generate an anti-trapping radial gradient, so it is important to adjust the geometry and current in the bars and coils such that the gradient due to the bars dominates along the radial direction. The resultant trapping field for a Ioffe trap is harmonic to lowest order, and can be expressed as :

$$|B| \propto B''_\rho \rho^2 + B''_z z^2 \qquad (8)$$

where B''_ρ is the radial curvature and B''_z is the axial curvature. The actual values depend on geometries and currents. Additional bias coils which apply a uniform field are often added to allow some choice over the value of the field at the center (and additional control of the gradients, since adding a uniform field also effects the curvatures). This allows adjustments of the stiffness and aspect ratio of the trap, which can be varied to produce anything from a spherical trap to a very

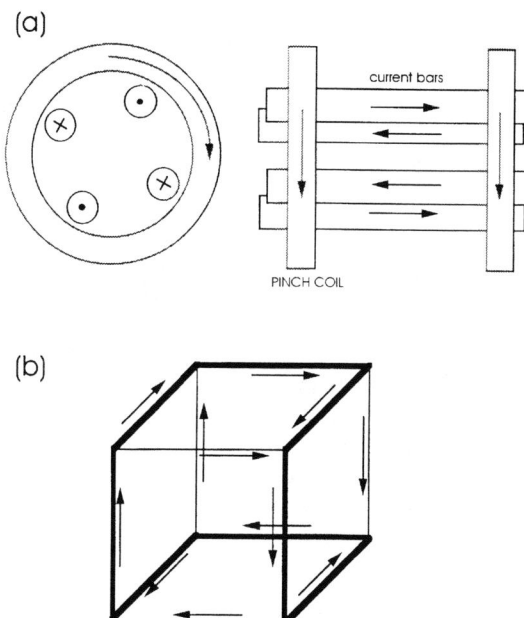

FIGURE 3. (a). The Ioffe trap consists of four current bars and a pair of pinch coils. (b). The baseball trap can be seen as a variation of the TOP trap, with some of the current segments removed.

anisotropic trap. Among the Ioffe trap variations used for BEC are the cloverleaf trap of MIT [12], the baseball trap used at JILA [13], and the 4-dee trap [14]. In each case, although there are often many more coils, they can be viewed as essentially Ioffe traps (see Fig. 3). Such traps typically use tens of turns and hundreds of amps of current, often dissipating 10 kW or more of electrical power.

Once we have a magnetic trap, it must be loaded with cold atoms, usually from a magneto-optical trap (MOT). Since BEC requires a large phase space density, it makes sense to try to match the stiffness of the magnetic trap to the size and temperature of the MOT, in order to preserve as much of the phase space density as possible. In 1-d the initial phase space density will be proportional to the initial size, σ_i, and thermal velocity, v_i, of the MOT. We will suddenly turn on a magnetic trap (assumed to be harmonic with a spring constant κ). Conservation of energy requires:

$$\frac{1}{2}mv_i^2 + \frac{1}{2}\kappa\sigma_i^2 = \frac{1}{2}mv_f^2 + \frac{1}{2}\kappa\sigma_f^2 \tag{9}$$

where we assume that the final velocity and size are measured after the atoms have come to equilibrium through elastic collisions. Using the equipartition theorem, we can write

$$v_f^2 \sigma_f^2 = \frac{1}{2}\left[\frac{m}{\kappa}v_i^4 + 2v_i^2\sigma_i^2 + \frac{\kappa}{m}\sigma_i^4\right]. \tag{10}$$

By differentiating this equation with respect to κ we find the optimum value of $\kappa = \frac{mv_i^2}{\sigma_i^2}$. Inserting this back into the expression for the phase space yields

$$v_f \sigma_f = v_i \sigma_i. \tag{11}$$

Phase space density is conserved! It may seem surprising that suddenly switching on a potential does not inevitably lead to heating and a decrease in phase space density. In fact, the total energy of the system does increase, but half of the energy goes into potential energy (equipartition theorem). Phase space density only depends on kinetic energy, and is conserved. In practice, there is usually some loss of phase space density upon transfer to a magnetic trap, simply because the aspect ratio of the dense MOT will not be exactly the same as that of the magnetic trap, which would be required to achieve absolutely no loss of phase space in three dimensions.

III EVAPORATIVE COOLING

When atoms are transferred into a magnetic trap, their phase space density is generally still many orders of magnitude (4-6) away from what is needed for BEC. Since cooling with light is not allowed, a new technique must be employed. Evaporative cooling, first suggested by Hess [15] in 1986 has proven to be wildly successful. Just as in the cooling of a cup of coffee, evaporative cooling in a magnetic trap relies on the fact that the evaporating particles carry off more than the average energy per particle. The most energetic particles are the ones most likely to have enough energy to escape. This cooling method relies on loss of atoms to provide cooling, which may seem counterproductive when we are trying to get to BEC, since high density is a necessity. Nonetheless it has proven to be quite effective, and in a trap evaporative cooling can actually lead to an increase in density, even though there are fewer atoms. A review article [16] on evaporative cooling contains much more information than I will present here.

The basic process can most easily be explained in a stepwise fashion, although it is usually implemented in a continuous way. Consider a gas with a Boltzmann distribution of energies in a trap, described by a temperature T. We then suddenly reduce the top of the trap in some way so that particles with an energy greater than a value E_{trap} escape. Just after this event the energy distribution will be a truncated Boltzmann distribution with no atoms with an energy higher than E_{trap}. We then wait for a long enough period of time such that elastic collisions between the trapped atoms rethermalizes the distribution. It will now have a lower temperature, at the expense of the loss of some atoms. This procedure can then be repeated, further reducing the lip of the trap, and cooling the atoms toward BEC.

For evaporative cooling to work, the density must be high enough so that elastic collisions can rethermalize the distribution in a reasonable time. What is reasonable is set by other inevitable loss processes, usually collisions between hot (room temperature) background gas atoms that eject cold atoms from the trap. For good UHV systems, this sets a time limit in the 10-100 s range. Because we are interested in low temperatures, it is reasonable to assume that the elastic collisions are predominantly s-wave collisions. The elastic collision cross section in this limit is:

$$\sigma_{el} = 8\pi a_s^2, \tag{12}$$

where a_s is the scattering length. This parameter can be viewed as the effective hard-sphere radius for the low energy collisions and fully characterizes elastic scattering in this limit. (The scattering length can actually be negative, so the hard sphere analogy is not complete). The thermalization rate is proportional to the elastic collision rate, and can be written as :

$$\Gamma_{therm} = \gamma n \sigma_{el} v_{th} \tag{13}$$

where n is the density, v_{th} is the thermal velocity, and γ is a constant linking the thermalization rate to the elastic collision rate. Monte Carlo simulations have found that $\gamma = 2.7$ [17]. The scattering lengths have been determined for most of the species that have been studied for BEC, and are 0.065, −1.5, 2.75, 5.5, and $|>14|$ nm for H,Li,Na,^{87}Rb, and Cs in the hyperfine states that have been Bose condensed [18] (cesium has yet to be condensed and the sign of the scattering length is undetermined).

The evaporative cooling process can be optimized as the temperature and density change as the cooling proceeds [16]. The process is governed by the parameter $\eta = E_{trap}/kT$, the ratio of the height of the trap lip to the temperature. Recall that the goal is to optimize phase space density ($\propto nT^{-\frac{3}{2}}$), which implies that we should try to minimize the loss of atoms. The limiting case would be to set $\eta = N$, so that one atom carries off all of the energy. Of course this would take an impractically long time. The length of time of the evaporation process will depend on the relative rates between "good" (elastic) and "bad" (inelastic loss) collisions. The presence of inelastic collisions argues for going as fast as possible, while the goal of conserving atoms argues for going as slow as possible. In practice the compromise between these competing requirements ends up with $\eta = 4-6$ in most experiments. In addition the shape of the trap will influence the evaporation method. If, for example, we tried to condense in a square well trap, where the volume of the cloud is independent of the energy, the thermalization rate will scale as $E^{\frac{1}{2}}$ due to the velocity term in the expression for the rate. Thermalization will take longer and longer as the sample cools, eventually turning off cooling. For a harmonic trap, the volume scales as $E^{\frac{3}{2}}$ so that the collision rate scales as E^{-1}. This produces what is known as runaway evaporation: the density actually increases so that the thermalization rate accelerates as the sample cools.

The method to produce a lip in the trap is usually to use an rf field that is tuned to drive a transition between a trapped and an untrapped Zeeman sublevel at a particular value of the field (which corresponds then to a particular energy) [19]. In a dressed picture the potential acquires a lip, and the problem can be addressed as a Landau-Zener curve crossing problem, although in general it is a multilevel problem. Evaporation using the circle of death (the rotating field zero in the TOP trap) has also been demonstrated [10], although the technique is less flexible, because the position of the circle of death is coupled to the parameters that determine the trap depth.

Evaporative cooling has many parameters that can be varied: trap shape, trap depth, η, rate of change of the position of the edge, and the functional time dependence of the position of the edge. The best strategy is of course to maximize the phase space density for each step taken. Typical results end up with a decrease in atom number of $10^3 - 10^4$, with an increase in phase space density of $10^4 - 10^6$, when BEC is achieved. The procedure typically takes 10-30 s.

IV BOSE EINSTEIN CONDENSATION

In 1924 Einstein predicted the existence of a phase transition in a gas of non-interacting Bose particles. The amazing feature of this transition was that at a finite temperature, the atoms would condense in the lowest energy state of the system, even though they did not interact with each other. While the phenomenon of BEC is in part responsible for some of the properties of superfluid liquid helium, it was not until the observation of BEC in a weakly-interacting gas of rubidium [20], that a system very close to that envisioned by Einstein could be studied.

Any standard statistical mechanics textbook will have a derivation of the conditions necessary for BEC, so I will not repeat that here, but simply highlight some of the results. From statistical mechanics the distribution of bosons in states with energy ϵ can be written as

$$n_\epsilon = \frac{1}{e^{\frac{\epsilon-\mu}{kT}} - 1} \quad (14)$$

where μ is the chemical potential (a measure of the extra energy added to the system with the addition of a particle). As the temperature decreases, a phase transition will occur if the phase space density is larger than a critical value. This corresponds to a phase space density where the deBroglie waves of the individual atoms begin to overlap one another. The "size" of the atoms is given by the thermal deBroglie wavelength:

$$\lambda = \frac{h}{(2\pi mkT)^{1/2}}. \quad (15)$$

When the condition

$$\lambda^3\left(\frac{N}{V}\right) > \zeta\left(\frac{3}{2}\right) = 2.612 \tag{16}$$

is fulfilled, a macroscopic occupation of the ground state of the system occurs. If we consider the homogeneous case where the ground state ϵ_0 has zero energy, at the phase transition the chemical potential goes to zero, and the population in the ground state gets macroscopically large. Although Einstein's theory was for a homogeneous system, it still applies in a trap. In a harmonic trap, the critical phase space density, and critical temperature are the same as for the homogeneous case, but the number in the ground state goes as

$$\frac{N_0}{N} = 1 - \left(\frac{T}{T_c}\right)^3 \tag{17}$$

where T_c is the critical temperature, defined as

$$T_c = \frac{h^2}{2\pi m k}\left(\frac{N}{2.612V}\right)^{\frac{2}{3}}. \tag{18}$$

For the alkali systems, BEC occurs in the range where $T \approx 1\mu K$ and the density is $\approx 10^{14} cm^{-3}$. While Einstein's theory was for non-interacting Bosons, the alkali system does have interactions with a strength determined by the size of the scattering length, a_s. Since the atoms are prevented from physically occupying the same space, and each atom has an effective volume of a_s^3, we can estimate the amount of the trap volume that must be excluded from the condensate wave function (depletion of the condensate). For the system to be weakly interacting, we want this to be small (unlike liquid helium). Explicitly, we require that $na_s^3 \ll 1$. For $a_s \approx 5$ nm, and a density of 10^{15} cm^{-3}, this is of order 10^{-4}. From this it is safe to conclude that the alkali systems are weakly-interacting over the entire experimental range of densities and temperatures that are achievable.

The condensates that are formed are of course not the true ground state of the system, which would be a lump of alkali metal. They are metastable, however, and can live for hundreds of seconds. In a magnetic trap a metal will never be formed, because the building blocks for the metal will be molecules formed in inelastic collisions. These will leave the trap long before anything approaching a metal can be formed. In general the collision rates are quite small. The two-body loss mechanism, which entails a spin-flip during a collision (dipolar loss) has a rate constant in the 10^{-15} to $10^{-17} cm^{-3}s^{-1}$ range, yielding lifetimes of 10-1000 s. The three-body loss rates are typically in the range of 10^{-29} to $10^{-30} cm^6 s^{-1}$ and are only significant at the high end of the densities achieved. One notable exception to these rate constants is cesium, which apparently has a low energy resonance that greatly increases the loss rates, and has to this date made cesium uncondensable [21]. Quite often the actual lifetime of the BEC is determined by collisions with background gas atoms due to imperfect vacuum.

Once we have evaporatively cooled a sample of alkali atoms to the point where the phase space density has reached or exceeded the critical value, the question remains as to how to detect that a BEC exists. In a trap, it is quite easy. Below the critical temperature, a macroscopic number of atoms will be in the ground state. This occurs while the population is still spread among many excited states. (For a trap with a 100 Hz oscillation frequency, the critical temperature might be 1μK, which corresponds to a frequency of 10kHz.) Since the ground state wave function has a small spatial extent, we would expect to see a sharp peak in the density profile corresponding to ground state atoms on top of a broader thermal background. (For the moment we will neglect interactions.) This is challenging to measure, since for typical trap parameters the size of the trap ground state will be of order 1μm, which is difficult to image. What is typically done instead is to shut off the trapping potentials, and let the cloud ballistically expand. In this case the rms of the distribution x_0 evolves to $\sqrt{(x_0^2 + v_{th}^2 \tau^2)}$ where τ is the expansion time. In the limit where the cloud has expanded to many times its original size the cloud size is simply proportional to the thermal velocity. An image of the cloud at this point directly measures the *momentum* distribution.

The experimental method used to actually measure the distribution is absorptive [20] or phase contrast imaging [22]. The expanded cloud is illuminated with a collimated laser beam, tuned far enough off resonance so that the sample is not very optically thick. An optical system images the position of the BEC onto the focal plane of a CCD camera, and the light scattered from the condensate produces a shadow in the laser beam, which is also hitting the camera. This image is then divided by an image with no BEC to extract the optical absorption from the atomic sample. The operative relationship is

$$\ln(\frac{I(x,y)}{I_0}) = -\sigma \int n(x,y,z) dz \qquad (19)$$

where I_0 is the normalizing intensity, and σ is the optical absorption cross section. From this, with some assumptions about the cloud shape in the z direction (which the imaging averages over) one can obtain density, number of atoms, and size, which is proportional to the momentum. Phase contrast imaging [22] works in a similar manner, except that the relative phase between the laser and the scattered light is adjusted to produce maximum destructive interference, even for the case of a large detuning where the BEC would act mainly like a real index of refraction with little absorption.

As mentioned earlier, the first observation of BEC in a weakly-interacting gas was reported by the JILA group in 1995 [20]. I will summarize their approach and evidence for BEC, but of course it is best to refer to the original reference. Their experiment consisted of loading ^{87}Rb atoms into a magneto-optical trap. The atoms were cooled to $\approx 20\mu$K with a phase of optical molasses, and loaded into a TOP trap in the state F=1, m=-1. The trapped sample was evaporatively cooled down to BEC. The trap was adiabatically lowered to a small spring constant (necessary because they were unable to rapidly shut off the magnetic fields when they were operating at maximum). The trap was then suddenly shut off, and after 60 ms of free expansion the cloud was measured by absorption imaging.

In the absorptive images they observed the formation of a high density component of the cloud when the rf frequency of the evaporation decreased below a certain value. This was of course suggestive of a phase transition. A critical piece of evidence was that the high density component was not isotropic, as was the cloud above the critical value. Since the BEC is a single wave function, the momentum spread in different directions is proportional to $\omega^{\frac{1}{2}}$ in those directions. In particular, in the TOP trap the oscillation frequencies in the two directions visible in the image differ by $\sqrt{8}$. They observed that the cloud had an anisotropy consistent with $8^{\frac{1}{4}}$, the large dimension corresponding to the tight direction of the trap, as expected.

That first experiment produced condensates with only a few thousand atoms, so that the effects of interactions between the condensate atoms were quite small. Since then it has become routine to form condensates of greater than 10^6 atoms, where although the system is still weakly interacting, the interactions cannot be neglected. The effects of the interactions can be accounted for by using a mean field approach. The result of such an approximation (for which the justification and derivation can be found in [23] and references therein) is a non-linear Schrödinger equation for the many-body condensate wave function:

$$i\hbar \frac{\partial \Phi}{\partial t} = (-\frac{\hbar^2 \nabla^2}{2m} + V_{\text{trap}} + \frac{4\pi \hbar^2 a_s}{m}|\Phi|^2)\Phi \qquad (20)$$

The non-linear term in the Hamiltonian describes the effects of collisions between atoms through a mean field that is proportional to the scattering length and the density ($|\Phi|^2$). This term is not necessarily small. Expressing it in terms of the characteristic scale length of the harmonic oscillator, $a_{\text{ho}} = (\frac{\hbar}{m\omega})^{1/2}$, this mean field energy is $\approx N^2 a_s/a_{\text{ho}}^3$, while the kinetic energy (proportional to $N\hbar\omega$) is $\approx N/a_{\text{ho}}^2$. The ratio of internal energy to kinetic energy then scales as $\approx Na_s/a_{\text{ho}}$ which is usually $\gg 1$ for $N > 10^4$. In this case the wave function is dominated by the mean field and the shape of the condensate is not that of the ground state of the harmonic oscillator. The many-body condensate wave function is no longer simply a product state of N harmonic oscillator ground state wave functions, but is a complicated entangled many-body wave function.

Most of the theoretical work that describes the behavior of these weakly-interacting condensates is based on the Gross-Pitaevskii (GP) equation (see the review article by Stringari [23]). This is derived from the non-linear Schrödinger equation by replacing the wave function Φ with:

$$\Phi = \phi e^{-i\mu t/\hbar} \qquad (21)$$

where μ is the chemical potential and

$$\int \phi^2 d^3 r = N. \qquad (22)$$

This leads to the GP equation:

$$\left(-\frac{\hbar^2 \nabla^2}{2m} + V_{\text{trap}} + \frac{4\pi \hbar^2 a_s}{m}|\phi|^2\right)\phi = \mu\phi. \tag{23}$$

When N gets to be large so that the internal energy is much larger than the kinetic energy, the GP equation can be further simplified (Thomas-Fermi limit) by dropping the kinetic energy term from the Hamiltonian, yielding:

$$\left(V_{\text{trap}} + \frac{4\pi \hbar^2 a_s}{m}|\phi|^2\right)\phi = \mu\phi, \tag{24}$$

which can easily be solved to give the condensate wave function in the Thomas-Fermi limit:

$$\Phi_{\text{TF}} = \left(m\frac{\mu - V_{\text{trap}}}{4\pi \hbar^2 a_s}\right)^{\frac{1}{2}} e^{-i\mu t/\hbar}. \tag{25}$$

For a harmonic trap the density profile of the BEC is simply an inverted parabola. The Thomas-Fermi solution has unphysical sharp edges, which are of course properly accounted for in the full solution of the GP equation. From the Thomas-Fermi approximation one can derive a number of useful expressions:

$$\mu_{\text{TF}} = \frac{\hbar\omega}{2}\chi^{\frac{2}{5}} \tag{26}$$

$$R_{\text{TF}} = a_{\text{ho}}\chi^{\frac{1}{5}} \tag{27}$$

$$n_{\text{TF}}(0) = n_{\text{ho}}(0)\frac{15^{\frac{2}{5}}\pi^{\frac{1}{2}}}{8}\chi^{-\frac{3}{5}} \tag{28}$$

$$\text{where } \chi = \frac{15Na_s}{a_{\text{ho}}}. \tag{29}$$

As mentioned earlier it is possible to have an atomic system with a negative scattering length. This is in fact the case for lithium, which has a scattering length of -1.45 nm in the $m_J = \frac{1}{2}, m_I = \frac{3}{2}$ state (their trap is in the high field region where F is no longer a good quantum number) that has been trapped by the Rice group [24]. In a homogeneous system condensation would not be possible, as the energy could always be lowered by increasing the density, effectively a "negative" pressure. The system would implode. In a trap the situation can be a little different. The zero-point energy can exert a positive "pressure" for small numbers of atoms which prevents collapse, until N get too large and the mean field dominates. Solutions of the GP equation [25] for the Rice parameters predicted a maximum number of atoms of 1400 before collapse. This is borne out in the experiment which cannot produce condensates with more than $\approx$ 1500 atoms. The dynamics of the collapse process are not yet understood and are under study by the Rice group.

V EXPERIMENTS WITH BEC

In my lectures in Mexico City I reviewed a number of experiments on BECs in some detail. I will not do that here, but instead will give a guide to the literature.

I will not attempt to be complete, nor do I mean to suggest that these are the only important experiments that have been performed with BEC. I will try to give some flavor for the variety and the diversity of what has been accomplished in the three years since BEC was observed at JILA.

Condensate Fraction: The groups at JILA and MIT made careful studies of the fraction of condensed atoms [26,12] as a function of temperature. The experiments consisted of measuring the temperature by carefully fitting the momentum distribution of the thermal component, and measuring the number of atoms in the condensate. They found excellent agreement with the $1 - (T/T_c)^3$ dependence, and could observe the effects of the mean field on the exact position of T_c.

Mean Field: The mean field approximation has been tested by observing the expansion of a condensate, which is dominated by a mean field "explosion" for large N. Using the Thomas-Fermi expressions, one can show that the energy released in such an expansion is $E_{\rm rel} = (2/7)\mu$. Since μ is proportional to $N^{2/5}$ the release energy should be as well, and this has been verified by [12]. Using a simple model that describes the size of a BEC in the Thomas-Fermi limit for any arbitrary time-dependent harmonic potential [27], the mean field was further tested by observing the change in aspect ratio from an expanding, initially anisotropic condensate. Again there was excellent agreement with the theory. It appears that the GP equation, and even the Thomas-Fermi limit, are very applicable to the experimental realizations so far produced.

Thermodynamics: One of the most dramatic aspects of superfluid liquid helium is the exceedingly sharp lambda-shaped phase transition that occurs at 2.2 K, when plotting specific heat vs temperature. An experiment [26] at JILA tried to replicate this, by measuring the release energy as a function of temperature. While they did not observe the dramatic shape evident in liquid helium, they did observe a discontinuous change in the slope of the release energy vs. temperature. Once again the data agreed with the GP equation [23]. Note that since this just measures the release energy it is not exactly a specific heat measurement.

Collective Modes: When the interaction between particles is included, the modes of excitations of the BEC are phonons with frequency $\nu = c/\lambda$. The speed of sound in the BEC is given by

$$c = \sqrt{\frac{n 4\pi \hbar^2 a_s}{m^2}}. \tag{30}$$

The time-dependent GP equation can be used to find the low-lying collective modes of oscillation. We assume a solution

$$\Phi(r,t) = e^{-i\mu t/\hbar}[\phi(r) + u(r)e^{-i\omega t} + v^*(r)e^{i\omega t}]. \tag{31}$$

Inserting this into the GP equation results in a set of coupled equations for $u(r)$ and $v(r)$ which can be solved numerically [28]. In the hydrodynamic limit, the modes can be classified by quantum numbers n,l,m. The eigenfrequencies [23] are then given by

$$\omega(n,l,m) = \omega_{\text{ho}}(2n^2 + 2nl + 3n + l)^{1/2} \qquad (32)$$

Note that the dipole mode ($n = 0, l = 1$) has an eigenfrequency of ω_{ho} and just corresponds to the whole condensate moving along a single-particle trajectory. The experiments [29,30] that have observed the collective modes operate by first exciting the mode by either varying the shape and/or depth of the trap (the coupling to the modes will be different for different perturbations), and then observing the periodic motion with imaging after some period of free oscillations.

The MIT group directly measured [31] the speed of sound by creating a small perturbation of the potential with a focussed laser beam in the middle of a long skinny BEC. They suddenly switched off the perturbation, and could see the sound wave propagate down the condensate.

Coherence Properties: One of the special properties of BEC is that it should have coherence properties similar to that of the laser. A distinct difference between a laser and a filtered thermal source is the second-order correlation function (intensity correlation). For a thermal source the normalized $g^{(2)}(\tau)$ starts at 2 at $\tau = 0$ and eventually falls to 1 at long times, while for a laser it is 1 at all times. The same is true for the spatial correlation function for a BEC. The mean field energy (effectively proportional to the two-body correlation function at $r = 0$) is twice as large for thermal atoms as it is in the condensate. This has been observed in a manner of speaking, in that the mean field estimates using independently measured scattering lengths are in good agreement with BEC measurements including the lack of the factor of two. More dramatic evidence of the coherence of BEC has been seen looking at third order coherence, which can be probed at $r = 0$ by three-body collisions. The JILA group measured [32] the difference in three-body rates for a thermal cloud and for a BEC and found that the rate slowed by a factor of 7 ± 2 in the condensate, in good agreement with the prediction of 6 (3 factorial). This demonstrates that the BEC does indeed have the coherence characteristics that make a laser such a special device.

First order coherence has been demonstrated [33] by the MIT group where they observed interference fringes between two condensates. In this experiment they formed two independent condensates in a magnetic trap by bisecting the trap with a laser-produced barrier. They then let the BECs expand and after some time imaged the atomic sample, and observed 20μm fringes from the interference of the condensates. Because there was no phase relationship between the independent condensates, on each successive measurement the position of the fringes would move randomly. An interesting experiment yet to be done would be to lock the phases of the two BECs together with tunneling and measure the phase of the fringe pattern.

Atom lasers: Since BEC shares similar coherence properties with a laser, it is natural to explore the possibility of producing what has become known as an "atom laser" (a clever acronym has yet to be found). One can develop a rather complete analogy between the two: the laser cavity is replaced by the trap; the gain medium for the atom laser is the thermal cloud; the pump is evaporative cooling; and the

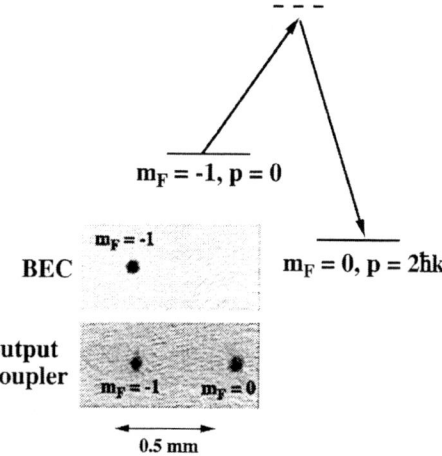

FIGURE 4. Raman output coupler: A stimulated Raman transition is used to transfer 2 $\hbar k$ of momenta, and change the magnetic sublevel to $m = 0$. The pictures are images of optical depth taken after a time-of-flight period.

output coupler must be developed for the atom laser. An often raised objection in this analogy is that atoms cannot be created or destroyed, unlike photons. In fact, the analogs to photons are *condensate* atoms, and they can be created and destroyed by collisions - the analogy is quite firm. One notable difference is the mean field interactions between atoms. While there are sometimes non-linear interactions in laser media that may be analogous, they are the exception, not the rule.

The MIT group demonstrated a simple output coupler [34], using pulses of rf to flip the spins of a portion or all of the condensate. The atoms are then in untrapped states and fall due to gravity (and are expelled by the trapping fields). This produced spreading pulses of presumably coherent matter waves. Since the rf transition is a coherent process, it is reasonable to assume that there was coherence between the pulses, although this was not measured. The mean field repulsion and spreading due to the trapping potential would introduce significant phase variations across the output wave, so even though it would be coherent it still might be difficult to use. Our group at NIST has developed a directional output coupler [35]. In this case we used stimulated Raman pulses instead of rf pulses. The advantage is that with the appropriate choice of frequencies and directions the atoms can be coupled out with 2 or more units of photon recoil momentum. We have demonstrated the output of a well-collimated beam of atoms in the $m = 0$ state where they no longer feel the trapping potential, with a longitudinal velocity of 6 cm/s and a corresponding deBroglie wavelength of 300 nm (see Fig. 4).

Two-component condensates: The JILA group has demonstrated [13] simultaneous trapping, cooling and condensing of the $F = 2, m = 2$ and $F = 1, m = -1$ states of ^{87}Rb. They were able to demonstrate sympathetic cooling, in that both

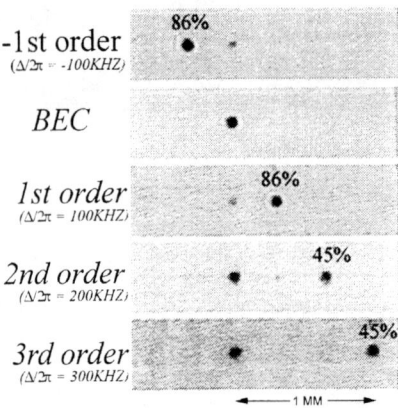

FIGURE 5. Bragg diffraction of a BEC: By applying a moving standing wave (whose velocity is determined by the frequency difference of the two waves comprising the standing wave) we can Bragg diffract a portion of the condensate into a well-defined momentum state.

spin states condensed when only one was being evaporative cooled. They observed a repulsion between the condensates so that in equilibrium they separate. This work, along with recent work at JILA and MIT, show that there will be a rich arena of study involving multiple-species condensates.

Atom Optics with BECs: Our group at NIST has concentrated on using our sodium BEC as a source for atom optics experiments. As mentioned above, we have developed a directional output coupler based on a Raman transition that transfers momentum. This evolved from an experiment (see Fig. 5) where we demonstrated Bragg diffraction of a BEC. We exposed the BEC to a moving optical standing wave, whose velocity was chosen so that energy and momentum were conserved. This process can be viewed as a stimulated Raman transition between an initial state with no momentum, and a final state with $2\hbar k$ of momentum. We demonstrated that this process could produce coherent momentum transfer with 100% efficiency. We observed higher order diffraction, including 15% efficiency at 6th order, which created an atomic velocity of 36 cm/s. We also studied simple diffraction of the condensate (see Fig. 6) from a short pulse of a stationary optical standing wave. The standing wave writes a periodic phase variation onto the BEC wave function. After some time evolution, the BEC splits into three components, corresponding to zeroth and first-order diffracted waves. We were able to demonstrate in a simple three pulse atom interferometer that these pieces of the condensate were in fact coherent. The separation between the coherent components of the wave function can be macroscopic, on the order of 1 mm in this experiment. BEC is the ideal source for atom optics, just as the laser is for optics.

Hydrogen: A discussion about BEC would not be complete without a mention of spin-polarized hydrogen. Work commenced in the late 1970's to Bose condense

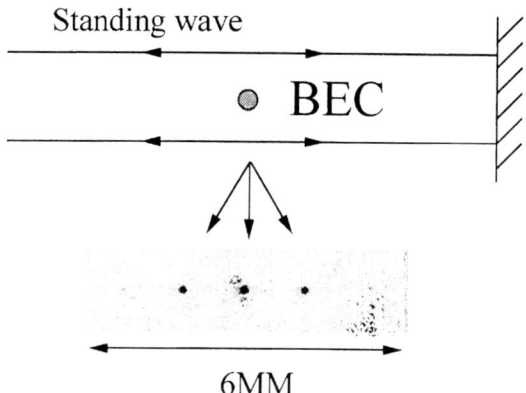

FIGURE 6. Diffraction of a BEC by an optical standing wave: Subjecting a BEC to a short standing wave generates diffraction into $\pm\hbar k$ modes. This represents a coherent wave function with a macroscopic distance between its pieces.

hydrogen. The use of a magnetic trap and evaporative cooling were first developed in that system. It was particularly challenging, because the lack of a $Ly - \alpha$ laser meant that the initial cooling required complicated cryogenics and dilution refrigerators. After two decades of effort, the MIT group lead by Kleppner and Greytak finally succeeded in creating a hydrogen BEC [36]. Although the condensed fraction is small (5 %) the number of atoms is the largest yet achieved in any system (10^9).

As is clear from this impressive, but partial, list of experiments that have been performed with weakly-interacting Bose condensates, we can expect many new and interesting results to develop in this field.

REFERENCES

1. Jessen, P. S., and Deutsch, I. H., *Adv. Atom. Mol. Opt. Phys.* **37**, 95 (1996).
2. Rolston, S. L., Phys. World **11**, 27 (1998).
3. Chu, S., Cohen-Tannoudji, C., and Phillips, W. D., *Rev. Mod. Phys.* **70**, 685,707, and 721 (1998).
4. Walker,T., Sesko, D., and Wieman, C., Phys. Rev. Lett. **64**, 408 (1990).
5. Reichel, J., et al., *Phys. Rev. Lett.* **75**, 4575 (1995), and Lawall, J., et al., *Phys. Rev. Lett.* **75**, 4195 (1995), for example.
6. Migdall, A., et al., *Phys. Rev. Lett.* **54**, 2596 (1985).
7. Davis, K. B., et al., *Phys. Rev. Lett.* **74**, 5202 (1995).
8. Davis, K. B., et al., *Phys. Rev. Lett.* **75**, 3969 (1995).

9. Petrich, W., et al., *Phys. Rev. Lett.* **74**, 3352 (1995).
10. Lutwak, R., Deng, L., Hagley, E., Kozuma, M., Ochinnikov, Y., Wen, J., Helemrson, K., Rolston, S., and Phillips, W., to be published.
11. Bergeman, T., Erez, G., and Metcalf, H., *Phys. Rev. A* **35**, 1535 (1987).
12. Mewes, M.O., et al., *Phys. Rev. Lett.* **77**, 416 (1996).
13. Myatt, C. J., et al., *Phys. Rev. Lett.* **78**, 586 (1997).
14. Hau, L. V., et al., *Phys. Rev. A* **58**, R54 (1998).
15. Hess, H. F., *Phys. Rev. B* **34**, 3476 (1986).
16. Ketterle, W., and VanDruten, N. J., *Adv. Atom. Mol. Opt. Phys.* **37**, 181 (1996).
17. Monroe, C., et al., *Phys. Rev. Lett.* **70**, 414 (1993).
18. Weiner, J., et al., to appear in *Rev. of Mod. Phys.* (1999).
19. Davis, K. B., et al., *Phys. Rev. Lett.* **74**, 5202 (1995).
20. Anderson, M. H., et al., *Science* **269**, 198 (1995).
21. Arndt, M., et al., *Phys. Rev. Lett.* **79** 625 (1997).
22. Andrews, M. R., et al. , *Science* **273**, 84 (1996).
23. Dalfovo, F., Giorgini, S., Pitaevskii, L. P., and Stringari, S., to appear in *Rev. Mod. Phys.*
24. Bradley, C. C., et al., *Phys. Rev. Lett.* **78**, 985 (1997).
25. Dodd, R. J., et al., *Phys. Rev. A* **54**, 661 (1996).
26. Ensher, J. R., et al., *Phys. Rev. Lett.* **77**,4984 (1996).
27. Castin, Y., and Dum, R., *Phys. Rev. Lett.* **77**, 5315 (1996).
28. Edwards, M., et al., *Phys. Rev. Lett.* **77**, 1671 (1996).
29. Jin, D. S., et al., *Phys. Rev. Lett.* **77**, 3331 (1996).
30. Mewes, M. O., et al., *Phys. Rev. Lett.* **77**, 988 (1996).
31. Andrews, M. R., et al., *Phys. Rev. Lett.* **79**, 553 (1997).
32. Burt, E. A., et al., *Phys. Rev. Lett.* **79**, 337 (1997).
33. Andrews, M. R., et al., *Science* **275**, 637 (1997).
34. Mewes, M. O., et al., *Phys. Rev. Lett.* **78**, 582 (1997).
35. Hagely, E., Deng, L., Kozuma, M., Wen, J., Edwards, M., Helmerson, K., Rolston, S., and Phillips, W., to be published.
36. Fried, D., Killian, T., Williams, L., Landhuis, D., Moss, S., Kleppner, D., and Greytak, T., to appear in *Phys. Rev. Lett.*

Quantum optics with trapped ions

Richard Thompson

*Blackett Laboratory, Imperial College,
Prince Consort Rd., London SW7 2BZ, UK
Email: r.c.thompson@ic.ac.uk*

Abstract. In these lecture notes, I describe the use of ion traps in experimental investigations of quantum optics. Ion traps are well suited to this type of investigation because of the well-controlled conditions under which ions are held in traps and because they are well isolated from the environment. The notes start with an account of the way that ion traps work, concentrating on the radiofrequency or Paul trap. The techniques of laser cooling in ion traps are then discussed. The rest of the notes deal with various experimental studies undertaken in quantum optics with trapped ions, including observations of quantum jumps; the quantum Zeno effect; cavity quantum electrodynamics; frequency standards; nonclassical states; and quantum logic gates. These notes do not attempt to give a full account of the theory of these phenomena, but rather to give an idea of the very wide range of investigations that have been undertaken with ion traps and the potential they show for future investigations in quantum optics and other fields.

I INTRODUCTION

Ion traps have now been in use for about forty years. In that time they have moved from being a spin-off from work with linear mass analysers to constituting a whole research field, with applications in mass spectrometry, measurements of fundamental constants, atomic physics, spectroscopy, quantum optics, frequency standards, cavity quantum electrodynamics and quantum computing. For many of these applications (though by no means all) it was the advent of laser cooling which allowed the technique to develop so effectively. Laser cooling allows signals to be observed even from single atomic ions and enables scientists to prepare ions in specific internal states and well-defined states of motion in the trap.

These lectures will concentrate on the use of ion traps for studies in quantum optics. The applications covered here complement those treated in the lectures by Gabrielse [1] at this school. There are also many similarities between work with neutral atoms and that with atomic ions so readers should also refer to the lectures by Rolston [2].

For a general review of work with trapped ions see, for example, the review by

Thompson [3] or the book by Ghosh [4] and for applications in fundamental studies in physics see Horvath et al. [5].

The lecture notes are organised as follows. First, we deal with the theory of ion traps, concentrating mainly on the radiofrequency or Paul trap (Section II). Then we look at how laser cooling is applied to ions in traps (Section III). The rest of the lecture notes is taken up with various applications of trapped ions: quantum jumps (Section IV); the quantum Zeno effect (Section V); cavity quantum electrodynamics (Section VI); frequency standards (Section VII); nonclassical states (Section VIII); and logic gates (Section IX).

II BASIC THEORY OF ION TRAPS

II.A The Penning trap

Penning and Paul traps use the same basic set of electrodes, shown in Fig. 1. This consists of a *ring* electrode, shaped like the inner part of a doughnut, and two *endcap* electrodes, which are similar to hemispheres. Actually these electrodes follow the shapes of the equipotential surfaces corresponding to a pure quadrupole potential of the form:

$$\phi(r,z) = -\frac{U_0}{R_0^2}(2z^2 - r^2) \qquad (1)$$

where $r^2 = x^2 + y^2$, $R_0^2 = r_0^2 + 2z_0^2$ is a geometrical constant depending on the size of the trap (with r_0 and z_0 defined in Fig. 1 and U_0 is the potential of the ring with respect to the endcaps. U_0 is made negative in order to generate a potential well in the axial (z) direction. This traps positively charged ions in one dimension. However, as discussed by Gabrielse in this volume [1], this potential cannot trap in three dimensions on its own, but needs some additional feature to give three-dimensional stability. In the Penning trap [1,6] this comes from an additional static axial magnetic field. Then an ion attracted out towards the ring electrode is forced into a combination of circular orbits in the radial (x,y) plane. The two resulting motions (at the *modified cyclotron* frequency (ω_c') and the lower *magnetron* frequency (ω_m) can perhaps best be seen by transforming into the frame rotating at half the cyclotron frequency (given by $\omega_c = eB/m$), in which the magnetic field is effectively cancelled out [7]. In this frame the two radial motions are seen to be a rotation in the positive or negative sense around the centre of the trap at a frequency ω_1 given by

$$\omega_1 = \sqrt{\omega_c^2/4 - \omega_z^2/2} \qquad (2)$$

where ω_z is the axial oscillation frequency given by

$$\omega_z^2 = 4e(-U_0)/mR_0^2. \qquad (3)$$

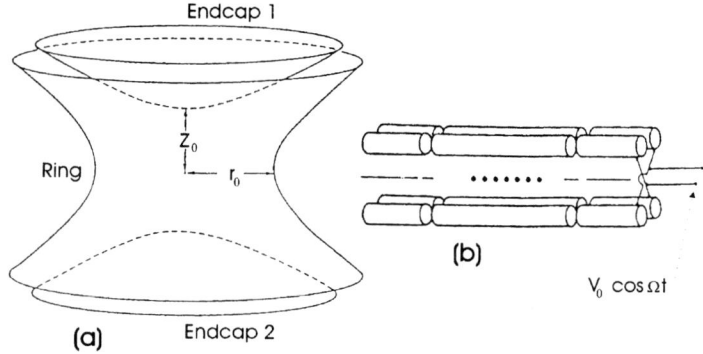

FIGURE 1. Electrodes of ion traps: (a) Paul and Penning traps; (b) linear RF trap.

U_0 is negative for trapping positive ions. The two radial frequencies are then found by transforming back into the laboratory frame and are found to be

$$\omega'_c = \omega_c/2 + \omega_1 \tag{4}$$

and

$$\omega_m = \omega_c/2 - \omega_1. \tag{5}$$

The Penning trap is a static trap but it has the disadvantage that the magnetron motion is unstable, meaning that the total energy associated with the magnetron motion is negative. Therefore in the presence of collisions there will always be a tendency for the magnetron orbit to increase in size, necessitating work under ultra-high vacuum (UHV) conditions (say less than 10^{-9} mbar).

II.B The radiofrequency or Paul trap

The Paul trap [8] uses an alternative method for generating an effective three-dimensional trapping potential. In this case it is a *dynamic* trap. It works by using a potential which has an AC component, that is, we use a potential given by $U_0 + V_0 \cos \Omega t$. For half of the cycle this gives a potential which is stable (i.e. has a minimum) in the z direction but is unstable (i.e. has a maximum) in the radial plane. For the other half of the cycle it is unstable in the z direction but is stable in the radial plane. The magic of the Paul trap is that the parameters can be chosen such that overall the motion can be stable in *both* directions. To see this, we need first to write down the equation of motion of the ion (say for the z motion):

$$\ddot{z} = (U_0 + V_0 \cos \Omega t)(4z)(e/mR_0^2). \tag{6}$$

This can be written in a standard form called a Mathieu equation:

$$\frac{d^2z}{d\tau^2} + (a_z - 2q_z \cos 2\tau)z = 0 \tag{7}$$

where $\tau = \Omega t/2$ and a_z and q_z are given by

$$a_z = -\frac{16eU_0}{m\Omega^2 R_0^2} \tag{8}$$

$$q_z = \frac{8eV_0}{m\Omega^2 R_0^2}. \tag{9}$$

This equation has stable solutions for certain ranges of values of a_z and q_z [9]. The final motion is quite complicated in general, but consists of a fast driven motion (the *micromotion*) at the applied frequency (Ω) and also motion at lower frequencies (the *macromotion* or *secular motion*). For small values of a_z and q_z we can represent the slow motion as resulting from an effective potential set up by the driven motion. This effective potential arises because the driven motion takes place in an inhomogeneous field so that the average force on the ion does not cancel to zero over a complete cycle of the driving field. The effective potential energy of the ion (also called a *pseudopotential*) is given by

$$V_{eff} = \frac{1}{8}m(a_z + \frac{1}{2}q_z^2)\Omega^2 z^2 \tag{10}$$

so that the oscillation frequency in this potential is

$$\omega_z = \frac{1}{2}\beta_z \Omega \tag{11}$$

where

$$\beta_z^2 = a_z + \frac{1}{2}q_z^2. \tag{12}$$

Similarly we can do the same analysis for the radial motion, and here we find that the values of a_r and q_r are $(-1/2)$ times the corresponding values for the axial motion. Therefore the stability conditions for the radial motion are different from those for the axial motion. The final situation is summarised in the *stability diagram* shown in Fig. 2. This shows that there are regions in the (a_z, q_z) plane where both motions are simultaneously stable. Although there are several such stability regions, in fact work has only been performed in the first and largest stability region, close to the origin. It can be seen that this has q_z ranging up to

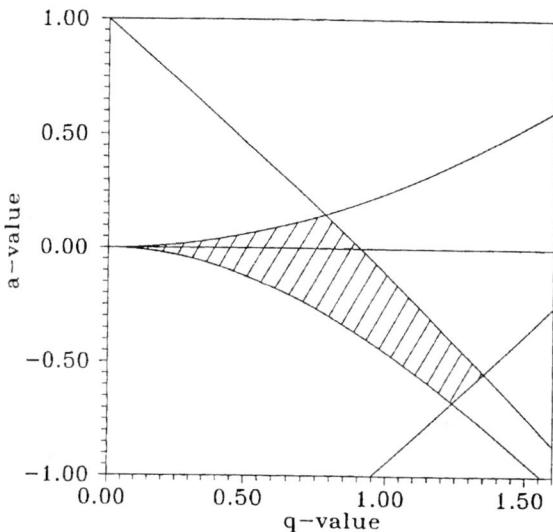

FIGURE 2. Stability diagram of a Paul trap. The motion is stable in both axial and radial directions within the shaded area. See the text for the definitions of $a\ (=a_z)$ and $q\ (=q_z)$.

a value of approaching unity, whereas a_z is typically much less, and is often set at zero (i.e. there is no applied static potential).

Since the a and q parameters depend on the charge to mass ratio of the ions, the Paul trap has the potential to be used as a mass-selective device. Indeed, the Paul trap was originally developed in Paul's group in Bonn out of work on linear mass filters [8]. If the trap parameters are chosen such that $a=0$, then the trap is stable for all masses down to a particular value (determined by the maximum value of q_z, which is around 0.9). On the other hand, the parameters can be chosen to put the operating point in the corner of the stability diagram and then the trap will be stable only for a very narrow range of masses. In this way it will act as a selective device, only trapping in a narrow range of charge-to-mass ratio. Furthermore, since the oscillation frequencies in the trap are mass-dependent, it is possible to selectively excite the motion of one type of ion in isolation. These motionally excited ions can then be detected using time-of-flight methods, for example. Commercial devices are available for use in mass spectrometry applications where the mass-selective nature of the Paul trap is exploited [10].

A useful concept for an ion trap is the trap depth, i.e. the potential energy which an ion must have in order to escape from the confining potential. In a Paul trap this is the magnitude of the pseudopotential at the edge of the electrode. In the axial direction this has the value $eD_z = \frac{1}{2}m\omega_z^2 z_0^2$. Typically this has the value of a few electron volts.

Since the oscillation frequency in the trap depends on (V/R_0^2), the applied voltage has to rise as the square of the size of the trap if the oscillation frequency is to be kept constant. If a high oscillation frequency is required (especially if the Lamb-Dicke regime is to be achieved —see Section III.C) then the applied voltages will have to be high and the trap also needs to be small, so that prohibitively large applied RF voltages can be avoided. This has led to the design and construction of miniature Paul traps (Section II.D).

II.C Motion in the Paul trap

The motion of ions in the Paul trap can become quite complicated as there are in general three components: first, the driving field at angular frequency Ω; second the radial secular motion at frequency ω_r (there are in fact two independent components to this at the same frequency which can be considered as the x and y amplitudes or equivalently as the clockwise and anticlockwise amplitudes); and third the axial secular motion at frequency ω_z. In a perfect trap the motion in a given direction (say the z direction) can be written (for low values of a_z and q_z) as:

$$z(t) = A \cos \omega_z t [1 + \frac{1}{2} q_z \cos \Omega t]. \tag{13}$$

This shows that it is not possible to eliminate the micromotion (at angular frequency Ω) unless the ion is stationary ($A=0$). This is because it is the motion at Ω that sets up the pseudopotential in which the ion is moving. Thus a single ion can in principle reside at the centre of the trap without moving. However, if there are more ions, they repel each other because of the Coulomb interaction so that even if they have no secular motion, they do not reside at the centre of the trap. This means that they always have some residual micromotion, which limits the minimum value of their mean kinetic energy, i.e., the minimum temperature to which they can be cooled.

In a real trap there are always small perturbations (e.g., contact potentials on the electrodes) that give rise to extra static fields which offset the equilibrium position of the ions from the centre of the trap. This leads to an extra term in the motion at frequency Ω, unrelated to the secular motion. The result of this is that even a single ion will not be free of the micromotion when it is at the equilibrium position. In many experiments, where a single ion is required to be stationary at trap centre, measures have to be taken to cancel out any of these stray fields. This is generally done by minimising the amplitude of the residual micromotion.

II.D Practical Paul traps

Many ion traps correspond quite closely to the diagram of the electrodes presented here (Fig. 1). However, for many experiments it is necessary to confine ions very tightly, and in this case miniature traps are constructed, which are difficult to manufacture with electrodes curved accurately in three dimensions. In this case the electrodes are generally constructed with simpler shapes, designed such that the leading term in the potential is the desired quadrupole term with all other higher-order terms being sufficiently small that they do not affect the potential appreciably, especially near the centre of the trap. Examples of traps with simplified electrode designs can be found in several papers, for example [11–13]. These traps are all of the order of 1 mm or less in diameter. Such traps are designed mainly for use with one ion (or at most a few ions). The applied frequency is typically tens of MHz and the voltage may need to be several hundred volts, giving secular frequencies of a few MHz. Some of these miniature traps are designed as a quarter-wave RF resonator giving a large voltage amplification factor, thus eliminating the need to generate a large RF voltage externally [14]. Larger traps designed for trapping large numbers of ions generally do not need to create such a steep trapping potential, and therefore they can be run with lower frequencies (hundreds of kHz) and lower voltages (of the order of 100-200 V) [15]. Such traps may be 10-40 mm in diameter. In all cases the secular frequencies are typically 0.1 to 0.2 times the applied frequency (i.e., β is typically 0.2 - 0.4).

II.E Linear RF trap

A development of the Paul trap is the linear RF trap. This is made from four rods and an RF potential is applied between the two opposite pairs of rods. This makes a two-dimensional trap (Fig. 1), and some means has to be provided to keep the ions in at both ends as otherwise they will escape. This can be done, for example, by adding endcap rods at both ends on which a DC potential is placed [16]. The reason for the interest in the linear trap is that instead of having a single point where an ion is free of micromotion, there is now a line that has the same property. Therefore it is possible to have a number of ions in a linear trap, *all* of which are free of micromotion. A number of applications are seen for this approach, but especially for frequency standards (where a larger signal-to-noise ratio can be obtained for a given maximum second-order Doppler shift) and quantum computing, where several ions are required to be free of micromotion simultaneously.

An alternative way to get a similar result is to have a very long set of electrodes which are bent round to form a ring trap [17]. Then no axial confinement is necessary as the ions are allowed to go round the whole circumference of the ring trap. The ring trap has been used for studies of crystallisation of ions at low temperatures [18]. In this experiment it was possible to form an ion crystal of up to one million ions.

II.F Operation of ion traps

There is not enough space here to go into details of the operation of ion traps but we need to discuss the subject briefly before moving on. More details can be found in reviews or in the book by Ghosh [4].

First, we consider the loading of traps. This is generally achieved using a weak atomic beam crossed with an electron beam. The atomic beam comes from a small oven which is heated electrically, and the electron beam comes from an electron gun or just a small filament that is heated so that it emits electrons which are then accelerated towards the electrodes. In a Penning trap the electrons have to travel along the magnetic field lines into the trap but in a Paul trap there is no such restriction on the location of the filament. The atoms will generally have thermal energies, whereas the electrons need to have sufficient energy to ionise the atoms (generally 10-20 eV for the creation of singly-charged ions). Once created, the ions will be attracted towards the centre of the trap and will generally have total energies of several eV (before cooling). It is always necessary for the base pressure in the vacuum system to to be low (say $< 10^{-9}$ mbar) for successful operation of traps. However, in the case of a Paul trap, a low mass buffer gas may be introduced at low pressure for ccoling purposes (Section III.A).

Then we need to be able to detect ions. For large clouds, electronic detection techniques are available, which are often similar to those used for the detection of electrons in traps (see [1]). There are also destructive means of detection which can be used, based, for example, on time-of-flight techniques for ions ejected from the trap. However, for small clouds of ions and continuous, non-destructive observation, there is no real alternative to optical means of detection. This is based on the detection of light from a laser beam which has been resonantly scattered by ions in the trap. This is very sensitive (Section IV.A) and does not perturb the ion cloud strongly. It is also selective and can even distinguish between isotopes of the same element. If laser cooling is being used (see the next section) then we simply need to detect the scattered laser cooling light, and this can give us information on the progress of the laser cooling. It is necessary to use a sensitive detector which can detect single photons, such as a photomultiplier or a high sensitivity imaging detector.

III LASER COOLING OF TRAPPED IONS

III.A The need for laser cooling

When ions are loaded into ion traps they have energies which are typically of the order of electron volts, much larger than normal thermal energies. This is because they are not generally created at the centre of the trap, so they gain energy as they fall towards the trap centre in the trap potential. The hot ions will have large amplitude motions and it will be possible for them to be ejected from the trap. In fact, in the Paul trap the energy tied up in the micromotion can also be coupled into the secular motion of the ions through ion-ion collisions [19]. This can also lead to loss of ions from the trap.

In order to localise the ions near the centre of the trap and to increase the density of the ion cloud, it is necessary to cool them. By this we mean a reduction in the mean kinetic energy of the ions, so the concept of temperature can still be applied even in the case of a single ion. For the magnetron motion in a Penning trap, the *total* energy associated with this motion actually increases as cooling takes place.

Cooling the ions has several other advantages. First, it reduces the chances of ions being lost through collisions. Second, it reduces the Doppler effect (to all orders) so that higher resolution can be obtained in spectroscopic measurements. Third, it also increases the interaction time between ions and any other radiation with which the ions are interacting (e.g., a laser beam or microwave radiation) thus reducing transit time broadening effects.

Cooling may be accomplished by several different means. For large clouds of ions in a Paul trap (but not in a Penning trap because of the unstable nature of the magnetron motion) a buffer gas can be used to reduce the temperature to something approaching room temperature [20]. This can be very effective and it is very simple to do: it just involves letting buffer gas (generally helium) into the vacuum system at a low pressure (typically 10^{-6} mbar). The buffer gas needs to be light and unreactive in order to be effective, so helium is usually the best choice.

However, buffer gas cooling can at best bring the temperature down to room temperature and generally it is not this low. In order to take the temperature lower it is necessary to use laser cooling. This can be applied in all cases except large clouds of ions in a Paul trap, where the RF heating prevents laser cooling being effective.

III.B Doppler cooling

Laser cooling has already been discussed elsewhere in these lectures in the context of laser cooling of neutral atoms [2]. For ions the most important process is Doppler cooling which was first suggested for ions by Wineland and Dehmelt in 1975 [21]. Advanced techniques such as polarisation gradient cooling are not generally applicable in ion traps due to the applied fields (but see [22]). There have been just one or two demonstrations of this sort of laser cooling in an ion trap [23].

Doppler cooling works in ion traps in a very similar way to neutral atoms, but there are some changes (see [24] for a full treatment). In particular, since the ions are bound by electromagnetic forces and move in closed orbits, they only need one laser beam to cool all degrees of freedom, so long as the beam is inclined at an angle to all the axes of the trap so that it has a component along all three directions. This is a big simplification compared to the case of neutral atoms.

There are a number of different limits which need to be treated slightly differently, depending on the relative values of the relevant frequencies in the problem. If the ion oscillation frequencies (ω_z, ω_r) are less than the linewidth (γ) of the laser cooling transition (generally the resonance transition of the ion) (i.e., $\omega_z, \omega_r \ll \gamma$), then this is called the weak binding limit and this is where laser cooling works in a very similar manner to the way it works for neutral atoms. We also assume here that the recoil energy, R, given by

$$R = \hbar^2 k^2 / 2m \tag{14}$$

is much less than $\hbar\gamma$. For optimal laser cooling the laser is tuned to $(\gamma/2)$ below the centre of the transition and in this case the limiting temperature is the same as for neutral atoms:

$$E_{min} = \frac{1}{2} k_B T_{min} \approx \frac{1}{4} \hbar \gamma. \tag{15}$$

where E_{min} refers to the kinetic energy of the ion. This gives a minimum temperature $T_{min} \approx \frac{1}{2} \hbar\gamma/k_B$ which for magnesium is of the order of 1 mK.

If, on the other hand, the oscillation frequencies are larger than the linewidth $(\omega_z, \omega_r \gg \gamma)$ then we have the tight binding limit and this is best thought of in terms of the vibrational quantum numbers of the ion in the trap potential well. In this case the Doppler effect gives rise to a carrier at the resonance frequency with a set of sidebands spaced at the trap oscillation frequency (Section III.C). Now the optimal cooling is obtained by tuning the laser to the position of the first sideband below resonance. Each time a photon is absorbed we now lose on average one vibrational quantum. This is referred to as *sideband cooling*. In this case the limiting mean vibrational quantum number, $\langle n_{min} \rangle$, is given by:

$$\langle n_{min} \rangle = \frac{5}{16} \frac{\gamma^2}{\Omega^2} \ll 1 \tag{16}$$

In the weak binding limit we can think of a photon as being absorbed at a specific time and place in the motion of the ion. In the strong binding case, the absorption is best thought of as being a transition from one state of vibrational motion to another, so the absorption is in some sense spread over the entire orbit.

For ions in Penning traps the cooling process is complicated by the fact that the laser beam needs to be offset from the centre of the trap in order to cool both radial motions (magnetron and cyclotron) at the same time. The beam needs to be offset to the side where the magnetron motion causes the ions to travel in the same direction as the laser beam [25]. This is related to the unstable nature of the magnetron motion. As a result of this, the temperature associated with the magnetron motion cannot be reduced as much as that associated with the modified cyclotron motion. The magnetron temperature is typically more like 1K rather than the Doppler limit of typically 1 mK. The cooling can be improved by making use of two laser beams rather than one [26].

III.C Lamb-Dicke regime

For ions bound in the simple harmonic potential of an ion trap, the radiated light can be thought of as coming from a frequency-modulated source, as a result of the oscillatory motion of the ions. Due to this motion, the ions do not radiate with the conventional Gaussian Doppler line shape, which is a continuous function of frequency, but rather they radiate at the unmodulated carrier frequency with sidebands spaced at integer multiples of the trap oscillation frequency each side of the carrier. The amplitudes of the various sidebands are determined by the distribution of amplitudes of the ion oscillation. In the case where $\omega \ll \gamma$ these sidebands merge into each other and we retrieve the normal Gaussian distribution. If, however, we have $\omega \gg \gamma$ then the sidebands are well resolved.

There is a special case when the amplitude of the motion (generally for a single ion) is so small that the amplitudes of the various sidebands become small enough to be ignored. This is referred to as the *Lamb-Dicke regime* [27]. In this case the motion has an amplitude which is much less than $(\lambda/2\pi)$ where λ is the wavelength of the radiated light. This corresponds to the phase of the radiated light never deviating from the unmodulated phase by more than one radian. We have given a classical description of this effect here but a similar result follows from a full quantum mechanical treatment. The effect is the same as that referred to as *Dicke narrowing* in the context of collisions in atomic gases.

The reason that this is important is that now all the radiation (or equivalently, all the absorption) is concentrated in the carrier, which is free of the first-order Doppler effect. Thus, the first order Doppler effect can be overcome, not by selecting particular atoms, but by forcing all the radiation to be at a frequency which is unaffected by the first order Doppler effect. In fact the second order Doppler effect (relativistic time dilation) is still present but this is of course also reduced dramatically because of the reduction in the average energy of the ion. The fractional second order Doppler effect, $\delta\nu/\nu$, is given by $\gamma - 1$ where

$$\gamma - 1 = E_{kin}/mc^2 = 3kT/2mc^2 \tag{17}$$

For laser cooled ions this can be reduced to one part in 10^{15} or so (at a temperature of the order of 1K).

The first demonstration of the achievement of the Lamb-Dicke regime for an optical transition was by Bergquist *et al.* [28] using mercury ions. They cooled a mercury ion initially using light tuned to the resonance line at 194 nm and then probed it at 282 nm on a narrow transition to a metastable level. Two sidebands are seen on each side of the carrier, corresponding to the Doppler temperature of a few mK (the vibrational frequency of the ion is about 1.5 MHz). Then the ion is cooled by sideband cooling on the 282 nm transition itself (this has a lower Doppler limit as it is a narrow transition). This is achieved by tuning a narrow band laser to the first lower (red) sideband corresponding to the $\Delta n = -1$ transition. Then we are left with just one sideband on either side of the carrier, and the amplitudes of the sidebands become asymmetrical. This is because in the quantum mechanical treatment of the problem it is clear that we can go from $n = 1$ (or above) in the ground state to either $n+1$ or $n-1$ in the excited atomic state, but from $n = 0$ we can only go to $n+1$ as a vibrational quantum number of $n = -1$ is not possible. Thus the asymmetry of the sidebands tells us how close the ion is to the lowest vibrational level ($n = 0$). In this case an equivalent temperature of about $50\mu K$ is achieved, though probably the best way to talk about this is in terms of the mean vibrational quantum number, $\langle n \rangle$, which had a value of about 0.05 in this experiment. This means that the system is in its ground state 95 % of the time. In this case a simple calculation of the temperature from the mean energy (ignoring the zero-point energy) gives a false result:

$$\langle E \rangle \approx \langle n \rangle \hbar \omega = k_B T \tag{18}$$

is not appropriate if $\langle n \rangle < 1$. Instead we must use the full Boltzmann distribution to find the relation between $\langle E \rangle$, $\langle n \rangle$ and T:

$$\langle E \rangle = \langle n \rangle \hbar \omega = \frac{\hbar \omega}{\exp(\hbar \omega / k_B T) - 1} \tag{19}$$

which, for $k_B T \ll \hbar \omega$, becomes

$$\langle E \rangle = \langle n \rangle \hbar \omega = \hbar \omega \exp(-\hbar \omega / k_B T). \tag{20}$$

The *Lamb-Dicke parameter*, η, gives an idea of how large the ground state wavefunction is compared to the wavelength of the radiation, and the Lamb-Dicke regime cannot be achieved unless $\eta \ll 1$. It is defined by

$$\eta = k x_0 = k \sqrt{\hbar / 2 m \omega} \tag{21}$$

where $k = 2\pi/\lambda$ and $x_0 = \sqrt{\hbar/2m\omega}$ is the spread of the ground state wavefunction in the potential well, which has oscillation frequency ω. In the above experiment, η has a value of approximately 0.09.

III.D Raman cooling

The above work is unusual in that standard Doppler cooling is used to achieve a very low temperature, and the Lamb-Dicke limit, by making use of a very narrow transition. One alternative route to very low temperatures in an ion trap is to use Raman cooling. Here we rely on a *driven* transition between two specific energy levels of the ion in the trap, using a pair of relatively intense laser beams whose difference frequency corresponds to the difference in energy between the levels. The intermediate virtual level is chosen to lie close to an excited state of the ion so that the rate of the Raman transition is enhanced.

Raman cooling has been used by the NIST group in Boulder to cool single Be ions to the lowest vibrational state in the trap [29]. It works in the following manner. First, the ion is cooled using Doppler cooling to the region of a few mK (the Doppler limit for Be is 0.5 mK, corresponding to a mean vibrational quantum number of about 0.5 in this trap, which has a vibrational frequency of 11.2 MHz (x direction). Then the Raman beams are used to transfer the ion from the electronic ground state (with vibrational quantum number n) to a different hyperfine level (with vibrational quantum number $n-1$). The Raman pulse is timed to be a π pulse —i.e., to transfer the ion with close to 100% probability. The ion is then excited resonantly to a state from which it decays back spontaneously to the ground state, probably remaining in the vibrational state $n-1$ (so long as $\eta \ll 1$; in fact the value is roughly 0.2 in this experiment). Thus the overall effect of this cycle is to reduce n by 1. The cycle is then repeated several times, and this should reduce n to zero with high probability. In fact Wineland's group are able to achieve an average value of n of about 0.014 in one dimension. This means that the ground state is occupied 98 % of the time. With cooling applied in all three dimensions, this figure becomes 92 %. Thus the ion can effectively be prepared in the ground vibrational state of the system. This forms the basis of many other experiments (Sections VIII and IX).

IV QUANTUM JUMPS

IV.A Single ion techniques

We have already discussed laser cooling of single ions, but at this stage it is worth thinking about how it is possible to work at all with single ions in ion traps. In particular, we consider how to prepare such an ion in a trap and what signal levels may then be expected.

The first experiments with single ions started only a few years after laser cooling was first introduced for trapped ions. The first laser cooling was reported in 1978 [11,30] and a photograph of a single ion was published in 1980 [31], with a spectrum of a single laser cooled ion the next year [32]. Preparation of a single ion in a trap is generally a case of turning the intensity of the atomic beam and the electron beam down to very low values so that on average only very few ions are loaded on each loading attempt. If the average number of ions loaded is less than one, then if a signal is seen, it is likely to be from a single ion. The technique of quantum jumps (Section IV.B) can then be used to confirm whether it is really a single ion or not.

An alternative is to load a small number of ions and then to attempt to lose all the ions except one. In some experiments it is possible to see the signal level reduce in definite steps as each ion is lost. This may be achieved through charge-exchange collisions with background gas or by taking the trap parameters to the edge of the stability region so that the trap is only just stable for a single ion and not for a cloud of ions.

It is worth considering how much optical signal can be seen with a single ion in a trap, because it is surprisingly large. The point is that a single cold ion can remain in the laser beam all the time and can also remain in resonance with the laser all the time, because it is no longer affected by the Doppler effect once it is cold. Then the ion can scatter of the order of $A/2$ photons per second, where A is the Einstein A-coefficient of the resonance transition, if the transition is saturated (this generally requires of the order of 100 μW or less in a focussed laser beam). Since A is typically 10^8 per second, we can collect of the order of 10^6 photons with a typical solid angle for collection, given the presence of the electrodes which generally limit the available solid angle. Of these, we may be able to detect 10^5 photons per second, with typical quantum efficiencies of photomultipliers and optical filter efficiencies. Thus in a real experiment we may expect to see 10 to 100 thousand counts per second detected from a single ion. This has indeed been seen in several different experiments [33–35] (this represents an overall efficiency of the detection system of the order of one part in a thousand). Indeed, for those experiments where the scattered light is in the visible region of the spectrum, experimenters have been able to see the ion with the naked eye [35]. Unfortunately, for many ions the resonance transition is in the UV region (since excitation energies of ions are generally larger than those for atoms, which are usually in the visible), so for these experiments one has to rely on other detectors than the eye. Many images of single ions or small crystals of ions in Paul or linear RF traps have now been published.

IV.B Setup for quantum jumps

It is very hard to observe directly a quantum mechanical system jumping between two levels, as the overall detection efficiency for the photon emitted in the transition is typically only 10^{-3} or less. However, if there are 3 levels in a suitable arrangement, the jumps on one transition can be used to turn the resonance fluorescence on another transition on and off. This is the basis for the experimental observation of quantum jumps in real time.

Quantum jumps therefore require an atomic system which has at least three levels arranged in a V-configuration with one ground state coupled to two excited states, one of which is a short-lived resonance level and the other of which is a long-lived metastable level. When the ion is in the ground state, it can be excited into the resonance level by the laser cooling radiation and it will then scatter resonance fluorescence at a high rate determined by the A-value for the transition. However, if the ion is excited in some manner into the metastable level (for instance, by an off-resonant transition, a low-probability decay from the resonance level, or a direct excitation from the ground state by a second laser) then it will no longer be able to scatter resonance light, and so the observed fluorescence signal will drop to zero. At some stage the ion will decay back to the ground state and then the process will start again. The times of switching between the *on* and *off* states are randomly distributed, giving a so-called *random telegraph signal*.

The theory of this process was first described by Cook and Kimble in 1985 [36]. The *on* and *off* times both have an exponential distribution (rather like radioactive decay) with mean values determined by the rate of excitation into the metastable state and the rate of decay out of it. These mean times can be several seconds (e.g., [35]) or they may be as short as 10-20 ms [33], depending on the atomic system and the various rates involved. The signal observed is of course quite different from what one observes if a larger number of ions is present in the trap, and before the first experimental observations were made, it was a matter of some debate as to whether one would really observe the random telegraph signal. This illustrates the way that quantum mechanics predicts manifestly different behaviour for a single atomic particle compared to that seen with an ensemble and highlights the importance and significance of experiments with single ions.

IV.C Observations of quantum jumps

The first observations of quantum jumps were made in 1986 in several different experiments with Ba^+ and Hg^+ ions. All these early experiments were performed in Paul traps, mostly miniature traps optimised for the trapping of single particles. The Ba experiment is particularly interesting as the main fluorescing transition is in the blue and this means that the ion can be observed undergoing quantum jumps in real time with the naked eye [35]. In this experiment two lasers are needed: one to drive this blue transition from the $^2S_{1/2}$ ground state to the $^2P_{1/2}$ resonance level, and one to recycle ions that have decayed from this level to a metastable $^2D_{3/2}$ level (this transition is in the red). The quantum jumps take place to a second metastable level, $^2D_{5/2}$, via an off-resonant excitation involving the $^2P_{3/2}$ state. The mean on and off times are of the order of several seconds, so the characteristic quantum jump signal can be clearly seen (Fig. 3). Another interesting feature of this experiment is that cooperative quantum jumps were seen in runs where 2 or 3 ions were trapped at the same time. Of course, a certain number of chance coincidences would be expected to be seen, but the observation in this experiment seems to indicate that these were more frequent than expected.

Since these early observations, many other experiments have been performed, and these are well summarised in reviews by Blatt and Zoller [37] and Cook [38]. The statistics of the quantum jumps have been verified in different systems and we now have a very good understanding of the way that single particles behave in these conditions. Since normal quantum mechanics deals with ensembles, it is not always the most appropriate way to treat the behaviour of single particles and new treatments of quantum mechanics which are more suited to this situation have been developed. See [33] and [39] for a discussion of these approaches to quantum mechanics.

Quantum jumps may be observed even in a two level system if a magnetic field is

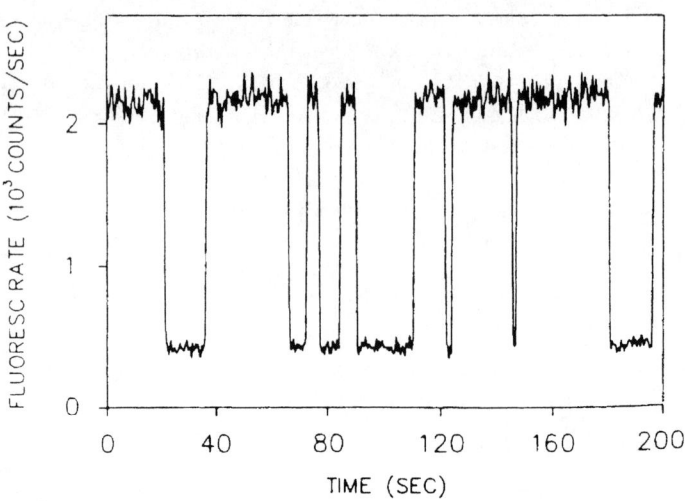

FIGURE 3. Quantum jumps in a single Ba$^+$ ion (from [35]). When the ion is in the *off* state there is still a small fluorescence signal due to scattered light off the electrodes of the miniature Paul trap.

used to split the ground state into its Zeeman components. In this way it has been possible to observe quantum jumps in Mg ions [33,34]. Then one of the ground state Zeeman levels acts as the metastable level and in the absence of laser radiation its lifetime is essentially infinite. The experiment is ideally suited to a Penning trap as the magnetic field is already there and has the right order of magnitude. In this experiment one laser (tuned to a strong transition out of one of the ground state Zeeman levels) can be used to serve several purposes: it acts as the laser source for laser cooling; it is used for detection of the ions; it drives the ions into the metastable state; and it drives them out of the metastable state back into the ground state. The spontaneous Raman transitions which are driven by the laser take place at a rate which is determined by the laser intensity and the detunings of the various levels, which is itself proportional to the strength of the magnetic field. In the Imperial College experiment the mean *off* time was of the order of 15 ms and theory shows that the ratio of *on* to *off* times is (for a perfect set-up) 16:1. A plot of quantum jumps from this experiment is shown in Fig. 4.

IV.D Shelving technique

Quantum jumps are not merely an interesting quantum mechanical phenomenon. They have use in the construction of frequency standards using ion traps. Here the

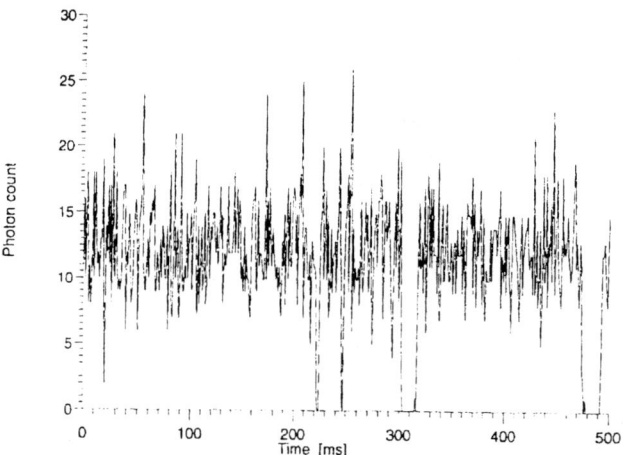

FIGURE 4. Quantum jumps in a single Mg$^+$ ion (from [33]). This experiment was performed in a Penning trap at a magnetic field of about 1 T. The fluorescence photons were counted in bins of 1 ms duration.

problem is that in order to obtain the greatest benefit from the use of trapped ions in frequency standards one would like to use just a single ion, but then the signal level is going to be very low, especially on a narrow transition such as would be useful for a standard. The solution is to use the ion as an *atomic amplifier* to increase the observed signal, in the manner first suggested by Dehmelt [40]. He points out that when an ion absorbs a photon taking it to a metastable level this results in the loss of many photons on the resonance transition which would otherwise have been detected. Thus although one is unable to detect the single absorbed photon with high probability, one can detect the fact that it has been absorbed with very nearly unit probability by recording the resonance fluorescence of the ion. This is called *electron shelving* and is used in all single ion frequency standards to obtain a high detection efficiency for transitions to the metastable state (Section VII.D).

V   QUANTUM ZENO EFFECT

V.A   Original proposal

The quantum Zeno effect was first introduced by Misra and Sudarshan in 1977 [41]. These authors pointed out that, in a quantum mechanical system that decays from one state to another, a measurement taking place during the decay collapses the wave function into either the initial state or the final state. Now in the early stages of the decay the probability of such a measurement resulting in the particle ending up in the final (decayed) state rises *quadratically* with time, even though for longer times the decay proceeds exponentially (which can be approximated as a *linear* decay for short times). Since the effect of a measurement is to destroy all the coherences in the system, in effect it restarts the decay process. Now if the decay is continually put back onto the quadratic part of the curve, the end result is that the overall rate of decay is slowed down from the rate expected from the exponential curve.

This was termed the *quantum Zeno effect* by analogy with the classic paradox due to Zeno about an arrow in flight. The point is that if we make many measurements on a decaying system (such that they probe this early quadratic region) we can slow down the decay. In the limit of continuous observation, we can expect the decay to slow to zero. This is a classic example of the measurement of a quantum mechanical system having an effect on the subsequent behaviour of the system.

The trouble with the observation of this phenomenon for real systems is that the time period over which the decay is quadratic is extremely short and inaccessible experimentally. The length of this period is related to the bandwidth of the available states in the decay, which is extremely large for a normal decay.

V.B Experimental observation

Itano et al. [42] performed an experiment in 1990 to observe the quantum Zeno effect. In their experiment they worked on a *driven* radiofrequency transition between hyperfine Zeeman sublevels in a cloud of Be ions in a Penning trap. In this way they avoided the problem of the short length of time available when the probability of a transition rises quadratically. In this case the quadratic period depends on the rate at which the transition is driven, which is under experimental control. Of course the effect is now not quite the same as in the original papers, but the principle is very similar.

Itano *et al.* used a π-pulse of radiofrequency radiation to drive the transition, so in the absence of any measurements the probability of the transition taking place is nearly equal to unity. However, if the coherent transition is interrupted by short measurements, they show that the overall transition probability is reduced in both theory and experiment [42]. Here a measurement consists of a brief laser pulse which is tuned to a transition out of *one* of the two hyperfine Zeeman levels involved and could in principle be used to detect whether the ions were in that state (by the observation of a photon) or the other one (by the failure to observe a photon). The laser pulse has to be sufficiently long and intense to ensure that an ion in that state has a high probability of being excited by the laser.

The wave function at time t is given by

$$|\psi(t)\rangle = \cos(\Omega t/2)|i\rangle - i\sin(\Omega t/2)|f\rangle \qquad (22)$$

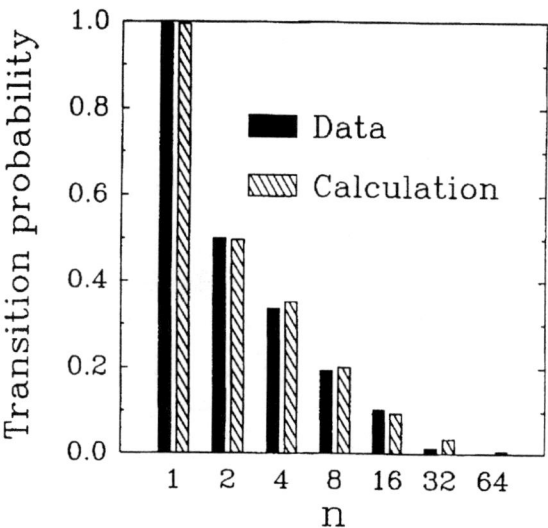

FIGURE 5. Results of the quantum Zeno experiment (from [42]). Here n is the number of times the π-pulse of microwave radiation is interrupted by the brief measurement pulses. The transition probability is only determined at the end of the π-pulse.

where Ω is the Rabi frequency, and i and f are the initial and final states. For a π-pulse, $\Omega t = \pi$, and this transfers all the population from i to f. If the π-pulse is interrupted at a time $t_1 = \pi/n\Omega$, then we find that the probabilities of the ion being found in the two states are:

$$P_i(t_1) = \cos^2(\pi/n) \tag{23}$$

$$P_f(t_1) = \sin^2(\pi/n). \tag{24}$$

At this point the decay restarts and the calculation can be repeated for each of the $(n-1)$ remaining measurement pulses. In practice the state of the ions is only determined in this experiment at the end of the π-pulse, when the fraction of ions that have made the transition is determined by measuring the amount of fluorescence on a cycling transition which yields a signal of many photons per ion in the first few ms of irradiation. The results obtained in this experiment are shown in Fig. 5.

V.C Detailed theoretical models

Itano *et al.* used a simple wavefunction collapse model to predict the fraction of ions to make the transition as a function of the number of measurement pulses, as discussed above (with some modifications) and the experiment verified these predictions to within experimental error. They concluded that this wavefunction collapse model gave a good description of the process.

However, many theorists were unhappy with this approach, and the publication of these experimental results gave rise to much theoretical work which aimed to give a more detailed and more rigorous description of what was going on in this experiment. A good treatment of this problem is given by Frerichs and Schenzle [43] (see also [44]). They treat the system of 3-level ion, radiofrequency radiation and laser radiation as a single system using a Bloch equation approach. This gives predictions for how intense the laser pulse has to be for it to count as a measurement. This was something that had to be introduced in an *ad hoc* manner in the treatment of Itano *et al.* The Bloch equation approach shows that the effect of the laser pulse is to destroy the coherences (i.e., the off-diagonal elements of the density matrix) which are built up by the radiofrequency radiation. This is the mechanism for the collapse of the wavefunction in this case.

V.D Proposed experiment

At Imperial College we are in the process of building an experiment to test the predictions of these theoretical models. We will perform an experiment which is similar to that of Itano *et al.* [42] but we need to perform the experiment on a single ion so that we can have a closer control of the experimental conditions. This will clearly reduce the signal level and the experiment will therefore need to run in a stable manner for a long time to collect data. We will study the variation of the effect of the measurement pulses with the strength of the pulses in an attempt to verify the theoretical predictions for what constitutes a measurement. We will also look at variations in the experiment which should allow us to introduce some decay by coupling to an excited state of the ion in the manner proposed by Plenio *et al.* [45]. This will take the experiment closer to the scheme originally proposed by Misra and Sudarshan [41]. Another possibility is to perform a quasi-continuous measurement, as proposed by [46]. The trap for this experiment has been constructed and tested and the experiment is now in the process of being put together.

VI CAVITY QUANTUM ELECTRODYNAMICS

VI.A Advantages of ion traps

The subject of quantum electrodynamics in a cavity (cavity QED) has been covered in detail elsewhere in these lecture notes [47]. Here we deal only with the special advantages which are offered by ion traps for experiments in cavity QED, with a brief account of the types of experiment in progress at present in this area.

The main advantage offered by the use of ion traps is the ability to have a definite number of ions in the cavity, and for this number to remain constant as a function of time. In other schemes, the atoms come from an atomic beam and so there is a *distribution* of atom number. Only the mean number can be controlled, and not the exact number of atoms at any time. Also the atoms fly through the cavity, giving a limited interaction time of the order of 100 μs. Although this has advantages in terms of gaining information about the state of the atom on leaving the cavity (the Rydberg states which are commonly employed can be detected with almost unity efficiency), it means that the number of atoms in the cavity is not constant, which is a distinct disadvantage.

With an ion trap we can be sure that there is zero, one, two etc ions in the cavity and this will not change with time, as the ions are confined by the trap potential. The number of ions present can be verified using quantum jumps or by calibrating the average fluorescence level observed per ion.

It is also possible to locate a single trapped ion accurately at the position of the mode waist in the optical cavity. This allows the use of a concentric optical cavity, which has a very small beam waist. In other experiments this is not possible as the position of atoms in an atomic beam cannot be controlled well enough, so confocal cavities are then used, which are very sensitive to aberrations.

Against these advantages has to be set the inconvenience of using a trap: the electrodes get in the way of the cavity and it is therefore difficult to make a cavity suitably small around the position of the trapped ion(s). Also since most ion resonance wavelengths are in the UV whereas those for neutral atoms are generally in the visible, we may have the added inconvenience of having to generate narrow-band, tunable laser radiation in the UV. This generally involves frequency-doubling techniques which yield much lower laser powers than are available in the visible. Finally, the ultra-high reflectance coatings needed for the construction of ultra-high finesse optical cavities are much harder to manufacture for UV wavelengths than for visible wavelengths, due to the increased absorption of the dielectrics used for the coatings. There are some ions which have visible transitions (e.g., Ba^+ and Ca^+) but there the atomic structure is such that more than one laser wavelength is required.

VI.B Experiments in progress

The experiments being set up at present are looking either for the modification of the spontaneous emission rate of an ion located at the centre of a cavity or for evidence of the *single atom laser* where a single ion generates coherent optical radiation in a cavity [48].

An excited atom in a cavity has its spontaneous decay rate modified because the radiation which goes into the optical cavity is either enhanced (if the cavity is resonant with the frequency of the radiation emitted by the ion) or suppressed (if the cavity resonance frequency is well away from that of the ion). Thus if the rate of decay of excited ions in the cavity is measured, this should be dependent on the tuning of the cavity resonance. The experiment is difficult for several reasons, but in particular because the emission is only modified for the radiation which goes into the cavity. Radiation which is outside the solid angle subtended by the cavity mode at the position of the ion is not affected by the position of the cavity resonance frequency. Thus the maximum modification to the spontaneous emission rate is limited by the solid angle of the cavity. If the cavity is not resonant with the atomic frequency the reduction in spontaneous emission rate will therefore be quite small, but if it is resonant the enhancement can still be quite large.

There are several different ways of observing the effect of the cavity on the spontaneous emission rate of a transition. One way of detecting this for a weak transition is by looking for a change in the rate of quantum jumps when the cavity is brought into resonance with the transition down from a metastable state. Alternatively, it is possible to observe the light emitted directly into the cavity mode (through a partially transmitting cavity mirror) or out of the cavity. A different approach would be to measure the change in the natural linewidth of the transition. These different approaches are suited to different experimental conditions.

The single atom laser is a particularly interesting experiment in cavity QED with a single trapped ion [48]. Under the right experimental conditions, a single ion can be made to lase, and the properties of this system are quite unusual, and different from those of the micromaser, for instance, where single atoms fly through the apparatus one at a time. This system has been studied theoretically by several authors but its practical implementation is difficult due to constraints on the atomic structure and practical difficulties in the construction of the cavity. A concrete proposal for its implementation from Walther's group makes specific predictions for the Ca^+ ion [48]. This work shows that two lasing thresholds are expected: one where the laser turns on and another where it turns off again. For certain parameters the light is expected to show sub-Poissonian statistics.

VII FREQUENCY STANDARDS

VII.A Background to frequency standards

Standards are needed for all of the physical units in use. In earlier times, length standards were based on physical artifacts such as a standard metre but this became difficult because it could not be reproduced at will at different locations and its accuracy was limited. Therefore this was replaced by an optical wavelength standard based on a highly reproducible optical transition in krypton gas. This was reproducible to about a part in 10^9 but in the last 15 years even this became inadequate. However, instead of then defining a new length standard based on a well-characterised laser wavelength it was decided to base the unit of length on that for time and frequency, which was already reproducible to something like a part in 10^{13}. This was done by defining the velocity of light to be a fixed number

(299 792 458 m/s). Then a measurement of the optical frequency of a stable laser determines its wavelength to the same precision.

The unit of time (and frequency) is defined using the caesium atomic clock, which runs on a microwave transition in caesium atoms at a frequency of around 9 GHz. The frequency of this transition is defined to have a certain fixed value and if the clock is run under certain well-defined conditions the frequency is guaranteed to have this value. The clock can then be used to generate a standard frequency against which other frequencies or time intervals can be calibrated. Standard laser frequencies can then be used as practical wavelength (or optical frequency) standards once their frequency has been determined in a measurement using a so-called frequency chain starting from the Cs clock.

In recent years the current generation of atomic clocks have been pushed to their limits of reproducibility and accuracy, and there is a need for new and more accurate clocks. The pressure for these improvements comes from pure science (e.g., the measurement of fundamental constants and studies in astronomy) and also from technology such as navigation.

VII.B Advantages of ion traps

Ion traps have been seen as having potential for ultra-high resolution spectroscopy and frequency standards for many years (see, for example, [21]). There are several advantages offered by the use of ion traps for frequency standards. First, the ions are located in a very well isolated environment with no collisions with foreign gas molecules or with walls, so it is possible to realise their inherent transition frequencies with high precision and reproducibility. The transitions are not broadened or shifted by external perturbations.

Second, because ions can be cooled in traps, we can drastically reduce the broadening and shift due to the Doppler effect. This is unlike any other Doppler-free techniques, where only the first-order effect is eliminated. The second order effect is only reduced if there is a genuine reduction in the kinetic energy of the particles. The use of the Lamb-Dicke regime (Section III.C) is a further help here as the carrier frequency is completely free of the first-order Doppler effect.

Third, the interaction time between the ions and the radiation (either laser light or microwaves) can be made very long with trapped ions, and this enables much narrower linewidths to be obtained. As an illustration, note that in a caesium atomic clock the time of flight of an atom through the atomic beam apparatus is of the order of 10 ms, and this sets a limit on the maximum interaction time. The corresponding transit time linewidth is of the order of 100 Hz. In the caesium clock the centre of the transition has to be found to about 1 mHz which is one part in 10^5 of the linewidth. In an ion trap, the interaction time can be as long as 500 s, giving a linewidth of the order of 1 mHz. Thus even if the centre of this transition can only be found to an accuracy of say 1 % of the linewidth (for a similar transition frequency), this represents a significant improvement over the caesium clock, and systematic errors will be much less of a problem in this case.

There has therefore been much interest in the construction of new frequency standards using ion traps, and this is the driving force behind many of the laboratories involved in ion trap research. Much progress has been made in the development of standards both in the microwave region of the spectrum (especially Hg^+, 40 GHz and Be^+, 300 MHz) and the optical region (e.g., Ba^+, 2 μm and Hg^+, 282 nm). There is a significant prospect that the next generation of frequency standards will be based on transitions observed in trapped ions. It is desirable for these standards to be based on transitions with as high a frequency as possible because then for a given linewidth $\delta\nu$ (limited in general by the interaction time) the *fractional* linewidth will be low, leading to a high quality factor (Q), defined by

$$Q = \frac{\nu_0}{\delta\nu} \tag{25}$$

where ν_0 is the frequency of the transition.

VII.C Microwave standards

Microwave frequency standards are typically in the region of 1-40 GHz, corresponding to ground state hyperfine transitions in ions having an alkali-like atomic structure (perhaps in the presence of a magnetic field) and the wavelength of these transitions lie in the range 1-30 cm. Therefore the ions are automatically within the Lamb-Dicke regime so long as the size of the ion cloud is of the order of a few mm or less. Therefore the need for laser cooling is not as strong as in the case of optical transitions (see below). However, in order to eliminate the second-order Doppler effect it is still advantageous to employ laser cooling. This is difficult for a large cloud of ions in an RF trap as RF heating prevents low temperatures being achieved. However, in a Penning trap there is no problem in cooling a large cloud of ions.

Mercury is a favourable candidate as it has a large mass, leading to low velocities, and it has a large ground state hyperfine structure splitting in $^{199}Hg^+$ (roughly 40.5 GHz), leading to a good fractional frequency stability. In fact prototype frequency standards using mercury were being actively worked on very early in the days of ion traps [49]. Experiments have recently been performed by the NIST group [50] and by the JPL group [51]. The NIST experiment [50] uses a linear ion trap with a string of a few ions which are laser cooled using radiation at 193 nm (eight ions were used in this experiment). It achieves a linewidth of 250 mHz (using a Ramsey-type interrogation scheme with an interrogation time of 1.8 s). The most recent results with this experiment report a Ramsey interrogation time of 100 s and, when run as a standard, the stability is reported to be about 4×10^{-15} over a 2 hour period [52]. This trap is run at cryogenic temperatures to keep the effects of collisions with background gas as small as possible. The other experiment [51] uses ions which are not laser cooled, but the signal is high as there is a large number of

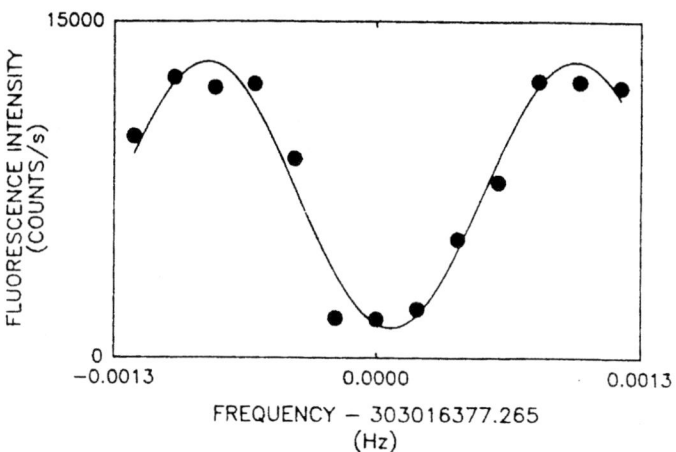

FIGURE 6. Be$^+$ microwave resonance at 303 MHz measured on a cloud of ions in a Penning trap (from [54]). The interaction time here was of the order of 500 s.

ions. The second order Doppler shift is significant (of the order of 10^{-12}) but is kept at a constant value by keeping the experimental conditions constant (especially the number of ions in the trap, which is maintained to approximately 0.1 %). The linear trap is able to achieve a smaller second order Doppler shift than a standard RF trap for a fixed number of ions. This experiment achieves a stability of 2×10^{-15} over a period of 24 000 s and is claimed to have an inherent stability of at least 10^{-15}. The linewidth of the microwave resonance is 160 mHz.

Beryllium is another element which has been worked with over a long period. The NIST group has recently run their Be$^+$ experiment as a prototype standard with very impressive performance figures. This experiment uses the technique of *sympathetic cooling* [53] where two different ionic species are held in the same trap at the same time (here Mg$^+$ and Be$^+$). The Mg$^+$ ions are laser cooled using light at 280 nm from a frequency-doubled dye laser. Through collisions, these ions sympathetically cool the Be$^+$ ions, and so the temperature of the Be$^+$ ions is reduced to very low values without them being directly laser cooled. This allows very long interaction times to be used when probing the clock transition, since the ions are not able to heat up during that period, as they otherwise would. As a result, in this experiment they are able to use an interaction time of typically 100 s, giving a linewidth of 5 mHz, and the standard has a stability of 3×10^{-14} over a period of 10 000 s [54]. The final limits on the accuracy of this standard are to do with the background pressure in the vacuum system. Figure 6 shows a lineshape of the microwave resonance used as a frequency standard in Be$^+$.

These various examples show that ion trap microwave standards are already at a similar level of performance to that of the caesium standard, and are still capable of improvement. Meanwhile, the Cs atomic clock has itself been improving due to the use of the techniques of optical pumping and laser cooling, so there is now a healthy competition between these different approaches to the production of new standards.

VII.D Optical standards

For an optical standard one requires an atomic structure which is very similar to that used in quantum jump experiments, and the standard arrangement is a three-level system with the resonance level and a metastable level both coupled to the ground state by a strong and a weak transition respectively. The lifetime of the metastable level has to be of the order of seconds so that the linewidth of the so-called *clock transition* down to the ground state (which is obviously limited by the natural linewidth of the metastable level) can be of the order of 1 Hz. It is also necessary to have an ultra-stable laser available with a comparable linewidth in order to take advantage of this low natural linewidth, and several laboratories around the world are working towards this level of stabilisation.

Detection is performed by observing the *absence* of many resonance transition photons when the ion is excited into the metastable level by a photon on the weak clock transition. Thus even with a single ion it is possible to detect nearly every transition made due to the use of this electron shelving technique (Section IV.D). The two lasers are applied at different times in order to avoid light shifts and broadening of the clock transition. Most of these experiments are performed with a single ion because of the need to achieve the Lamb-Dicke regime and the need to avoid RF heating effects. Thus a miniature RF trap with strong confinement and a single ion is the most common arrangement. Such experiments have been performed in Hg^+, Ba^+, Ca^+, Sr^+ and Yb^+. Most of these experiments have not progressed as far as the microwave standards experiments, due to the technical difficulties involved: tight confinement, strong laser cooling, ultra-stable laser sources etc. However, a linewidth of less than 100 Hz on an optical transition in Hg^+ has been observed by the NIST group [55], giving a Q-value of roughly 10^{13}, with a potential reproducibility of one part in 10^{15} or better.

An extreme example of a potentially very narrow transition is one in Yb^+ which has been studied by the group at NPL in the UK. This transition at 467 nm goes from the $S_{1/2}$ ground state of the ion to a very long lived $F_{7/2}$ state which has a lifetime of several years. The NPL group measured this transition in a single laser cooled ion in a miniature RF trap using a quantum jump technique [56] (Fig. 7). The laser with which they probed the ion had a linewidth of 350 kHz, and this is reflected in the linewidth of the observed transition, but the potential linewidth of this transition is very small indeed. The difficulty, of course, is obtaining a sufficient signal from such a weak transition to steer the frequency of an ultra-stable laser

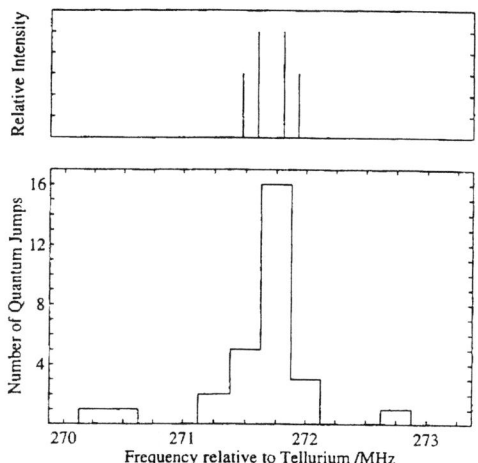

FIGURE 7. Spectrum of the 467 nm transition in a single Yb$^+$ ion (from [56]). The resolution is limited by the laser linewidth, which is 350 kHz. The laser frequency was changed in steps of 250 kHz between measurements.

source. The Yb$^+$ ion also has other transitions of interest for optical frequency standards and is being studied by several groups around the world.

Because the frequency of an optical transition is so much higher than that of a microwave transition, it is likely that in the end the transition used to define the second will be an optical one rather than a microwave one because the potential for high stability is so much better. However, it is likely to be a while before such standards are generally available and microwave ion trap standards are likely to be in use before then. There are of course other systems which may be used as well as ions in traps (particularly laser cooled atoms) but the work with trapped ions is very promising and has already come a long way towards the realisation of the ultimate frequency standard.

VIII NONCLASSICAL STATES

VIII.A Introduction

Quantum mechanics allows a much greater range of possible states of a system compared to those allowed by classical mechanics, as is well known. This is because a quantum mechanical system can exist in a *superposition* of different eigenstates, whereas a classical system has to be in a definite state at any time. One consequence of this is that it is possible to construct quantum mechanical superpositions that have noise levels which are lower than the standard quantum mechanical limits set by the uncertainty principle. In fact, in these so-called *squeezed states* the noise associated with one type of measurement is decreased while that associated with another is increased, so there is no violation of the uncertainty principle overall [57]. Furthermore, some quantum mechanical states have properties associated with them that cannot be realised in classical systems (for example, sub-Poissonian statistics of a light field).

All these features of quantum mechanical systems have been of great interest to physicists over the years as they probe the very foundations of quantum mechanics and shed light on the inherent statistical nature of the theory. There has therefore been a great deal of effort which has gone into experimental demonstrations of nonclassical states (especially of light) in order to see whether quantum mechanics correctly predicts these often counter-intuitive phenomena. With ion traps these experiments fall into two areas: demonstration that the resonance fluorescence emitted by a single ion is sub-Poissonian and antibunched; and generation of nonclassical states of motion in the vibrational state of a single trapped ion. These experiments are discussed in the following sections.

VIII.B Nonclassical light

It is well known that the light scattered by a single atomic particle on resonance has non-classical properties. This is because (in a simplified semiclassical view) after the particle has emitted a photon in decaying from the excited state to the ground state there has to be a delay while the particle is excited again before it can emit another photon. This means that there is a very low probability of two photons being emitted together: there will on average be a delay of the order of the spontaneous emission lifetime before the second photon is emitted. This is in contrast to classical light where two photons may be emitted together. The end result of this is that the resonance fluorescence is both *antibunched* and *sub-Poissonian*. Antibunched light has a low probability of two photons arriving at the detector close to each other in time, while sub-Poissonian light has a variance of photon counts which is less than that associated with normal counting processes having Poisson statistics (with the variance equal to the mean). These two properties are related but distinct.

Walther's group at Garching has made demonstrations of the non-classical nature of resonance fluorescence from a single ion. In the latest experiment [58] they study a single magnesium ion in a miniature end-cap trap [13]. They can choose to measure the particle nature of the light scattered by the ion or its wave nature. By counting photons and recording the statistics of the counts they are measuring the particle nature of the light, and this shows that the light is indeed antibunched, as expected [58]. This is done by measuring $g^{(2)}(\tau)$, the intensity correlation function,

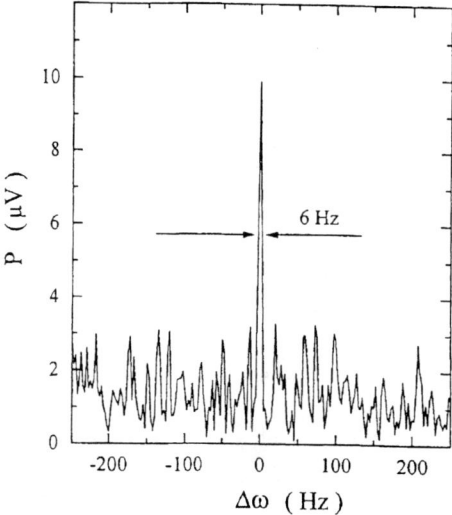

FIGURE 8. Resonance fluorescence heterodyne spectrum of a single Mg$^+$ ion in a miniature end-cap trap (from [58]). As expected, the elastically scattered component has a linewidth limited only by the resolution of the detection electronics.

which is the normalised probability of detecting a photon an interval τ after one photon has been detected. For a single ion, $g^{(2)}(\tau)$ is predicted to approach zero as τ approaches zero, and this is confirmed in the experiment (an earlier experiment by the same group shows plots of $g^{(2)}(\tau)$ for 1, 2 and 3 ions and also demonstrates that the light is sub-Poissonian [59]).

However, by measuring instead a beat signal obtained by mixing the resonance fluorescence from the ion with the laser light driving the transition, they are also able to see the spectrum of the resonance fluorescence, which probes the wave nature of the fluorescence. As expected, this spectrum is very narrow (remember the laser linewidth is eliminated by this heterodyne technique) and the measured linewidth of the elastically scattered light is less than 6 Hz (Fig. 8). This experiment is technically very difficult as the signal level available with only a single ion is very low.

VIII.C Nonclassical states of motion

Nonclassical properties of simple harmonic oscillators (SHOs) in quantum mechanics are generally treated in the context of modes of the radiation field, as radiation modes are a straightforward example of a quantum mechanical simple harmonic oscillator with the position and momentum becoming the electric field components of the mode [57]. Several different types of radiation can be defined: the vacuum state, where the mode is not excited at all; Fock states (or number states), which are states with a definite number of excitations (photons) in the mode; coherent states (with an amplitude and phase which are as definite as the uncertainty principle will allow - these are the states generated by lasers); thermal states (an incoherent superposition of number states such as is generated by a classical thermal source); and squeezed states (where the uncertainty in one quadrature (e.g., phase) is reduced at the expense of increased noise in another (e.g., amplitude). All these different types of state of the radiation field (and more!) have been generated in experiments on beams of light.

However, with a trapped ion we have available a system which also matches the requirements of a quantum mechanical simple harmonic oscillator if we concentrate on the vibrational state of the ion. This is directly a single oscillator mode, well isolated from the environment and, with the Raman techniques developed for laser cooling, it is possible to manipulate the state of the system at will. The NIST group has, in a remarkable series of experiments, demonstrated the main non-classical properties of quantum mechanical SHOs using a single Be^+ ion in a miniature RF trap [60].

The experiment builds on the initial preparation of the $n = 0$ state of the vibrational motion using the Raman techniques discussed earlier (Section III.D). Once the ion is in the $n = 0$ state, it can then be manipulated using similar techniques. For example, a Raman pulse tuned to the $\delta n = +1$ (blue) sideband will place the ion in the $n = 1$ state (in the other hyperfine level). If it is returned to the ground electronic state on the carrier, we have generated a Fock state with $n = 1$ and if it is returned on the (lower) sideband corresponding to $\delta n = +1$ again, we generate the Fock state with $n = 2$. In this way they are able to generate any Fock state with high purity and by plotting the probability of making a Raman transition as a function of the length of a subsequent Raman pulse on the blue sideband, they can directly see the Rabi oscillations between these two states. This Rabi frequency is itself a function of n, as expected for the theory of this system:

$$\Omega_{n,n+1} = \sqrt{n+1}\eta\Omega \qquad (26)$$

where Ω is the Raman coupling parameter. For this experiment, they find that η, the Lamb-Dicke parameter, has the value 0.202. A coherent state can be generated in several ways, for example by applying a short coherent drive to the ion in the trap using a weak electric field applied across the endcaps. This is of course a forced (or driven) SHO, and the result is a motion with a definite amplitude and phase, limited in uncertainty only by the uncertainty principle, as for any quantum mechanical system. This state can also be expressed as a coherent sum over Fock states of the oscillator, as was verified in the experiment, by the observation of collapses and revivals in the ground state occupation probability after a subsequent Raman pulse was applied.

The NIST group were able to generate a thermal state in simple manner too. All they did was to laser cool the ion *without* using the Raman cooling at all. The state generated by Doppler cooling is a thermal state with a temperature corresponding to the Doppler limit. In their case this gave an average value of $\langle n \rangle$ equal to 1.3 ± 0.1. A squeezed state was also generated, this time by driving the $n = 0$ state with Raman beams having a frequency difference corresponding to $\delta n = 2$. They were able to achieve a squeeze parameter β of 40 ± 10 in this manner, corresponding to a value of $\langle n \rangle$ of roughly 7.1.

Finally, the NIST group were able to generate a *Schrödinger cat state* using the vibrational state of the ion [61]. A Schrödinger cat state is a superposition of *macroscopically* different states of the system (in the original conception it was of course a superposition of alive and dead states of the cat). In this realisation a complicated series of operations is used to put the ion into a superposition of two electronic states, each having a large amplitude of motion but with a phase shift ϕ between them. We then have in effect two wavepackets in the trap oscillating with different phases. This can be observed by mixing the two states with another Raman pulse. Then these two wavepackets can overlap and interfere. The experiment consists of measuring these interferences as a function of the phase difference ϕ, and the results are striking (Fig. 9). The measurements show that the wavepackets are roughly 7 nm wide, and in the largest amplitude motion state generated, they have a maximum separation of 80 nm. In this way they really have been able to create a superposition of macroscopically different states of a single atomic particle in the trap.

IX ION TRAP LOGIC GATES

IX.A Quantum computing

As discussed earlier, quantum mechanical systems have the unique property that they can be prepared in a quantum superposition of different states. This is not the same as a statistical mixture of states, for which the system is in one state or another, but it is not known which until a measurement is made - only the probability of it being in either state is known. If a system is in a quantum superposition, it is in some sense in both states at the same time. However, when a measurement is made, the system will be projected into one state or the other with probabilities determined by the coefficients of the different states.

One application of this idea is that of *quantum computing*. This is a large subject on its own, and is treated, for example, by Steane [62]. Put simply, the advantage of making a computer out of quantum mechanical logic gates rather than classical ones is that they can then operate on input states which are quantum superpositions of different states, and therefore in effect perform many calculations at the same time, giving an output which is itself a superposition of different output states. However, this does not always help because any measurement of the output state can only yield one piece of information and the rest is lost. The trick is therefore to find problems which can be expressed in such a way that a single measurement of the output *can* be employed to give useful information.

There appear to be a few problems which fall into this category. One is the factorisation of very large numbers, which is an extremely difficult problem on classical computers, taking a very long time to complete. There exist algorithms (see [62]) which can find the factors of large numbers in a much smaller number of steps with a quantum computer than with a classical one. Another possible area of application is in algorithms for the sorting of lists.

However, there are many problems to be solved. One is that clean, well-isolated quantum systems have to be found which can be used to realise a quantum computer. These systems have to be able to exist in a quantum mechanical superposition of states with sufficiently low decoherence rates that many operations can be performed on the states while preserving the coherence. If the coherence is lost,

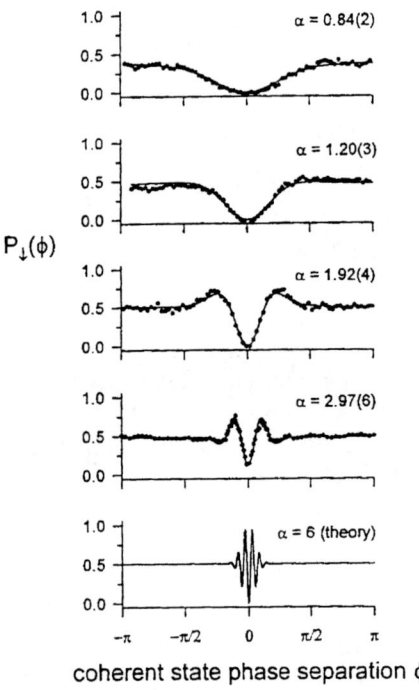

FIGURE 9. Realisation of a Schrödinger cat state using a single Be$^+$ ion (from [61]). The plots show the interference between the two macroscopically distinguishable components of the ion motion in the coherent superposition, as a function of the phase difference ϕ between them.

then the quantum superposition turns into a statistical mixture of states, and this loses all the advantages of having a quantum mechanical system. For practical systems this problem appeared to be insurmountable but recently the development of *quantum error correction codes* has meant that quantum computers can be made that correct for errors as they are generated through the sophisticated use of redundancy in the operations [62].

In a quantum computer the information is held in the form of *qubits*. A qubit is a single quantum mechanical piece of information, analogous to a bit in a classical computer, except that now a qubit may exist in any superposition of 1 and 0 rather than just one or the other. In terms of the density matrix of the system, this is represented by the presence of off-diagonal elements in the density matrix. The qubits are used as inputs to *quantum gates*, which operate on one or more qubits to give output states. These gates are similar to the gates in classical computers and complex operations are built up out of a small number of different types of elementary gates. One of the most important of these is the *controlled-NOT* or C-NOT gate. This reverses the state of the *target* qubit (i.e. $|1\rangle \to |0\rangle$ and $|0\rangle \to |1\rangle$) if and only if the value of the control qubit is $|1\rangle$.

It appears that there are two main ways of realising a quantum computer at present. One is using nuclear magnetic resonance (NMR) techniques on a solid sample. Elementary operations have been performed with this type of system [62], but there is a difficulty in extending the size of such a quantum computer because of signal to noise problems with large numbers of qubits. The other is using a string of laser cooled ions in a linear RF trap [63]. There may also be possibilities using cavity QED [47].

IX.B Single ion gates

The first suggestion that a quantum computer could be made from a string of ions in an RF linear trap was made by Cirac and Zoller [64]. The suggestion was that each qubit could be represented by the electronic state of an ion in the trap, and that coupling between them could be performed using the centre of mass motion of the string of ions. A procedure was given for the construction of a C-NOT gate using this idea. This used the useful feature that a 2π pulse of radiation on a transition between two states leaves the system in the same state as it started in, except that the *sign* of the wavefunction is reversed. Normally this has no effect, but in this scheme it transforms a superposition of the form $|0\rangle + |1\rangle$ to $|0\rangle - |1\rangle$ where the radiation only couples state $|1\rangle$ to a third state. This is used very ingeniously together with other operations in the gate to transform the target qubit if the control bit is set but to leave it alone if the control bit is not set.

Only one group has managed so far to build a gate using trapped ions [65]. In fact, although based loosely on the ideas in [64], this gate works with a single ion. The two qubits are represented by the electronic state of the ion (in fact, two of the Zeeman-split ground state hyperfine levels of a Be^+ ion - the target qubit) and

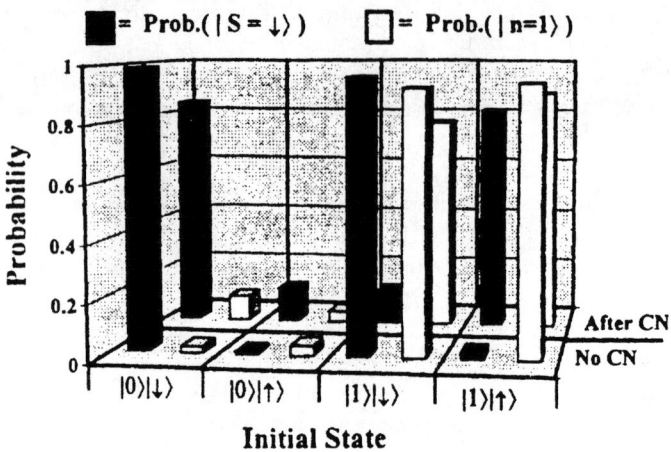

FIGURE 10. Demonstration of a controlled-NOT (C-NOT or CN) quantum gate using a single trapped ion (see the text for details) (from [65]).

the first two vibrational states of the ion in the trap potential well (the control qubit). A third atomic level is also needed at some point. With this system, the NIST group have been able to demonstrate that a C-NOT gate can be made to work with high efficiency. The results are shown in Fig. 10. In this figure, which is like a truth table for the gate, the black bars represent the target qubit while the white bars represent the control qubit. The state of the control bit is unchanged by the gate, as expected (at least 90% of the time) and if the control qubit is zero the target qubit is also unchanged. However, when the control qubit is one, the target qubit is reversed by the gate with high efficiency.

Although the efficiency of this gate is nothing like as good as it would need to be in order to be a part of a quantum computer, it constitutes a convincing demonstration that quantum gates can be built with trapped ions and that the experimental problems can be overcome.

IX.C Future prospects

In their initial experiment, the NIST group also looked at the decoherence rates in the system [65]. They found the rate to be a few kHz, to be compared to the switching speed of 20 kHz. The decoherence rates need to be made lower in order to improve the fidelity of the gate. The problems to be dealt with are discussed in great detail by Wineland *et al.* [66]. This paper shows that there are severe problems to be dealt with, and several unknowns, but that the prospects are good for demonstrating real computation with quantum mechanical systems in the next few years.

The NIST group has since begun to work on two-ion systems. A crystal of two ions in a trap has two modes of oscillation along the axis of the crystal: a centre of mass (COM) mode where the ions move together and a stretching mode where the ions move in opposite directions. Here it is of course necessary to cool both modes of motion in order to achieve the Lamb-Dicke regime for the two ions. The NIST group has managed to cool a two-ion crystal to $\langle n_{COM} \rangle = 0.11$ and $\langle n_{stretch} \rangle = 0.01$ [67]. This is an important first step towards the construction of a gate with two ions, which is itself a significant step towards the construction of many-ion gates.

The NIST group have made important measurements of the heating rates in a two-ion crystal. They find that the heating rate for the stretching mode is much lower than that for the centre of mass mode, and suggest that this mode would be the better one to use for quantum gates [67]. The measured rates in this experiment were $\delta \langle n \rangle / \delta t = 20$ ms^{-1} (COM mode) and < 0.18 ms^{-1} (stretch mode).

In future work it is likely that logic gates with more than one ion will soon be constructed but the use of such gates for real computation is a long way away. The most recent work reported by the NIST group is the deterministic entanglement of two trapped ions [68], which represents another important step. At present the most important aspect is the demonstration that the ideas which have been developed for computation with quantum mechanical systems can indeed be realised in practice. Other groups around the world are also working in this area, including Los Alamos [69] and Innsbruck [70].

X CONCLUSION

These lectures have attempted to demonstrate the wide range of problems in quantum optics to which ions in traps can be applied. It has not been possible to be comprehensive, and there are inevitably some areas which have had to be left out. However, I hope that it has been made clear that trapped ions have made many contributions to our understanding of quantum optics in the last twenty years since laser cooling of trapped ions was first demonstrated. Ion traps offer a uniquely well-controlled and isolated environment for small numbers of atomic particles and the applications, as we have seen, are numerous. I have every confidence that in the next twenty years they will continue to have a dramatic impact throughout quantum optics and other areas of physics.

Acknowledgements

I wish to thank my colleagues at Imperial College for their help in reading these notes, and the Engineering and Physical Sciences Research Council for their support of our work at Imperial College. I would also like to thank by colleagues at UNAM for their invitation to give these lectures and for their hospitality during my stay in Mexico.

REFERENCES

1. Brown, L. S., and Gabrielse, G., *Rev. Mod. Phys.* **58**, 233-311 (1986).
2. Rolston, S. L., these lecture notes (1998).
3. Thompson, R. C., in *Adv. At. Mol. Opt. Phys.* **31**, 63-136 (1993).
4. Ghosh, P. K., *Ion traps*, Oxford University Press (1995).
5. Horvath, G. Zs. K., Thompson, R. C., and Knight, P. L., *Contemporary Physics* **38**, 25-48 (1997).
6. Penning, F. M., *Physica* **3**, 873-94 (1936).
7. Thompson, R. C., and Wilson, D. C., *Z. Phys. D* **42**, 271-7 (1997).
8. Fischer, E., *Z. Phys.* **156** 1-26 (1959).
9. McClachlan, N. W., *Theory and application of Mathieu functions*, Oxford University Press (1947).
10. March, R. E., and Hughes, R. J., *Quadrupole storage mass spectrometry*, Wiley Interscience, New York (1989).
11. Neuhauser, W., Hohenstatt, M., Toschek, P. E., and Dehmelt, H., *Phys. Rev. Lett.* **41**, 233-6 (1978).
12. Beaty, E. C., *J. Appl. Phys.* **61**, 2118-122 (1987).
13. Schrama, C. A., Peik, E., Smith, W. W., and Walther, H., *Opt. Comm.* **101**, 21 (1993).
14. See e.g. Jefferts, S. R., Monroe, C., Barton, A. S., and Wineland, D. J., *IEEE Trans. Instrum. Meas.* **44**, 148-50 (1995).
15. See e.g. McGuire, M. D., Petsch, R., and Werth, G., *Phys. Rev. A* **17**, 1999-2004 (1978).
16. Prestage, J. D., Dick, G. J., and Maleki, L., *J. Appl. Phys.* **66**, 1013-7 (1989).
17. Waki, I., Kassner, S., Birkl, G., and Walther, H., *Phys. Rev. Lett.* **68**, 2007-2010 (1992).
18. Birkl, G., Kassner, S., and Walther, H., *Nature* **357**, 310-3 (1992).
19. Blümel, R., Kappler, C., Quint, W., and Walther, H., *Phys. Rev. A* **40**, 808-23 (1989).
20. See e.g. Plumelle, F., Desaintfuscien, M., Duchene, J. L., and Audoin, C., *Opt. Comm.*, **34**, 71-6 (1980).
21. Wineland, D. J., and Dehmelt, H., *Bull. Am. Phys. Soc.* **20**, 637 (1975).
22. Wineland, D. J., Dalibard, J., and Cohen-Tannoudji, C. *J. Opt. Soc. Am.* **9**, 32-42 (1992).
23. Birkl, G., Yeazell, J. A., Rückerl, R., and Walther, H., *Europhys. Lett.* **27**, 197 (1994).
24. Itano, W. M., and Wineland, D. J., *Phys. Rev. A* **25**, 35-54 (1982)

25. Itano, W. M., Brewer, L. R., Larson, D. J., and Wineland, D. J., *Phys. Rev. A* **38**, 5698-5706 (1988).
26. Heinzen, D. J., Bollinger, J. J., Moore, F. L., Itano, W. M., and Wineland, D. J., *Phys. Rev. Lett.* **66**, 2080-3 (1991).
27. Dicke, R. H., *Phys. Rev.* **89**, 472-3 (1953).
28. Bergquist, J. C., Itano, W. M., and Wineland, D. J., *Phys. Rev. A* **36**, 428-30 (1987).
29. Monroe, C., Meekhof, D. M., King, B. E., Jefferts, S. R., Itano, W. M., Wineland, D. J., and Gould, P., *Phys. Rev. Lett.* **75**, 4011-4 (1995).
30. Wineland, D. J., Drullinger, R. E., and Walls, F. L., *Phys. Rev. Lett.* **40**, 1639-41 (1978).
31. Neuhauser, W., Hohenstatt, M., Toschek, P. E., and Dehmelt, H., *Phys. Rev. A* **22**, 1137-40 (1980).
32. Wineland, D. J., and Itano, W. M., *Phys. Lett. A* **82**, 75-8 (1981).
33. Gisin, N., Knight, P. L., Percival, I . C., Thompson, R. C., and Wilson, D. C., *J. Mod. Opt.* **40**, 1663-71 (1993).
34. Hulet, R. G., Wineland, D. J., Bergquist, J. C., and Itano, W. M., *Phys. Rev. A* **37**, 4544-7 (1988).
35. Sauter, T., Blatt, R., Neuhauser, W., and Toschek, P. E., *Opt. Comm.* **60**, 287-92 (1986).
36. Cook, R. J., and Kimble, H. J., *Phys. Rev. Lett.* **54**, 1023-6 (1985).
37. Blatt, R., and Zoller, P., *Eur. J. Phys.* **9**, 250-6 (1988).
38. Cook, R. J., in *Progress in Optics XXVIII* (E Wolf, Ed.) Elsevier, pp. 361-416 (1990).
39. Garraway, B. M., and Knight, P. L., *Phys. Rev. A* **49**, 1266-74 (1994).
40. Dehmelt, H., *IEEE Trans. Instrum. Meas.* **IM-31**, 83-7 (1982).
41. Misra, B., and Sudarshan, E. C. G., *J. Math. Phys.* **18**, 756-63 (1977).
42. Itano, W. M., Heinzen, D. J., Bollinger, J. J., and Wineland, D. J., *Phys. Rev. A* **41**, 2295-300 (1990).
43. Frerichs, V., and Schenzle, A., *Phys. Rev. A* **44**, 1962-8 (1991).
44. Schenzle, A., *Contemp. Phys.* **37**, 303-20 (1996).
45. Plenio, M., Knight, P. L., and Thompson, R. C., *Opt. Comm.* **123**, 278-86 (1996).
46. Beige, A., and Hegerfeldt, G. C., *J. Phys. A* **30**, 1323-34 (1997).
47. Haroche, S., these lecture notes (1998).
48. Meyer, G. M., Briegel, H-J., and Walther, H., *Europhys. Lett.* **37**, 317-22 (1997).
49. Jardino, M., Desaintfuscien, M., Barrilet, R., Viennet, J., Petit, P., and Audoin, C., *Appl. Phys.* **24**, 107-12 (1981).
50. Raizen, M. G., Gilligan, J. M., Bergquist, J. C., Itano, W. M., and Wineland, D. J., *J. Mod. Opt.* **39**, 233-42 (1992).
51. Prestage, J. D., Tjoelker, R. L., Dick, G. J., and Maleki, L., *J. Mod. Opt.* **39**, 221-232 (1992).
52. Berkeland, D. J., Miller, J. D., Bergquist, J. C., Itano, W. M., and Wineland, D. J., *Phys. Rev. Lett.* **80**, 2089-92 (1998).
53. Larson, D. J., Bergquist, J. C., Bollinger, J. J., Itano, W. M., and Wineland, D. J., *Phys. Rev. Lett.* **57**, 70-3 (1986).
54. Bollinger, J. J., Heinzen, D. J., Itano, W. M., Gilbert, S. L., and Wineland, D. J., *IEEE Trans. Instrum. and Meas.*, **40**, 126-8 (1991).

55. Wineland, D. J., Bergquist, J. C., Bollinger, J. J., Itano, W. M., Heinzen, D. J., Gilbert, S. L., Manney, C. H., and Raizen, M. G., *IEEE Trans. Ultrasonics, Ferroelectrics and Freq. Control*, **37**, 515-23 (1990).
56. Roberts, M., Taylor, P., Barwood, G. P., Gill, P., Klein, H. A., and Rowley, W. R. C., *Phys. Rev. Lett.* **78** 1876-9 (1997).
57. Teich, M. C., and Saleh, B. E. A., *Quantum Opt.* **1**, 153-91 (1989).
58. Hoeffges, J. T., Baldauf, H. W., Eichler, T., Helmfrid., S. R., and Walther, H., *Opt. Comm.* **133**, 170-4 (1997).
59. Diedrich, F., and Walther, H., *Phys. Rev. Lett.* **59**, 2931-4 (1987).
60. Meekhof, D. M., Monroe, C., King, B. E., Itano, W. M., and Wineland, D. J., *Phys. Rev. Lett.* **76**, 1796-9 (1996).
61. Monroe, C., Meekhof, D. M., King, B. E., and Wineland, D. J., *Science* **272**, 1131 (1996).
62. Steane, A., *Rep. Prog. Phys.* **61**, 117-73 (1998).
63. Steane, A., *Appl. Phys.* **B 64**, 623-42 (1997).
64. Cirac, J. I., and Zoller, P., *Phys. Rev. Lett.* **74**, 4091 4 (1995).
65. Monroe, C., Meekhof, D. M., King, B. E., Itano, W. M., and Wineland, D. J., *Phys. Rev. Lett.* **75**, 4714-7 (1995).
66. Wineland, D. J., Monroe, C., Itano, W. M., Liebfried, D., King, B., and Meekhof, D. M., "Experimental issues in coherent quantum state manipulation of trapped atomic ions", preprint (1997) (quant-ph/9710025).
67. King, B. E., Wyatt, C. J., Turchette, Q. A., Liebfried, D., Itano, W. M., Monroe, C., and Wineland, D. J., *Phys. Rev. Lett.*, in press (1998)(quant-ph/9803023).
68. Turchette, Q. A., Wood, C. S., King, B. E., Myatt, C. J., Liebfried, D., Itano, W. M., Monroe, C., and Wineland, D. J., preprint (1998) (quant-ph/9806012).
69. Hughes, R. J., et al., *Fort der Physik–Prog. of Phys.* **46**, 329-61 (1998).
70. Nagerl, H. C., Bechter, W., Eschner, J., Schmidt-Kaler, F., and Blatt, A., *Appl. Phys. B* **66**, 603-608 (1998).

II. FOUNDATIONS OF QUANTUM MECHANICS

Linear Stochastic Electrodynamics: Looking for the Physics Behind Quantum Theory[*]

Luis de la Peña and Ana María Cetto

Instituto de Física, UNAM, México
luis@fenix.ifisicacu.unam.mx, ana@fenix.ifisicacu.unam.mx

Abstract. In this chapter, which covers part of the course given at ELAF, a straightforward procedure is presented that leads from the basic postulates of stochastic electrodynamics to the usual formalism of quantum theory. The theory thus developed is called linear stochastic electrodynamics, to underline that one of its basic features is the (asymptotic) linear response of atomic systems to the background field.

The chapter starts with a brief discussion of some open questions in quantum theory and of the possibility to find an answer to them by resorting to the zeropoint radiation field as the source of the quantum behaviour of matter. The basic properties of this field are discussed, and a brief enumeration is made of some of the positive results and vital shortcomings of standard stochastic electrodynamics. After identifying the source of these shortcomings in the assumption that the background field is not altered by its interaction with matter, linear stochastic electrodynamics is developed and shown to lead, under certain approximations, to a consistent picture of both matter and field quantization. In the concluding part, it is shown that also the electron spin can be considered to be generated by the interaction of the particle with the zeropoint field; in particular, the two-valuedness of the spin projection is associated with the existence of just two independent states of polarization of the field.

---

[*] This is an abridged version of the course given at ELAF98; the unabridged version can be obtained from the authors, under the title: *Stochastic Electrodynamics: Looking for the Physics Behind Quantum Theory.*

I BRIEF CONSIDERATIONS ABOUT QUANTUM MECHANICS AND STOCHASTIC ELECTRODYNAMICS

1. Is there any need to look for a fundamental theory of quantum mechanics?

For many physicists this seems a strange question, since quantum mechanics *is* the fundamental theory of matter. However, for others —like John Bell for instance—, quantum mechanics is no more than a *fapp* theory[1] about measurements and observables, but not about *beables* [1]. Of course Bell is not alone in his discomfort, as is testified by the thousands of papers and monographs that have been written in an attempt to get a more intuitive or, rather, a more realistic picture of quantum mechanics. There are also hundreds of popular books that try to make the strange paradoxes of quantum mechanics palatable to the beginner; only too often, however, their authors fail to distinguish science from ideology, and in some occasions even from magic. There are of course some very fortunate exceptions, such as the outstanding little book by David Wick [2], or the one by Baggott [3].

Let us recall some instances that illustrate what we have in mind.

a) What is a photon? If you believe you know the answer, the old words of Einstein [4] near the end of his life may be of some value: "All the fifty years of conscious brooding have brought me no closer to the answer to the question, 'What are light quanta?' Of course today anyone thinks he knows the answer, but he is deluding himself." It is true that a lot of new physics has been learned in the last fifty years, but it would be difficult to say that the new knowledge has led us significantly closer to the answer [5].

b) What is an electron? In other words (as suggested by the discussion in [6]: where does matter reside in quantum mechanics? For in the usual (Copenhagen) interpretation, the notion of corpuscle as a localized object disappears; yet the wave that takes its place cannot be considered a physical wave (unless strange ad hoc hypotheses are introduced), since its normalization is arbitrary and, moreover, it exists only in configuration space, which for the case of several interacting particles differs from ordinary three-dimensional space.

c) What are we really trying to say when we speak of an electron that interferes with itself? Quantum theory is supposed to deal only with observables, yet the fact is that halves of an electron have never been observed. Particle interference and other wave manifestations of matter seem to remain as mysterious as they were seven decades ago.

d) The very meaning of the wave function remains practically as obscure as it was at the onset of present-day quantum mechanics. It is true that virtually all textbooks declare to adhere to the single-particle interpretation of the Copenhagen school; but many of them are not particularly fine examples of consistency on this

---

[1]) fapp= *f*or *a*ll *p*ractical *p*urposes.

point. This is why the ensemble interpretation of the wave function has survived for decades; simply, it gives more satisfaction to many spirits than the orthodox one.

e) Even such an elementary and fundamental aspect as the stability of atomic states remains without a satisfactory *physical* explanation. Indeed the Schrödinger equation predicts it, but the physical mechanism that supports atomic stability remains unexplained.

f) In quantum mechanics we constantly stumble with such strange and unintelligible things as negative probabilities and the lack of precise recipes to construct correlations and the like. We do not even know how to interpret products of observables when they contain the same factors in a different order. Such deep problems provide the rationale for extreme approaches like the construction of "quantum logics", the mere existence of which is an indication of the discomfort produced by the usual rendering of the theory.

g) What is the physics behind entangled states? Entanglement means nonlocality at arbitrary distances; quantum mechanics predicts the phenomenon, but gives no hint as to a physical explanation or mechanism behind it. Are we forced to accept that quantum influences can act (instantaneously) over indefinite distances, with no physical element being exchanged?

h) More generally, quantum mechanics seems to tell us that we must give up the pretension of making a local realistic description of quantum systems, since we become inextricably intertwined with the observed (sub)system, and nonlocal (instantaneous) effects are bound to occur. In plain words, we become part of what we are studying. It is now fashionable to hear that experiment has confirmed the existence of such nonlocal influences taking place with the exchange of nothing physical, just information at superluminal velocities. Taken at its face value this means that quantum mechanics and relativity are mutually inconsistent. It is at least surprising (to say it gently) that this happens at the time of the almost heroic efforts of the superstring theoreticians to construct a quantum theory that incorporates gravity automatically, by its mere structure, to get a unified theory of all interactions; and when something similar occurs within quantum field theory, where important efforts have been made to explicitly demonstrate its causal properties, i.e., the compatibility of interactions and their effects with relativity (see, e.g., [7] and references therein). Our present day theoretical physics seems to be schizoid, at least in what refers to some major principles as locality and causality, and hence realism.[2]

---

[2] Contrary to a popular belief, it is not true that the experiments to demonstrate the violation of some Bell inequalities have already given a definitive verdict. The inequalities that have been probed are not the original ones, based on a couple of fundamental and very general principles, but alternate ones that are easier to test, the derivation of which requires extra (and unwarranted, even if plausible) hypotheses, and which are of a much less fundamental nature. It is unclear whether the results are an outcome of the violation of the fundamental premises, or merely of the auxiliary ones. The interested reader may find an analysis of these matters, and other criticisms, in [8,9].

2. Why should stochastic electrodynamics provide an answer to the questions left open by present day quantum mechanics?

For SED, the origin of quantum properties of matter constitutes a legitimate subject of physical scrutiny. Nobody knows to what extent SED can give a satisfactory answer, yet some of us who have been working in different aspects of the theory, believe that there is an ample possibility for this to be the case eventually, and this is the force than keeps us going. When some years ago it became clear that the treatment of non linear, multi-frequency sytems would require a theory different from the one that was being developed following the conventional approaches to the theory of stochastic processes, many interpreted this as an evidence that SED does not describe nature and abandoned the arena. We hope to demonstrate here that there are good and strong reasons to consider SED as a viable theory.

SED is built on the basic assumption that the electromagnetic vacuum, normally known as the zeropoint field, is a real field that pervades all space. Thus all matter is in permanent contact and interaction with this random field. Its ensuing irreducible stochastic motion is believed to be the source of quantum behaviour (even of that of the field itself, as we will have opportunity to see below). Hence, according to SED, the quantum properties are acquired by matter, rather than being intrinsic.

SED is not a particularly new theory. Indeed, the honour to be its first proponent belongs to Nernst (around 1916!). Its basic idea (explained in the paragraph above) is so simple and intuitive that it has been rediscovered many times since. The interested reader will find a detailed account of this story in the monograph *The Quantum Dice*,[3] to which we will refer in what follows as *The Dice*.

We believe that SED contains the seed for a physical approach to quantum mechanics because:

a) It contains stochastic physical elements that may explain the origin of the quantum stochasticity (indeterminism is a more usual term, but it is so contaminated with ideological elements that we prefer to avoid it).

b) It contains a physical field in interaction with matter, and thus is able, in principle, to explain the appearance of the wave behaviour of matter as something not intrinsic, but impressed by the field and revealed by the particle (as we briefly discuss below).

c) Under appropriate assumptions (to be disclosed below) it may explain the stability of certain (atomic) configurations, thus opening a door to an understanding of the origin of quantization.

d) It is able to explain, at least in principle, the apparent non localities of quantum mechanics as mediated by the field. Also, the possibility exists to get a physical explanation of the quantum (exchange) correlations among identical particles.

e) It allows to precise the physical meaning of quantum observables.

f) Those working on the optical version of SED, the theory called *stochastic*

---

[3] *The Quantum Dice. An introduction to stochastic electrodynamics* [10] is a detailed review of the work on SED up to 1995. An important part of the matters discussed in the present essay are presented there in more detail; it also contains an ample list of references.

optics, believe that the quantum behaviour of the field is due to interference and correlation effects between the macroscopic and the zeropoint fields. Indeed, a good deal of interesting results have been obtained by following this idea.[4]

It is true that, at least for the time being, SED does not make experimentally verifiable predictions different from standard quantum mechanics. However, the theory does set limits to the application of the present-day description of quantum systems. We will see that according to it, quantum mechanics describes the behaviour of systems only once they have reached the *quantum regime*, and we shall give a meaning to this concept.

II THE ZEROPOINT FIELD

Space is not empty but occupied by a collection of zeropoint fields, which together constitute the physical vacuum. According to SED, of all these vacua the electromagnetic component, the *zeropoint radiation field*, is crucial for an understanding of atomic and quantum physics; in other words, the quantum behaviour of matter is considered a consequence of the interaction of matter with the zeropoint radiation field.

Classical electrodynamics is built upon the assumption that in the absence of nearby sources there is no electromagnetic field; however, after a moment's reflection one perceives that this assumption is unnecessarily restrictive and poorly founded. A more natural boundary condition would be a random field at infinity, i. e., a nonzero homogeneous solution of Maxwell's equations [15,16]. However, as is well known it was only with the advent of quantum theory that the idea of a zeropoint field began to take shape. While elaborating his second theory of the blackbody law, Planck arrived at the expression

$$\mathcal{E}(\omega, T) = \tfrac{1}{2}\hbar\omega + \frac{\hbar\omega}{e^{\hbar\omega/kT} - 1} \quad (1)$$

for the average energy of an elementary oscillator of natural frequency ω in equilibrium with the radiation field at temperature T. The zeropoint energy appeared thus for the first time, represented by the term $\mathcal{E}(\omega, 0) = \tfrac{1}{2}\hbar\omega$. Contrary to classical physics, where all motions are assumed to freeze at absolute zero, this result shows that in the atomic world, fluctuations continue to take place even at $T = 0$. Now, Planck's prediction referred to the zeropoint energy of the *mechanical oscillators*, but for the electromagnetic field modes the spectral energy distribution was still given by (1) *without* the zeropoint contribution. It was Nernst (1916) who argued that the difference between field and matter oscillators is inadmissible if

---

[4]) Since there will be no opportunity to touch upon these matters in the present essay, we refer the interested reader to some of the original papers on the subject, as [11,12], and more recent ones as [13,14].

these systems are to attain statistical equilibrium when in thermal contact, and that equation (1) should therefore hold for both. The concept of the *zeropoint field* was thus born.

Even if physics went along a different course and Nernst's ideas were soon forgotten, eventually the idea of a zeropoint fluctuating radiation field has found support, both on the theoretical and on the experimental side. There are several phenomena widely considered to be caused by the interaction of matter with the zeropoint field. Some of the most frequently adduced ones are the van der Waals and Casimir forces between macroscopic bodies [17-20], and the natural atomic linewidth, due in equal parts to the zeropoint field and radiation reaction [21,22].[5]

An objection frequently raised against the existence of a real zeropoint field is that it implies huge effects —both electromagnetic and gravitational— which could not possibly pass unnoticed. It is well known that for this reason Pauli strongly opposed the notion of a real zeropoint field, a line that has been adhered to by many contemporary authors. A partial answer proposed by SED with respect to the electromagnetic effects is that they do not at all pass unnoticed, but quite on the contrary: they are being systematically observed in the form of quantum properties of matter —in fact so systematically, that we are not aware of it—. It is true that the zeropoint field should seemingly produce other effects due to its huge energy. This problem, which is common to all quantum field theories, is usually set aside in the case of QED by considering the vacuum as virtual. Of course in a truly relativistic treatment the zero of the energy cannot be defined arbitrarily, so the problem remains open, both in QED and in SED [20,25].

In SED one wants to find out the extent to which the quantum properties of matter can be considered to arise from its interaction with the vacuum field. A program of this kind should not in principle start by assuming a *quantum* vacuum, which would be expected to be the outcome of the theory; the starting point, the original field of SED, should be a random field described in nonquantized terms. That the ultimate correspondence between the two theories *cannot* be exact follows immediately from the different views of both theories as regard realism, locality, causality and so on, without the need to enter into details. Quite apart from the differences in physical insight afforded by the two theories, approximations are necessarily introduced in making the transition from SED to the quantum description, and the predicted differences should therefore lead eventually to the design of empirical tests that allow to discriminate between them..

As an example of the possibilities that SED offers for a better understanding of the behaviour of matter, let us consider the problem of atomic stability, which finds no solution in classical physics, and only a formal one in quantum mechanics.[6] In SED atomic stability is conceived as a result of the balance between the aver-

---

[5] It can even be argued that the zeropoint field is required for the formal consistency of QED. For instance, in absence of this field the canonical commutation relation $[\hat{q}(t), \hat{p}(t)] = i\hbar$ for a dipole oscillator or a free particle would decay exponentially with time due to radiation reaction [23,24].
[6] We have in mind the fact that quantum theory *predicts* a stable ground state despite the presence of radiation reaction, but the mechanism of such stability remains hidden.

age rates of absorption and emission of energy by the atomic electrons. Indeed a semiquantitative analysis for the circular Kepler hydrogenic orbits of energy $\mathcal{E}$ and radius r shows that the ratio of the power absorbed from the field to the power radiated, assuming that the zeropoint field has a spectral density $\rho(\omega)$ proportional to ω^s, with s a positive integer, can be written in the form

$$\eta \equiv \frac{\dot{W}_a}{\dot{W}_r} = \left|\frac{\mathcal{E}}{\mathcal{E}_e}\right|^{(3s-8)/2} = \left(\frac{r_e}{r}\right)^{(3s-8)/2}, \qquad (2)$$

where $\mathcal{E}_e$, r_e are the equilibrium energy and radius respectively, defined by the condition $\dot{W}_r = \dot{W}_a$. It follows that when the electron is losing energy and falling towards the nucleus, $|\mathcal{E}| \to \infty$, and for stability to exist one should have $\eta > 1$, which demands that $3s > 8$, or $s \geq 3$. We verify that with this selection, when the electron gains energy ($|\mathcal{E}| \to 0$), $\eta < 1$, as is required to recover the equilibrium situation. The result was encouraging when it was first obtained (1976), since it was well understood that one must take $s = 3$ for the SED field to be Lorentz invariant, which appears thus as the simplest field able to stabilize the atom. When it became possible to make more detailed calculations, it turned out that this nice picture is spoiled by the elliptic orbits, for which the electron absorbs energy in excess, and escapes and ionizes the system. This result (obtained around 1980) opened a period of disenchantment and uncertitude about the viability of SED as a physical theory. The main scope of the present work is to offer an understanding of the reason of such failure and to show that there is a way to solve this complex problem.

One of the most important and distinctive properties of the zeropoint field is its frequency spectrum. This spectrum can be easily derived from the argument that the minimum energy $\mathcal{E}(\omega, T = 0)$ of a quantum oscillator is $\frac{1}{2}\hbar\omega$; this, combined with Planck's formula for the spectral energy density of the equilibrium field in terms of the average energy of its modes,

$$\rho(\omega, T) = \frac{\omega^2}{\pi^2 c^3} \mathcal{E}(\omega, T), \qquad (3)$$

gives the important result [15]

$$\rho_0(\omega) = \rho(\omega, T=0) = \frac{\hbar \omega^3}{2\pi^2 c^3}. \qquad (4)$$

Considering the central role played by this spectrum, it is rather unsatisfactory to use a quantum law to establish it. Therefore the problem of determining the spectral density of the vacuum field has been considered in the most general possible terms. In particular this same spectrum can be recovered from Wien's displacement law $\rho(\omega, T) = \omega^3 \Phi(T/\omega)$ (with Φ a *universal* function, which remains undetermined within classical physics), reformulated so as to become applicable at

$T = 0$, when $\Phi(0)$ becomes a universal constant [33]; another and much more appealing argument, starts from the observation that all inertial observers move freely, i.e., with no friction, through the zeropoint field. This demand can be shown to be equivalent to that of the isotropy of the zeropoint field in all inertial frames of reference and, of course, it leads to the same spectrum $\sim \omega^3$. [15,10,24,26–33]

The free zeropoint field is usually considered as maximally disordered, under the natural assumption that it has been generated by a huge number of independent sources. The simplest representation of this field in free space is made by considering a finite parallelopiped of volume V and performing a discrete Fourier expansion in terms of plane waves, to write

$$\mathbf{A}(\mathbf{x},t) = \sum_{n,\lambda} \sqrt{\frac{2\pi c^2 \mathcal{E}_n}{\omega_n^2 V}} \mathbf{e}_n^\lambda \left(a_{n\lambda} e^{-i\omega_n t + i\mathbf{k}\cdot\mathbf{x}} + a_{n\lambda}^* e^{i\omega_n t - i\mathbf{k}\cdot\mathbf{x}} \right). \qquad (5)$$

The vector $\mathbf{k} = \mathbf{k_n}$ has components

$$k_i = \frac{2\pi}{L} n_i, \quad i = x, y, z \qquad (6)$$

and each n_i can assume any integer value, including zero, so that $\mathbf{A}(\mathbf{x_i}) = \mathbf{A}(\mathbf{x_i + L_i})$ along each coordinate axis. The mode frequencies are given by

$$\omega_\mathbf{n} = ck_\mathbf{n} = 2\pi \sqrt{\frac{n_1^2}{L_1^2} + \frac{n_2^2}{L_2^2} + \frac{n_3^2}{L_3^2}}. \qquad (7)$$

The contribution from each polarization has been explicitly separated, so that $\mathcal{E}_n$ refers to the energy of the mode with wave vector $\mathbf{k}$ and polarization λ. For each wave vector $\mathbf{k_n}$ there are two orthogonal directions, distinguished by the polarization index $\lambda = 1, 2$. In other words, for every $\mathbf{n} = (\mathbf{n_1}, \mathbf{n_2}, \mathbf{n_3})$ the $\mathbf{e}_n^\lambda$ represent two polarization vectors which along with $\hat{\mathbf{k}}_\mathbf{n} = \mathbf{k_n}/k_\mathbf{n}$ form a triplet of orthogonal unit vectors, so that, with the Coulomb gauge $\nabla \cdot \mathbf{A} = \mathbf{0}$, one gets

$$\mathbf{k_n} \cdot \mathbf{e}_n^\lambda = \mathbf{0}, \quad \mathbf{e}_n^\lambda \cdot \mathbf{e}_n^{\lambda'} = \delta_{\lambda\lambda'}, \qquad (8)$$

$$\hat{\mathbf{k}}_\mathbf{n} = \mathbf{e}_n^1 \times \mathbf{e}_n^2, \quad \mathbf{e}_n^1 = -\hat{\mathbf{k}}_\mathbf{n} \times \mathbf{e}_n^2, \quad \mathbf{e}_n^2 = \hat{\mathbf{k}}_\mathbf{n} \times \mathbf{e}_n^1. \qquad (9)$$

The electric field is then

$$\mathbf{E}(\mathbf{x},t) = i\sum_{n,\lambda} \sqrt{\frac{2\pi \mathcal{E}_n}{V}} \mathbf{e}_n^\lambda \left(a_{n\lambda} e^{-i\omega_n t + i\mathbf{k}\cdot\mathbf{x}} - a_{n\lambda}^* e^{i\omega_n t - i\mathbf{k}\cdot\mathbf{x}} \right), \qquad (10)$$

and a similar expression applies for the magnetic field, with $\hat{\mathbf{k}}_\mathbf{n} \times \mathbf{e}_n^\lambda$ instead of $\mathbf{e}_n^\lambda$. (When there is no risk of confusion we write $\mathbf{k}$ (or $\mathbf{n}$) instead of $\mathbf{k_n}$ in the indices.)

The integrals of motion of the free field can be easily calculated; one gets for the Hamiltonian and momentum

$$H = \frac{1}{8\pi} \int d^3x \left(\mathbf{E}^2 + \mathbf{B}^2 \right) = \sum_{\mathbf{n},\lambda} H_{\mathbf{k}\lambda} = \sum_{\mathbf{n},\lambda} \mathcal{E}_n a^*_{\mathbf{n}\lambda} a_{\mathbf{n}\lambda}, \tag{11}$$

$$\mathbf{P} = \frac{1}{4\pi c} \int d^3\mathbf{x} \mathbf{E} \times \mathbf{B} = \sum_{\mathbf{n},\lambda} \frac{\mathcal{E}_n}{c} \hat{\mathbf{k}}_n a^*_{\mathbf{n}\lambda} \mathbf{a}_{\mathbf{n}\lambda}. \tag{12}$$

When these expressions are applied to the zeropoint field, the amplitudes $a_{\mathbf{n}\lambda}$, $a^*_{\mathbf{n}\lambda}$ are taken to be random variables; all the stochasticity of the field is contained in the set $\{a_{\mathbf{n}\lambda}, a^*_{\mathbf{n}\lambda}\}$, which are constant complex numbers. According to (11), the average energy of a mode is given by $\langle H_{\mathbf{n}\lambda}\rangle = \mathcal{E}_k \langle |a_{\mathbf{n}\lambda}|^2\rangle = \mathcal{E}_n$, with the scale of the $a_{\mathbf{n}\lambda}$ chosen such that

$$\langle |a_{\mathbf{n}\lambda}|^2 \rangle = 1. \tag{13}$$

(The symbol $\langle f \rangle$ means average over all realizations of the field, i.e., of the amplitudes $\{a_{\mathbf{k}\lambda}, a^*_{\mathbf{k}\lambda}\}$). With $\mathcal{E}_k = \frac{1}{2}\hbar\omega_k$ for the zeropoint field, we get

$$\mathbf{A}(\mathbf{x},t) = c \sum_{\mathbf{k},\lambda} \sqrt{\frac{\pi\hbar}{\omega_k V}} \mathbf{e}^\lambda_\mathbf{k} \left(\mathbf{a}_{\mathbf{k}\lambda} e^{-i\omega_k t + i\mathbf{k}\cdot\mathbf{x}} + \mathbf{a}^*_{\mathbf{k}\lambda} e^{i\omega_k t - i\mathbf{k}\cdot\mathbf{x}} \right). \tag{14}$$

In the continuous description the vector potential of the vacuum field takes the form

$$\mathbf{A}(\mathbf{x},t) = c \sum_\lambda \int d^3\mathbf{k} \sqrt{\frac{\hbar}{8\pi^2\omega_\mathbf{k}}} \mathbf{e}^\lambda(\mathbf{k}) \left(\mathbf{a}_\lambda(\mathbf{k}) e^{-i\omega_k t + i\mathbf{k}\cdot\mathbf{x}} + \mathbf{a}^*_\lambda(\mathbf{k}) e^{i\omega_k t - i\mathbf{k}\cdot\mathbf{x}} \right). \tag{15}$$

Since the field (14) describes the vacuum, its average must be zero, which requires that

$$\langle a_{\mathbf{k}\lambda} \rangle = 0, \quad \langle a^*_{\mathbf{k}\lambda} \rangle = 0. \tag{16}$$

The amplitudes $a_{\mathbf{k}\lambda}$ corresponding to different modes of the free field are assumed to be statistically independent, so in consistency with (13) we must write

$$\langle a^*_{\mathbf{k}\lambda} a_{\mathbf{k}'\lambda'} \rangle = \delta_{\lambda\lambda'} \delta_{\mathbf{k}\mathbf{k}'}, \tag{17}$$

and

$$\langle a_{\mathbf{k}\lambda} a_{\mathbf{k}'\lambda'} \rangle = 0, \quad \langle a^*_{\mathbf{k}\lambda} a^*_{\mathbf{k}'\lambda'} \rangle = 0. \tag{18}$$

In the limit $V \to \infty$ the random amplitudes transform into $a_{\mathbf{k}\lambda} \to a_\lambda(\mathbf{k})$, with

$$\langle a_\lambda^*(\mathbf{k})\mathbf{a}_{\lambda'}(\mathbf{k}')\rangle = \delta_{\lambda\lambda'}\delta(\mathbf{k}-\mathbf{k}'). \tag{19}$$

It is often convenient to use the set of canonical variables for the field modes $\{q_{\mathbf{k}\lambda}, p_{\mathbf{k}\lambda}\}$ defined by

$$q_{\mathbf{k}\lambda} = i\sqrt{\frac{\mathcal{E}_k}{2\omega_k^2}}\left(a_{\mathbf{k}\lambda} - a_{\mathbf{k}\lambda}^*\right), \quad p_{\mathbf{k}\lambda} = \sqrt{\frac{\mathcal{E}_k}{2}}\left(a_{\mathbf{k}\lambda} + a_{\mathbf{k}\lambda}^*\right). \tag{20}$$

From (16)-(18) it follows that the random variables $q_{\mathbf{k}\lambda}$, $p_{\mathbf{k}\lambda}$ have zero average

$$\langle q_{\mathbf{k}\lambda}\rangle = 0, \quad \langle p_{\mathbf{k}\lambda}\rangle = 0, \tag{21}$$

and are uncorrelated

$$\langle q_{\mathbf{k}\lambda} p_{\mathbf{k}\lambda}\rangle = 0, \tag{22}$$

whereas their second-order moments, or variances, are given by

$$\sigma_{p_k}^2 = \sigma_{p_{\mathbf{k}\lambda}}^2 \equiv \langle p_{\mathbf{k}\lambda}^2\rangle = \mathcal{E}_k = \tfrac{1}{2}\hbar\omega_k, \quad \sigma_{q_k}^2 = \sigma_{q_{\mathbf{k}\lambda}}^2 \equiv \langle q_{\mathbf{k}\lambda}^2\rangle = \frac{\mathcal{E}_k}{\omega_k^2} = \frac{\hbar}{2\omega_k}. \tag{23}$$

One can alternatively express the complex amplitudes $a_{\mathbf{k}\lambda}$ in terms of a new pair of real, independent random variables $r_{\mathbf{k}\lambda}$, $\varphi_{\mathbf{k}\lambda}$ in the form

$$a_{\mathbf{k}\lambda} = \frac{1}{\sqrt{2\mathcal{E}_k}}\left(p_{\mathbf{k}\lambda} - i\omega_k q_{\mathbf{k}\lambda}\right) = r_{\mathbf{k}\lambda} e^{i\varphi_{\mathbf{k}\lambda}}, \tag{24}$$

where $r_{\mathbf{k}\lambda} = |a_{\mathbf{k}\lambda}|$. One gets

$$\omega_k q_{\mathbf{k}\lambda} = -\sqrt{2\mathcal{E}_k}\, r_{\mathbf{k}\lambda} \sin\varphi_{\mathbf{k}\lambda}, \quad p_{\mathbf{k}\lambda} = \sqrt{2\mathcal{E}_k}\, r_{\mathbf{k}\lambda} \cos\varphi_{\mathbf{k}\lambda} \tag{25}$$

and the energy of the mode assumes the simple form

$$H_{\mathbf{k}\lambda} = \mathcal{E}_k |a_{\mathbf{k}\lambda}|^2 = \mathcal{E}_k r_{\mathbf{k}\lambda}^2. \tag{26}$$

From (16)-(18) we see that

$$\langle \sin\varphi_{\mathbf{k}\lambda}\rangle = 0, \quad \langle \cos\varphi_{\mathbf{k}\lambda}\rangle = 0, \tag{27}$$

which means that the phases $\varphi_{\mathbf{k}\lambda}$ are uniformly distributed in the interval $(0, 2\pi)$, and that

$$\langle r_{\mathbf{k}\lambda}^2\rangle = 1. \tag{28}$$

For the two-point covariances of the electric field components one gets using (15)-(19), after performing the summation over λ

$$\langle E_i(\mathbf{x},t)\mathbf{E_j}(\mathbf{x}',\mathbf{t}')\rangle = \frac{2\pi\hbar}{V}\sum_{\mathbf{k}}\omega_k\left(\delta_{ij}-\hat{k}_i\hat{k}_j\right)\cos(\omega_k s - \mathbf{k_n}\cdot\mathbf{r}), \qquad (29)$$

with $s \equiv t - t'$ and $\mathbf{r} \equiv \mathbf{x} - \mathbf{x}'$. For the explicit evaluation it is convenient to take the continuum limit $V \to \infty$, in which the triple sums transform into a triple integral; one then obtains

$$\langle \mathbf{E}(\mathbf{x},t)\cdot\mathbf{E}(\mathbf{x}',t')\rangle = \frac{\hbar c}{2\pi^2}\int d^3k\, k\, \cos(\omega_k s - \mathbf{k_n}\cdot\mathbf{r}). \qquad (30)$$

An important expression is obtained by taking $\mathbf{x} = \mathbf{x}'$ in equation (30) (we will not require the more general expressions in this work); the result can be written in the form

$$\langle E_i(\mathbf{x},t)\mathbf{E_j}(\mathbf{x},t')\rangle = \delta_{ij}\int_0^\infty d\omega S(\omega)\cos\omega(t-t'), \qquad (31)$$

where the *power spectrum* $S(\omega)$ is related to the spectral density (4) by

$$S(\omega) = \frac{4\pi}{3}\rho(\omega), \qquad (32)$$

as is readily seen by considering the case $t = t'$. The Fourier transform of equation (31) is the SED version of the Wiener-Khintchine theorem, which gives the power spectrum in terms of the Fourier transform of the two-time covariance [34]. In terms of the Fourier transform of the field,

$$\tilde{E}_i(\mathbf{x},\omega) = (2\pi)^{-1}\int_{-\infty}^\infty \mathbf{dt E_i}(\mathbf{x},\mathbf{t})e^{i\omega t}, \qquad (33)$$

equation (31) gives the useful result

$$\left\langle \tilde{E}_i(\mathbf{x},\omega)\tilde{\mathbf{E}}_\mathbf{j}^*(\mathbf{x},\omega')\right\rangle = \delta_{ij}S(\omega)\delta(\omega-\omega'). \qquad (34)$$

In the case of the zeropoint field it is necessary to introduce a cutoff to give a definite meaning to expressions involving an integration over all frequencies, such as (31).

To construct all higher-order correlations it is necessary to have a full knowledge of the distributions. A most frequent assumption in the SED literature is that the amplitudes of the (statistically independent) field modes are normally distributed random variables. This means in particular that each $q_{\mathbf{k}\lambda}$ and $p_{\mathbf{k}\lambda}$ have a Gaussian distribution with the respective variances given by equations (23). Since this postulate will require revision in what follows, in attention to space limitations we do not elaborate further on it here, referring the interested reader to the detailed explanations in *The Dice*.

The theory just explained was applied to a number of problems. In the linear and single-frequency cases the results were very satisfactory, in that the predicted behaviour was either in full correspondence with the quantum one, or in general agreement with it, except that no sharp values for dynamical variables can be predicted with a theory as the present one. To be more explicit, we recall the examples of the Casimir and van der Waals forces, and the harmonic oscillator. In the first case, a complete agreement is reached between the QED and SED predictions for all temperatures and different shapes and constitution of the bodies involved (macroscopic or microscopic, as the case may be) [10,38,35–37,18]. The harmonic oscillator provides a unique opportunity to study with simple mathematics the properties acquired by a mechanical system in permanent contact with the zeropoint field, and to assess with a specific example the merit of the assumption that the classical particle becomes through this interaction a quantum object. Of course, due to its linearity the oscillator is far from being a fair representative of the general quantum system, but nevertheless the importance of getting a concrete expression of the relationship between quantum theory and SED can hardly be overestimated. Even such fine properties of the harmonic oscillator as the Lamb shift and lifetimes can be calculated, with results in close agreement with those of QM and QED [39,40]. In the present theory, though, the notion of sharp values for the dynamical variables must be sustituted by that of distributed values; this could seem at first glance as the end of the theory, since quantum mechanics and quantum levels seem to be inseparable notions. However, in the quantum mechanical description of stationary states in terms of the Wigner distribution (or any other appropriate phase space distribution) [41]) the 'sharp levels' become distributed, with a non zero variance. It is quite remarkable that this latter description when restricted to states with non negative probabilities, and the one afforded by SED, are in full agreement for the harmonic oscillator.

The results just mentioned and others, such as the prediction of diamagnetism [42], a theory of the Compton effect [43], the general determination of the Lamb shift and the atomic lifetimes, as well as some cavity effects [39], clearly indicate that matter acquires quantum properties through its interaction with the background field. However, the theory as developed up to this point proved several years ago to be limited to the treatment of single-frequency systems. In particular, specific calculations for the hydrogenic atom and the anharmonic oscillator gave wrong results [10,32,44–50]. A reason for this failure is that nonlinear problems cannot be treated in the Markovian approximation (as was assumed), at least due to the long tails of the correlation times to which the coloured noise of the zeropoint field leads. A more general argument refers to the condition of *detailed (energy) balance,* obtained when the energy equilibrium condition is Fourier-analysed into a relation that should hold for every separate (relevant) frequency. This is the case, for instance, in the derivation of the Einstein A and B coefficients in quantum theory, for every particular transition frequency. When detailed energy balance does not hold, the mechanical system keeps a constant average energy, but there remains a sustained periodic energy exchange between some modes of the field; in other words, although a steady state is reached, it is not an equilibrium state. Since we want the SED system to reach a real equilibrium state, so that the background field remains unchanged (once the quantum regime has been attained), we must find a way to guarantee that detailed energy balance holds.

Now, as is well known, an ensemble of classical multiply periodic mechanical systems in equilibrium with an electromagnetic field satisfies the Maxwell-Boltzmann distribution and the field obeys the Rayleigh-Jeans law; under these circumstances detailed balance holds [10,51].[7] However, since these properties do not correspond to the quantum behaviour of multiply periodic systems, it becomes crucial in SED to find a way to circumvent them if one wants to succeed in making contact with quantum mechanics in the general case. This means that the above theory requires some revision, a task in which we now engage.

III LINEAR STOCHASTIC ELECTRODYNAMICS

What one needs, then, is a theory that is consistent with detailed energy balance, but that does not lead to M-B statistics for matter, nor R-J for the field. To meet these strong requirements we need to revise some of the fundamental tenets of SED as it has been developed up to this point. The ensuing theory is what we call *linear stochastic electrodynamics*, or linear SED for short, for reasons that will be apparent below (but that do *not* imply any linearization of the nonlinear dynamics).

A natural candidate to start this revision is the background field that is in permanent interaction with matter, since by the very effect of this interaction, there is no reason to expect that this field preserves the properties of the free vacuum. We therefore start by reconsidering the statistical properties that have been assumed for the amplitudes of the background field.

III.A Field correlations

As before, we write the field in the form

$$\mathbf{A}(\mathbf{x},\mathbf{t}) = \sum_\lambda \int d^3k \widetilde{A}(\mathbf{k},\omega_\mathbf{k}) e^\lambda(\mathbf{k}) \left(a_\lambda(\mathbf{k}) e^{-i\omega_\mathbf{k} t + i\mathbf{k}\cdot\mathbf{x}} + a_\lambda^*(\mathbf{k}) e^{i\omega_\mathbf{k} t - i\mathbf{k}\cdot\mathbf{x}} \right), \quad (35)$$

$$\widetilde{A}(\mathbf{k},\omega_\mathbf{k}) = \sqrt{\frac{hc^2}{8\pi^2 \omega_\mathbf{k}}},$$

and assume that each component averages to zero,

$$\langle a_{\mathbf{k}\lambda} \rangle = 0, \quad \langle a_{\mathbf{k}\lambda}^* \rangle = 0. \quad (36)$$

---

[7] The condition $\rho(\omega) \sim \omega^2$ (which leads to the Rayleigh-Jeans distribution) is required to isolate the terms pertaining to different frequencies in the Fourier-analysed condition of energy equilibrium. The remaining condition is then frequency-independent and leads to the Maxwell-Boltzmann distribution for the phase-space density of the multiply periodic system, regardless of specific details.

We had also assumed that the field amplitudes corresponding to different modes are statistically independent,

$$\langle a^*_{\mathbf{k}\lambda} a_{\mathbf{k'}\lambda'}\rangle = \delta_{\lambda\lambda'}\delta_{\mathbf{kk'}}, \tag{37}$$

$$\langle a_{\mathbf{k}\lambda} a_{\mathbf{k'}\lambda'}\rangle = 0, \quad \langle a^*_{\mathbf{k}\lambda} a^*_{\mathbf{k'}\lambda'}\rangle = 0. \tag{38}$$

At this point, we define a set of new partially averaged amplitudes

$$\mathbf{a}(\omega_{\mathbf{k}}, \mathbf{x}) = \frac{\omega_{\mathbf{k}}^2}{c^2} \sum_\lambda \int_{\Omega_{\mathbf{k}}} d\Omega_{\mathbf{k}} \frac{\widetilde{A}(\mathbf{k},\omega_{\mathbf{k}})}{\widetilde{A}(\omega_{\mathbf{k}})} \mathbf{e}^\lambda(\mathbf{k}) a_\lambda(\omega, \mathbf{k}) e^{i\mathbf{k}.\mathbf{x}}; \tag{39}$$

the quantity $\widetilde{A}(\omega)$ will be selected below and the integration is performed over the solid angle in the reciprocal space $\mathbf{k}$, for fixed magnitude (fixed frequency) $k = |\mathbf{k}| = \omega/c$. The resulting (partially averaged) stochastic fields $\mathbf{a}(\omega_{\mathbf{k}}, \mathbf{x})$ contain the contribution at a given point $\mathbf{x}$ of all the field modes of frequency ω. Due to the partial averaging, the $\mathbf{a}(\omega_{\mathbf{k}}, \mathbf{x})$ should be expected to be much less random than the original amplitudes $a_\lambda(\omega, \mathbf{k})$. Now the field is expressed as

$$\mathbf{A}(\mathbf{x}, t) = \int d\omega \widetilde{A}(\omega) \mathbf{a}(\omega_{\mathbf{k}}, \mathbf{x}) e^{-i\omega_{\mathbf{k}} t} + \text{c.c.} \tag{40}$$

From (39) it is immediate that $\langle \mathbf{a}(\omega_{\mathbf{k}}, \mathbf{x})\rangle = 0$ everywhere. For the correlations we get

$$\langle a_i(\omega_{\mathbf{k}}, \mathbf{x}) a_j^*(\omega'_{\mathbf{k}}, \mathbf{x}')\rangle = \frac{\omega_k^2}{c^2} \frac{\omega_{k'}^2}{c^2} \sum_{\lambda,\lambda'} \int_{\Omega_k} d\Omega_k \int_{\Omega_{k'}} d\Omega_{k'} e_i^\lambda(\mathbf{k}) e_j^{\lambda'}(\mathbf{k}')$$

$$\times \frac{\widetilde{A}(\mathbf{k},\omega_k)}{\widetilde{A}(\omega_k)} \frac{\widetilde{A}^*(\mathbf{k}',\omega_{k'})}{\widetilde{A}^*(\omega_{k'})} e^{i(\mathbf{k}.\mathbf{x} - \mathbf{k}'.\mathbf{x}')} \langle a_\lambda(\omega,\mathbf{k}) a^*_{\lambda'}(\omega',\mathbf{k}')\rangle . \tag{41}$$

From equation (37) it follows that

$$\langle a_\lambda(\omega,\mathbf{k}) a^*_{\lambda'}(\omega',\mathbf{k}')\rangle = \delta_{\lambda\lambda'} \frac{1}{k^2}\delta^2(\Omega_{\mathbf{k}} - \Omega_{\mathbf{k'}})\delta(\omega - \omega'),$$

so that one obtains

$$\langle a_i(\omega_{\mathbf{k}}, \mathbf{x}) a_j^*(\omega'_{\mathbf{k}}, \mathbf{x}')\rangle = \delta(\omega_k - \omega_{k'})$$

$$\times \int_{\Omega_{\mathbf{k}}} d\Omega_{\mathbf{k}} \left(\delta_{ij} - \frac{k_i k_j}{k^2}\right) \frac{\omega_k^2}{c^2} \left|\frac{\widetilde{A}(\mathbf{k},\omega_k)}{\widetilde{A}(\omega_k)}\right|^2 e^{i\mathbf{k}.(\mathbf{x}-\mathbf{x}')} \tag{42}$$

since

$$\sum_\lambda e_i^\lambda(\mathbf{k}) e_j^\lambda(\mathbf{k}) = \delta_{ij} - \frac{k_i k_j}{k^2}.$$

Here we make an approximation. We assume that in our applications the quantities $\mathbf{x} - \mathbf{x}'$ of interest remain inside or in the neighbourhood of the atom; for such distances of the order of atomic sizes, and frequencies in the visible region of the spectrum, or even higher, the magnitude of the exponent in (42) is much smaller than 1 and can be neglected (this assumption corresponds to the usual long wavelength approximation).[8] Under these conditions and with an appropriate selection of the value of $\widetilde{A}(\omega_k)$ we obtain from (42)

$$\left\langle a_i(\omega_\mathbf{k},\mathbf{x})\mathbf{a}_\mathbf{j}^*(\omega'_\mathbf{k},\mathbf{x}')\right\rangle = \left\langle |a(\omega_\mathbf{k})|^2\right\rangle \delta_{ij}\delta(\omega_k - \omega_{k'}). \tag{43}$$

The coefficient $\left\langle |a(\omega_\mathbf{k})|^2\right\rangle$ is to be selected for normalization purposes, so that the amplitudes $a_i(\omega_\mathbf{k})$ become dimensionless numbers (usually this factor is independent of ω_k). Of course we also have that

$$\left\langle a_i(\omega_\mathbf{k},\mathbf{x})\mathbf{a}_\mathbf{j}(\omega'_\mathbf{k},\mathbf{x}')\right\rangle = 0, \quad \left\langle a_i^*(\omega_\mathbf{k},\mathbf{x})\mathbf{a}_\mathbf{j}^*(\omega'_\mathbf{k},\mathbf{x}')\right\rangle = 0. \tag{44}$$

To carry out general calculations we require a knowledge of all moments of the field amplitudes. According to the usual rules, one assumes a Gaussian distribution for the modulus of the amplitudes, so that one would write, for example (recalling that the $\mathbf{x}$ dependence has now disappeared)

$$\langle a_i(\omega_\mathbf{k})\mathbf{a_j}(\omega'_\mathbf{k})\mathbf{a_k^*}(\omega''_\mathbf{k}) \rangle = 0 , \tag{45}$$

$$\langle a_i(\omega_\mathbf{k})\mathbf{a_j}(\omega'_\mathbf{k})\mathbf{a_k^*}(\omega''_\mathbf{k})\mathbf{a_l^*}(\omega'''_\mathbf{k}) \rangle = \left\langle |a(\omega_\mathbf{k})|^2\right\rangle \left\langle |a(\omega_{\mathbf{k}'})|^2\right\rangle$$
$$\times \left[\delta_{ik}\delta_{jl}\delta(\omega_k - \omega_{k''})\delta(\omega_{k'} - \omega_{k'''}) + \delta_{il}\delta_{jk}\delta(\omega_k - \omega_{k'''})\delta(\omega_{k'} - \omega_{k''})\right], \tag{46}$$

and so on. It is just here that we meet the limitations of the usual theory, because these rules and their extensions to products of more factors fail to capture some fundamental aspects of the behaviour of atomic systems. To see this and at the same time find the correct rules we resort to some simple observations.

The first remark of interest may seem almost obvious, but is nevertheless very important. We have in mind that once the whole SED system, namely field plus matter, has reached the quantum regime, i.e., is in equilibrium or very near to equilibrium, not only the field has modified the state of matter, but one should expect that also the matter has affected the nearby field. This trivial observation leads however to a most important conclusion, namely, that the field in interaction with matter in the stationary situation (or better, once the system has entered the quantum regime) is no more the free zeropoint field, but something else, that

---

[8] For distances $r = |\mathbf{x} - \mathbf{x}'| \lesssim a_B$, where $a_B = \hbar^2/me^2$ ($= 1$ au) is the Bohr radius and frequencies of the order of $\omega_H = |E_H^0|/\hbar = me^4/\hbar^3$ ($= 1$ au), where E_H^0 is the ground state energy of the hydrogen atom, one has $kr \lesssim (\omega_H/c)|\mathbf{x} - \mathbf{x}'| \approx e^2/\hbar c = \alpha$ ($= c^{-1}$ au), where α is the fine structure constant, $\alpha \simeq 1/137$. Of course, there could be situations for which the long wavelength approximation is too rough. It should be clear that we are not neglecting the spatial structure of the field (which may become essential in some circumstances), but merely assuming it to be uniform within the volume of the system under study.

may even depend on the specific state of the specific system. This means that its statistical properties are not necessarily anymore those of the *free zero point* field, as given by equation (45). For instance, we may ask whether it is true that under this more general context the product of three amplitudes as in equation (45) has always an average null value. We will certify that this postulate is indeed erroneous, and that it is just in postulating this and other similar (more general) statistical properties of the field amplitudes that the previous version of SED failed. It is not the fundamental ideas and concepts of old SED that went wrong, but the specific postulates that were arbitrarily assembled (as is always the case with postulates) to build and develop the theory. So now we must answer the question, what are the correct statistical properties of the background field in interaction with matter in the quantum regime?

In studying the harmonic oscillator one observes that there are modes of frequencies to which the mechanical system has a very sharp and intense resonant response. For the non resonant frequencies the response is relatively negligible. We will refer to the frequencies that generate a resonance-like sharp and intense response as the *relevant frequencies*. For a given system and state, the field can then be seen as consisting of two parts,

$$\mathbf{A} = \mathbf{A}_{\text{res}} + \mathbf{A}_{\text{noise}}. \tag{47}$$

In $\mathbf{A}_{\text{res}}$ we collect all terms involving relevant frequencies, whereas $\mathbf{A}_{\text{noise}}$ stands for the rest, which we will take as a noise. The fundamental part of the dynamics of the system is controlled by $\mathbf{A}_{\text{res}}$; the noise just produces small random corrections to it. Due to the high Q of the resonant behavior, in its turn a consequence of the very small value of the radiation reaction force coefficient τ —of the order 10^{-23} s for the electron—, the response of the system to each of these two components is usually orders of magnitude apart, so that for many important purposes it is acceptable to leave aside the effects of the noise, although it always introduces dispersion into the system.

Let us consider now a representative multiply periodic system. The nonlinear dynamics mixes the different fundamental frequencies and generates motions with combination frequencies of all orders, so that the analysis will contain an infinite number of frequencies, given by linear relations of the form

$$n_\alpha \omega_\alpha + n_\beta \omega_\beta + \ldots = 0, \tag{48}$$

where the n_λ are positive or negative integers. It seems only natural to consider that the modes of the field with frequencies so related cannot be statistically independent. Of course relations of the type given by equation (48) are of interest only for the field components that enter into $\mathbf{A}_{\text{res}}$, so that for those of $\mathbf{A}_{\text{noise}}$ we accept without further ado conditions as (43)-(46). Let us focus our attention on the components of $\mathbf{A}_{\text{res}}$ that are correlated. To simplify the notation and avoid complications that would add little to the reasoning that follows, we restrict our discussion to the one dimensional case and discrete representation of the field. The generalization to other descriptions is immediate.

Since the amplitudes a_α (that replace the $a(\omega)$ in the discrete representation) are complex numbers, we write them in the form

$$a_\alpha = r_\alpha e^{i\varphi_\alpha}, \qquad (49)$$

with $r_\alpha = |a_\alpha|$ and φ_α real numbers. As we said it is usual to assume that the magnitudes r_α follow a normal distribution centered around a common value $\bar{r}$, whereas the phases φ_α are uniformly distributed in the interval $(0, 2\pi)$ (although it will be more convenient to consider the interval $(-\pi, +\pi)$), and statistically independent. As was also said, owing to the partial averaging over the directions and polarization of the waves of the frequency under consideration, the fluctuations of the magnitudes r_α are expected to become greatly reduced with respect to those of the variables in the original description (we are averaging over an infinite number of components, all of them with the same central value $\frac{1}{2}\hbar\omega_\alpha$ for the energy). Thus one may safely assume that the r_α are so sharply peaked around their mean value, that they can be taken as constant (i.e., not distributed); this constant value r_α is the same for all frequencies and equal to 1, so that equation (49) takes the simpler form

$$a_\alpha = e^{i\varphi_\alpha}. \qquad (50)$$

In the limit in which the normalization box has an infinite volume, both representations of the field, with fixed or distributed modulus, become equivalent [10].

Now we can write, for example,

$$\langle a_\alpha^* a_{\alpha'} \rangle = \left\langle e^{i(\varphi_\alpha - \varphi_{\alpha'})} \right\rangle = \delta_{\alpha\alpha'}. \qquad (51)$$

The result is an immediate consequence of the fact that if two independent random phases φ_α and φ_β are uniformly distributed in the interval $(0, 2\pi)$, their difference $\varphi_\alpha - \varphi_\beta$ mod $(0, 2\pi)$ is also uniformly distributed in the same interval [34]. In what follows we take all resulting phases within the interval $(-\pi, \pi)$, adding if necessary an appropriate (and irrelevant) constant phase factor.

From (50) it follows that

$$a_\alpha a_\beta = e^{i(\varphi_\alpha + \varphi_\beta)} = a_{\alpha+\beta}. \qquad (52)$$

The quantity $a_{\alpha+\beta}$ is also an amplitude, referred to the modes of frequency $\alpha + \beta$ (in a notation that should be evident). The result can be extended to the product of any number of factors,

$$a_\alpha a_\beta \cdots a_\lambda = a_{\alpha+\beta+\ldots+\lambda}. \qquad (53)$$

From this follows a fundamental result, namely, that all functions of the amplitudes a_μ that can be expressed as a power series are reducible to linear functions of the amplitudes. The resulting product amplitude is associated with the combination frequency $\omega_{\alpha+\beta+\ldots+\lambda}$, and from equation (40) it follows that $e^{i(\omega_\alpha+\omega_\beta+\ldots+\omega_\lambda)t} = e^{i\omega_{\alpha+\beta+\ldots+\lambda}t}$, whence

$$\omega_{\alpha+\beta+\ldots+\lambda} = \omega_\alpha + \omega_\beta + \ldots + \omega_\lambda. \tag{54}$$

If the initial set of relevant frequencies is complete, then the new combination frequency belongs to that set and no new frequency appears; in other words, only a recombination of terms has taken place.

A simple example may illustrate the point. Assume that a certain calculation gives the quantity

$$Q_{123} = f_1 a_1 + f_2 a_2 + f_3 a_3 + g_{12} a_1 a_2 + g_{23} a_2 a_3;$$

applying (53) one obtains (the simplification $a_1 a_2 = a_3$ is of course arbitrary here)

$$Q_{123} = f_1 a_1 + f_2 a_2 + f_3 a_3 + g_{12} a_3 + g_{23} a_4 =$$
$$= f_1 a_1 + f_2 a_2 + (f_3 + g_{12}) a_3 + g_{23} a_4.$$

The manipulations have modified the coefficient of a_3 and added a contribution corresponding to frequency $\omega_4 = \omega_2 + \omega_3$ that was hidden.

To be more specific let us write the $\mathbf{A}_{\text{res}}$ part of the field in the form (here all frequencies are positive; we omit the subindex)

$$A(t) = \sum_\alpha \left(\widetilde{\mathbf{A}}_\alpha(\omega_\alpha) a_\alpha e^{-i\omega_\alpha t} + \widetilde{\mathbf{A}}^*_\alpha(\omega_\alpha) a^*_\alpha e^{i\omega_\alpha t} \right). \tag{55}$$

Let us consider the field squared

$$A^2(t) = \sum_{\alpha\beta} \left(\widetilde{\mathbf{A}}_\alpha \widetilde{\mathbf{A}}_\beta e^{-i(\omega_\alpha+\omega_\beta)t} \mathbf{a}_{\alpha+\beta} + \widetilde{\mathbf{A}}_\alpha \widetilde{\mathbf{A}}^*_\beta e^{-i(\omega_\alpha-\omega_\beta)t} \mathbf{a}_{\alpha-\beta} \right) + \text{c.c.} \tag{56}$$

With a small rearrangement of terms this reduces to an expression exactly of the same form as equation (55). Since the amplitudes $a_{\alpha\pm\beta}$ are statistically independent of their separate factors a_α and a_β (and of the rest of amplitudes), the field A^2 seems to be statistically independent from A ($\langle A(t) A^2(t') \rangle = 0$). But this is just what we do *not* want, since we have seen that $\mathbf{A}_{\text{res}}$ should contain some amplitudes that *are* correlated. In other words, we should take explicitly into consideration that the combination frequencies $\omega_\alpha \pm \omega_\beta$ are contained in the field $A(t)$ given by (55), and hence that A^2 contains amplitudes correlated with the $a_{\alpha\pm\beta}$ of A. Consider for example the terms of frequency $\omega_\gamma \equiv \omega_\alpha - \omega_\beta$. Then $a_{\alpha-\beta} = a_\gamma$, or

$$e^{i(\varphi_\alpha-\varphi_\beta)} = e^{i\varphi_\gamma}. \tag{57}$$

In plain words, to a combination of frequencies corresponds just the same combination of phases. Now the index γ has become indistinguishable from the composite index $\alpha - \beta$. This can be extended to the product of any number of amplitudes a_α or to the corresponding sum of any number of phases φ_α. However, all non-trivial cases (i.e., with more than one element involved) can be reduced to the case of two terms, since this contains the essence of the information about the correlations. Thus it suffices to denote every frequency and every phase with a pair of indices in place of only one, to take into account all possible results; this amounts to taking all relevant (resonant) frequencies as combination frequencies. In other words, in the previous example instead of writing φ_γ we write $\varphi_\alpha - \varphi_\beta$, and similarly for the frequencies, $\omega_{\alpha\beta}$ instead of ω_γ, whence equation (57) holds. For a consistent and systematic notation we should write

$$\varphi_{\alpha\beta} \equiv \varphi_\alpha - \varphi_\beta. \tag{58}$$

(with the resulting phase always within $(-\pi, \pi)$). With this procedure we have guaranteed that the amplitudes (or phases) containing a common index are correlated. For instance, for different α, β, γ we have

$$\langle (\varphi_\alpha - \varphi_\beta)(\varphi_\beta - \varphi_\gamma) \rangle = \langle \varphi_\alpha \varphi_\beta - \varphi_\alpha \varphi_\gamma - \varphi_\beta^2 + \varphi_\beta \varphi_\gamma \rangle$$
$$= -\langle \varphi_\beta^2 \rangle = -\sigma_\beta^2 = -\sigma^2, \tag{59}$$

where $\sigma^2 = \pi^2/3$ is the dispersion of the uniform distribution in $(-\pi, \pi)$.

The step we have just made is crucial and deserves some comments [10,52–54]. The need for a double index to denote the relevant frequencies might be seen as going back to the construction of matrix mechanics by Heisenberg and Born. However, the idea (at least in a latent form) can be considered to go further back to Bohr and his intuition to recast the empirical laws of the spectroscopists in the form $E_n - E_m = \hbar\omega_{nm}$. In other words, it corresponds to the empirical discovery by the spectroscopist of the 19th century that each frequency (or wave number) emitted by an atom can be written as the difference of two terms (the Ritz combination principle). This means that in making the transition $\omega_\gamma \to \omega_{\alpha\beta} = \omega_\alpha - \omega_\beta$ we are *not* introducing a quantum law, but a principle known to physics many decades before the birth of quantum theory, even in the form of Bohr's theory. Now we see the enormous strength and deep meaning of this frequency combination principle when it is used in combination with random phases.

By writing

$$a_{\alpha\beta} = e^{i\varphi_{\alpha\beta}} = e^{i(\varphi_\alpha - \varphi_\beta)} \tag{60}$$

we obtain, e.g.,

$$a_{\alpha\beta} a_{\beta\gamma} = e^{i(\varphi_\alpha - \varphi_\beta + \varphi_\beta - \varphi_\gamma)} = e^{i(\varphi_\alpha - \varphi_\gamma)} = a_{\alpha\gamma}, \tag{61}$$

where the repeated index β has been 'contracted', so to speak. This is the kind of algebra that leads to the linear dependence in the amplitudes for the field $\mathbf{A}_{\text{res}}$. Of course this reduction does not apply to the noise field $\mathbf{A}_{\text{noise}}$, since its components are uncorrelated. The reduction law in equation (60), namely

$$a_{\alpha\beta}a_{\beta\gamma} = a_{\alpha\gamma}, \tag{62}$$

is the central property of the amplitudes associated with the relevant frequencies. It is interesting to see what happens with products of three amplitudes, which in the conventional description average always to zero (see equation (45)). We have, for instance,

$$a_{\alpha\beta}a_{\beta\gamma}a_{\alpha\gamma}^* = a_{\alpha\gamma}a_{\alpha\gamma}^* = 1, \tag{63}$$

in open disagreement with (45).

It is just this disagreement between the present formulation of the theory and the previous (standard) form of SED that makes the difference. Now it is clear that the two theories will predict entirely different results for all quantities that involve powers higher than two of the amplitudes, in particular for multiply periodic (nonlinear) systems. The theory constructed around the principles just explained is what we call linear SED. When in the SED literature (from around 1979 on) it is said that SED fails for nonlinear systems (or for multiply periodic systems), it is the standard version of the theory what is being referred to, with a totally uncorrelated background field. Thus, it was not SED what failed, but the (unwarranted) selection of some of its postulates. Since the postulates of a physical theory are freely (but of course not arbitrarily) selected, there is ample room to experiment with different sets of them, until one is found that gives the closest representation of nature.

III.B Making Contact with the Quantum Formalism

The consequences of the above construction are immense, despite its simplicity. Let us discuss a simple but representative example. Consider a generic dynamic variable of the system under study, and call it $y(t)$. We write, taking into account that all functions of time reduce to a linear function in the relevant field amplitudes by a systematic application of the rule (62),

$$y(t) = \sum \widetilde{y}_{\alpha\beta} a_{\alpha\beta} e^{i\omega_{\alpha\beta}t} + \text{noise terms.} \tag{64}$$

Here we have introduced some new conventions that are very effective. The sum contains now both positive and negative frequencies, so this term represents itself for positive frequencies and its complex conjugate for negative frequencies. This amounts to writing

$$\widetilde{y}_{\alpha\beta} = \widetilde{y}_{\beta\alpha}^*, \tag{65}$$

for real dynamical variables, since in going from $\omega_{\alpha\beta} = \omega_\alpha - \omega_\beta$ to $-\omega_{\alpha\beta} = \omega_\beta - \omega_\alpha = \omega_{\beta\alpha}$ an interchange of the indices is performed, so that $\widetilde{y}_{\alpha\beta}^*$ goes into $\widetilde{y}_{\beta\alpha}$. In equation (64) we have also separated the set of terms containing only relevant frequencies (the purely resonant terms) from the rest, all of which contain noisy factors and have been grouped under the single term 'noise terms'. In what follows we focus our attention on the relevant terms (considered to be the ones that generate the interesting dynamics); the structure and details of the noisy terms will not be further explored here.

Now let us calculate a power of $y(t)$; for instance, we have

$$\begin{aligned}y_\alpha^3(t) &= \sum \left(\widetilde{y}_{\alpha\beta}\widetilde{y}_{\alpha'\beta'}\widetilde{y}_{\alpha''\beta''}\right) a_{\alpha\beta}\, a_{\alpha'\beta'}\, a_{\alpha''\beta''}\, e^{i\left(\omega_{\alpha\beta}+\omega_{\alpha'\beta'}+\omega_{\alpha''\beta''}\right)t} + \text{noise}_1 \\ &= \sum \left(\widetilde{y}_{\alpha\beta}\widetilde{y}_{\beta\beta'}\widetilde{y}_{\beta'\beta''}\right) a_{\alpha\beta}\, a_{\beta\beta'}\, a_{\beta'\beta''}\, e^{i\left(\omega_{\alpha\beta}+\omega_{\beta\beta'}+\omega_{\beta'\beta''}\right)t} + \text{noise}_2 \\ &= \sum \left(\widetilde{y}_{\alpha\beta}\widetilde{y}_{\beta\beta'}\widetilde{y}_{\beta'\beta''}\right) a_{\alpha\beta''}\, e^{i\omega_{\alpha\beta''}t} + \text{noise}_2 \\ &= \sum \left(\widetilde{y^3}\right)_{\alpha\beta''} a_{\alpha\beta''}\, e^{i\omega_{\alpha\beta''}t} + \text{noise}_2, \end{aligned} \qquad (66)$$

where we have introduced the definition

$$\left(\widetilde{y^3}\right)_{\mu\nu} = \sum_{\lambda\lambda'} \left(\widetilde{y}_{\mu\lambda}\widetilde{y}_{\lambda\lambda'}\widetilde{y}_{\lambda'\nu}\right). \qquad (67)$$

In performing the calculation the first step was to separate the relevant contributions (the correlated components) from the noisy terms and add the later to the noise; the rest of the calculation is a direct application of the reduction formula (62). Equation (67) is just a particular case of the product of three factors and can immediately be generalized to any number of factors. It is clear from such generalization that we arrive at the rule of matrix multiplication. In other words, the resonant part of the generic dynamical variable $y(t)$ is now represented by a matrix $\hat{y}(t)$ with matrix elements

$$(\hat{y}(t))_{\alpha\beta} \equiv \widetilde{y}_{\alpha\beta}(t) = \widetilde{y}_{\alpha\beta}\, e^{i\omega_{\alpha\beta}t}. \qquad (68)$$

Of course we also took into account that the frequencies combine according to equation (54) to give (in the above example; the rule can be extended to any number of terms)

$$\omega_{\alpha\lambda} + \omega_{\lambda\mu} + \omega_{\eta\beta} = \omega_{\alpha\beta}. \qquad (69)$$

This rule allows us to single out the terms that contribute to a given combination frequency. (Recall that these are the only resonant contributions; when the combination rule does not apply the resulting terms are part of the noise).

With this result we go back to the equation of motion, i.e., the Abraham-Lorentz equation,

$$m\,\ddot{x} = F(x) + m\tau\,\dddot{x} + e \sum \left(\widetilde{E}_{\alpha\beta} a_{\alpha\beta} e^{i\omega_{\alpha\beta}t} + \text{c.c.}\right) + \text{noise}_b. \qquad (70)$$

The term represented by noise_b refers to the incoherent part of the background field, which generates noise but no resonant response from the mechanical system.

Equations (64) and (68) can be used to expand x and $F(x)$; we write (again, each sum includes its complex conjugate)

$$x(t) = \sum \tilde{x}_{\alpha\beta} a_{\alpha\beta} e^{i\omega_{\alpha\beta}t} + \text{noise}_x = \sum \tilde{x}_{\alpha\beta}(t) a_{\alpha\beta} + \text{noise}_x, \qquad (71)$$

and similarly for the force, with time dependent matrix elements $\tilde{F}_{\alpha\beta}(t) = \tilde{F}_{\alpha\beta} e^{i\omega_{\alpha\beta}t}$; of course, we also write $\tilde{E}_{\alpha\beta}(t) = \tilde{E}_{\alpha\beta}(t) e^{i\omega_{\alpha\beta}t}$. We get, grouping all noisy terms under noise$_t$,

$$m \sum \ddot{\tilde{x}}_{\alpha\beta}(t) a_{\alpha\beta} = \sum \tilde{F}_{\alpha\beta}(t) a_{\alpha\beta} + m\tau \sum \dddot{\tilde{x}}_{\alpha\beta} a_{\alpha\beta} + e \sum \tilde{E}_{\alpha\beta}(t) a_{\alpha\beta} + \text{noise}_t. \qquad (72)$$

Multiplying by $a_{\alpha\beta}^*$ and taking the expectation value with respect to the realizations of the field we get

$$m \ddot{\tilde{x}}_{\alpha\beta}(t) = \tilde{F}_{\alpha\beta}(t) + m\tau \dddot{\tilde{x}}_{\alpha\beta} + e\tilde{E}_{\alpha\beta}(t). \qquad (73)$$

This is the $\alpha\beta$ element of the matrix equation

$$m \ddot{\hat{x}}(t) = \hat{F}(t) + m\tau \dddot{\hat{x}}(t) + e\hat{E}(t). \qquad (74)$$

Thus we conclude that the description of the resonant behaviour, i.e., of the most conspicuous part of the dynamics, excluding the purely noisy components of the motion, is given by the Heisenberg equation (74), in which both matter and field variables are represented by (square) matrices.[9] Equation (74), which is perhaps the main result of this essay, affords the Heisenberg description of the system according to non relativistic QED, including the radiation reaction force [23,24]. Let us further elaborate on these results.

III.B.1 *Quantum mechanics*

We have learned that by transforming into matrices the original dynamical variables that describe the atomic system, we are singling out the essentials of the dynamical behaviour impressed by the background field and discarding all purely noisy motions. This description becomes meaningful only after the background field has taken control over the mechanical system and the transient behaviour has dissapeared, which happens in a time of the order of $(\tau\Omega^2)^{-1}$, with Ω a lowest characteristic frequency of the atomic motions. To this we refer for short by saying that the system has entered the *quantum regime*. Under these circumstances, and if we are interested only in the essentials of the behaviour of matter, we may further approximate equation (74) by

$$m \ddot{\hat{x}}(t) = \hat{F}(x(t)). \qquad (75)$$

---

[9] Below we show that these matrices can be represented more generally by operators on a Hilbert space.

This implies leaving aside all radiative corrections to the motions. Now we have entered the quantum territory and it seems an appropriate place to make some comments on the meaning of the results with regard to the contents of quantum mechanics.

- We have got rid of the noise terms, because in the quantum regime their net effect is very small; however the background noise is essential for the system to attain the quantum regime and for the condition of detailed balance to be maintained, which defines the (relative) dynamical stability of quantum states.

- We got a simplified description, but we lost all effects due to the neglected terms ($m\tau\,\dddot{\hat{x}}\,(t)$ and $e\hat{E}(t)$), among them a very important one, namely the instability of the excited states. We lost even the possibility to explain the stability of the ground state, since equation (75) is *purely mechanical* and no trace of the field and of the mechanism that led to the stationary situation has been kept. This explains why it has been impossible to understand atomic stability *from within* quantum mechanics.

- Even causality has been lost, since the agent that produces quantum behaviour, the cause behind the description in terms of matrices, remains hidden for quantum mechanics. A part of the causal element is restituted in the transition from quantum mechanics to QED.

- It should not be a surprise to notice that the theory founded on the basis of equation (75), completed with the equation $\hat{p} = m\,\dot{\hat{x}}$ which is an immediate consequence of the formalism, is neither consistent with a stochastic description nor with a statistical description, if only because the equations refer only to that part of the dynamics which is more regular, but exclude the noise. Once more, we see that an analysis of the dynamics, now from the stochastic or statistical point of view but from *within* quantum mechanics, cannot provide the hint that leads to the solution of the puzzle.

- The above considerations also explain why, despite quantum mechanics being an essentially (unavoidably) statistical theory, it nevertheless is not a consistent statistical theory.

- From the previous analysis it follows that we should consider quantum mechanics as an *intrinsically* statistical theory, at least because the behaviour at each (transition) frequency depends on the *averaged* contributions of the (infinite) set of modes of the background field of that frequency, with random direction, polarization and phase. Further, because the initial conditions are erased by the partial averaging process, the description refers to all those systems with different initial conditions that lead through the interaction with the background field to the same final state. Thus the interpretation of the wave function in terms of individual events finds no support in the present theory: the wave function refers to a representative system.

- In making the transition from the original atomic dynamical variables to matrices, we are eliminating the notion of trajectory; this notion becomes impossible to recover from within the quantum description.

- The theory just developed gives a basis for an explanation of one of the most distinctive properties of quantum mechanics, namely, the superposition principle and wavelike behaviour of matter. For the linear response to the field means that where several fields superpose, also the quantum response functions superpose. Thus, the degree of coherence of the superposed fields will determine the coherence manifested by the 'guided' matter, so to speak.

- The present description explains another very obscure aspect of quantum systems, namely, the coexistence of stochasticity ('indeterminism') with dynamical variables having precise values (the eigenvalues). The point is that, as we have been insisting along the derivations above, the quantum theory neglects all the noisy contributions, keeping only the correlated terms. This means that the eigenvalues give only an estimate of the real values attained by the variables, but a very close estimate indeed, thanks to the resonant nature of the response.

- According to the present theory the corresponding dynamical variables of the classical and the quantum description of a given system are qualitatively different things, the latter referring to the *possibilities* of response rather than to a specific behaviour. Thus, there is no reason to expect a priori that the classical meaning and properties of a specific variable should hold in the quantum domain, without a deep modification or adjustment.

The Heisenberg equations of motion have been obtained, but this does not complete the quantum description. Let us pay attention to the two most important missing elements.

We know that the element that fixes the ground state of the atomic system is the zeropoint field. For example, the energy of the ground state of the harmonic oscillator is $\frac{1}{2}\hbar\omega$ just because this is the energy of the corresponding field oscillator. This is part of the information that has been lost in the reduction of the original description to the simplified Heisenberg equations (75), so we need to recover it for the theory to be complete. In usual quantum mechanics this is done with the help of the fundamental commutators, a procedure that also works pretty well here. Let us start by considering the following matrix element

$$(\hat{x}\hat{p})_{\alpha\alpha} = \sum_\lambda \tilde{x}_{\alpha\lambda}\tilde{p}_{\lambda\alpha} = \sum_\lambda \frac{\partial x}{\partial a_{\alpha\lambda}}\frac{\partial p}{\partial a_{\lambda\alpha}} = \sum_\lambda \frac{\partial x}{\partial a_{\alpha\lambda}}\frac{\partial p}{\partial a_{\alpha\lambda}^*} = \sum_\lambda \frac{\partial x_\alpha}{\partial a_{\alpha\lambda}}\frac{\partial p_\alpha}{\partial a_{\alpha\lambda}^*}. \qquad (76)$$

The oscillating time factors have been absorbed in the amplitudes, so that the $a_{\lambda\mu}$ are, using a more explicit notation,

$$a_{\lambda\mu}(t) = a_{\lambda\mu}e^{i\omega_{\lambda\mu}t}. \qquad (77)$$

However, no time dependence is contained in equation (76) because the time factors cancel out. The last step in that equation requires some explanation. To derivatives such as $(\partial x/\partial a_{\alpha\lambda})$, only terms explicitly containing the corresponding indices can contribute; since a sum is being performed over the index λ, the only terms that contribute to the result are those contained in the part of $x(t)$ given by

$$x_\alpha(t) = \sum_\beta \tilde{x}_{\alpha\beta} a_{\alpha\beta} e^{i\omega_{\alpha\beta} t} + \text{noise}_\alpha \tag{78}$$

(and similarly for $p_\alpha(t)$). The variable $x_\alpha(t)$ refers to the x that describes the *state* α, and so on, so that only transition coefficients $\tilde{x}_{\alpha\beta}, \tilde{x}^*_{\alpha\beta} = \tilde{x}_{\beta\alpha}$ connecting state α with any other state β can contribute to it.

A similar calculation leads to

$$(\hat{p}\hat{x})_{\alpha\alpha} = \sum_\lambda \tilde{p}_{\alpha\lambda}\tilde{x}_{\lambda\alpha} = \sum_\lambda \frac{\partial p_\alpha}{\partial a_{\alpha\lambda}} \frac{\partial x_\alpha}{\partial a^*_{\alpha\lambda}},$$

hence for the commutator we get

$$[\hat{x},\hat{p}]_{\alpha\alpha} = \sum_\lambda \left(\frac{\partial x_\alpha}{\partial a_{\alpha\lambda}} \frac{\partial p_\alpha}{\partial a^*_{\alpha\lambda}} - \frac{\partial p_\alpha}{\partial a_{\alpha\lambda}} \frac{\partial x_\alpha}{\partial a^*_{\alpha\lambda}} \right). \tag{79}$$

This result can be written in terms of the canonical variables of the field $q_{\alpha\beta}, p_{\alpha\beta}$ instead of $a^*_{\alpha\beta}, a_{\alpha\beta}$ (this is a linear transformation) using equations (20):

$$[\hat{x},\hat{p}]_{\alpha\alpha} = i\hbar \sum_\lambda \left(\frac{\partial x_\alpha}{\partial q_{\alpha\lambda}} \frac{\partial p_\alpha}{\partial p_{\alpha\lambda}} - \frac{\partial p_\alpha}{\partial q_{\alpha\lambda}} \frac{\partial x_\alpha}{\partial p_{\alpha\lambda}} \right). \tag{80}$$

The right hand side is the Poisson bracket of x_α and p_α with respect to the field variables. Since in the quantum regime x_α and p_α depend no more on atomic variables (for instance, on initial conditions), this is the full Poisson bracket, and we obtain

$$[\hat{x},\hat{p}] = i\hbar [x,p]_{PB} = i\hbar. \tag{81}$$

In writing this equation we have taken into account that $[x,p]_{PB} = 1$ and have eliminated the indices $\alpha\alpha$ to consider all matrix elements. This can be done because the above relation can be extended to the non diagonal elements when the noise terms are neglected (for the non diagonal elements, variables x_α, p_β belonging to different states are to be considered, and they have no correlated terms, assuming that there is no degeneracy).[10] Of course, equation (81) generalizes in the multidimensional case to the usual rule

---

[10] A non diagonal element is, e. g.,

$$[\hat{x},\hat{p}]_{\alpha\mu} = i\hbar \sum_\lambda \left(\frac{\partial x_\alpha}{\partial q_{\alpha\lambda}} \frac{\partial p_\mu}{\partial p_{\alpha\lambda}} - \frac{\partial p_\mu}{\partial q_{\alpha\lambda}} \frac{\partial x_\alpha}{\partial p_{\alpha\lambda}} \right).$$

Since p_μ depends on the set $\{a_{\mu\lambda}\}$ that is different from the set $\{a_{\alpha\lambda}\}$, the result is zero.

$$[\hat{x}_i, \hat{p}_j] = i\hbar \delta_{ij}. \tag{82}$$

Incidentally, the quantity in the right hand side of equation (79) is what we have termed the *Poissonian* (between x_α and p_α in this case). As can be seen in *The Dice*, the Poissonians are a useful tool within linear SED.

Let us now comment briefly on the Hilbert space description. Just as was the case with the random phases, the combination rule of frequencies (69) can be identically satisfied (as was noticed since the early days of matrix mechanics) by writing each frequency as the difference of two terms. At this point the step can be considered as merely formal, but afterwards it will be seen that a physical (and important!) sense can be given to the new quantities. Indeed, with

$$\omega_{\alpha\beta} = \Omega_\alpha - \Omega_\beta, \tag{83}$$

the equation (69) is satisfied automatically :

$$\omega_{\alpha\beta} + \omega_{\beta\gamma} + \omega_{\gamma\delta} = \Omega_\alpha - \Omega_\beta + \Omega_\beta - \Omega_\gamma + \Omega_\gamma - \Omega_\delta = \Omega_\alpha - \Omega_\delta = \omega_{\alpha\delta}.$$

Since this form works for any number of terms, it must hold in general. Even if this property is useful and physically meaningful ($\hbar\Omega_\alpha$ corresponds to the energy eigenvalue E_α, as we know), it is not essential in what follows.

We proceed to verify that the action of the background field on the atomic system carries it from its original phase space into the Hilbert space of states. For this purpose we recall that the generic dynamical variable $y(t)$ can be put in the form

$$y(t) = \sum \tilde{y}_{\alpha\beta} a_{\alpha\beta} e^{i\omega_{\alpha\beta}t} + \text{noise terms}$$
$$= \sum \tilde{y}_{\alpha\beta} e^{i(\varphi_\alpha - \varphi_\beta)} e^{i(\Omega_\alpha - \Omega_\beta)t} + \text{noise terms}$$
$$= \sum \tilde{y}_{\alpha\beta} e^{i(\Omega_\alpha t + \varphi_\alpha)} e^{-i(\Omega_\beta t + \varphi_\beta)} + \text{noise terms}$$

or

$$y(t) = \sum \tilde{y}_{\alpha\beta} a_\alpha(t) a_\beta^*(t) + \text{noise terms}, \tag{84}$$

where we have introduced the quantities

$$a_\alpha(t) = e^{i(\Omega_\alpha t + \varphi_\alpha)}, \tag{85}$$

so that

$$a_{\alpha\beta}(t) = a_\beta^*(t) a_\alpha(t). \tag{86}$$

The stochastic amplitudes $a_\alpha(t)$ can be taken as describing a set of elementary oscillators of frequency Ω_α, amplitude 1, and random phase. Let us use them to construct a set of (stochastic) vectors

$$|\alpha\rangle = \begin{pmatrix} 0 \\ 0 \\ \vdots \\ a_\alpha(t) \\ \vdots \\ 0 \end{pmatrix}, \tag{87}$$

with a_α in the α row and zeroes elsewhere; their components are

$$(|\alpha\rangle)_\lambda = a_\alpha(t)\delta_{\alpha\lambda}. \tag{88}$$

These vectors $|\alpha\rangle$ are orthonormal,

$$\langle \beta|\alpha\rangle = \sum_\lambda a_\beta^*(t) a_\alpha(t) \delta_{\beta\lambda}\delta_{\lambda\alpha} = \delta_{\alpha\beta}, \tag{89}$$

and constitute a complete basis of a Hilbert space

$$\left(\sum_\lambda |\lambda\rangle\langle\lambda|\right)_{\mu\nu} = \sum_\lambda |a_\lambda(t)|^2 \delta_{\lambda\mu}\delta_{\lambda\nu} = \delta_{\mu\nu}. \tag{90}$$

With the help of these vectors and noting that $|\alpha\rangle\langle\beta| = a_\beta^*(t)a_\alpha(t)\hat{M}_{\beta\alpha}$, where the square matrix $\hat{M}$ has zeroes everywhere and a 1 in the position (α,β), one can recast equation (84), leaving aside the noise, into the form

$$\hat{y}(t) = \sum_{\alpha\beta} \tilde{y}_{\alpha\beta} |\alpha\rangle\langle\beta|, \quad \tilde{y}_{\alpha\beta} = \langle\alpha|\hat{y}(t)|\beta\rangle. \tag{91}$$

Thus, what we are using may be called the a representation of the Hilbert space of the system. With the help of a canonical transformation, which amounts to performing a unitary transformation of the basis, the present description can be translated to any other equivalent one. The point is that with equation (91) we have identified the coefficients $\tilde{y}_{\alpha\beta}$ of the dynamical variables as matrix elements of an operator $\hat{y}(t)$ in the Hilbert space of the problem. For instance, the Schrödinger theory corresponds to constructing this Hilbert space with a set of square integrable functions in configuration (or any other pertinent) space.

III.B.2 Quantum electrodynamics

We have seen that the part of the background field denoted by $\mathbf{A}_{\text{res}}$ in (47) transforms simultaneously with the atomic variables into an operator (a matrix) as soon as the whole system enters the quantum regime. It is easy to verify that the new field is also represented by a matrix, since this property is impressed by the correlated field amplitudes to both parts of the system simultaneously. Specifically, writing a cartesian component of the field $\mathbf{A}_{\text{res}}$ in the schematic form

$$A = \sum \tilde{A}_{\alpha\beta} a_{\alpha\beta}(t) + \text{noise}_1 \tag{92}$$

one has, for example,

$$A^2 = \sum \tilde{A}_{\alpha\beta} \tilde{A}_{\beta\gamma} a_{\alpha\gamma}(t) + \text{noise}_2, \tag{93}$$

where, as before, only the correlated terms have been written explicitly. This result shows that one should write

$$\left(A^2\right)_{\alpha\gamma} = \sum_\beta \tilde{A}_{\alpha\beta} \tilde{A}_{\beta\gamma} \tag{94}$$

and so on, in agreement with the matrix structure of the field.

Since $\mathbf{A}_{\text{res}}$ is represented by a superposition of harmonic oscillators, the set of canonical field variables q_k, p_k refer just to quantum oscillators when the field is written as a matrix, with the usual quantum rules derived from (20):

$$\hat{a} = \sqrt{\frac{1}{2\hbar\omega}}\left(\hat{p} - i\omega\hat{q}\right), \quad \hat{a}^\dagger = \sqrt{\frac{1}{2\hbar\omega}}\left(\hat{p} + i\omega\hat{q}\right), \tag{95}$$

and the usual commutation rules. This shows that the operators appearing in the description of the quantized background field are the usual creation and annihilation operators of quantum field theory, and that the use of the (second) quantized field is needed for the consistency of the description of the whole system, to the extent to which this can be achieved with the present rules. We confirm a well known old result: quantization of matter implies field quantization and viceversa. One cannot go consistently without the other; they go hand in hand.[11]

Using the inverse of equations (95) we write

$$\tfrac{1}{2}\hat{p}^2 + \tfrac{1}{2}\omega^2\hat{q}^2 = \hbar\omega\hat{a}^\dagger\hat{a} + E_0, \tag{96}$$

from which it follows that the Hamiltonian of the field should be written in the usual form

$$\hat{H} = \sum \hbar\omega_{\alpha\beta} \left(\hat{a}^\dagger_{\alpha\beta}\hat{a}_{\alpha\beta} + \tfrac{1}{2}\right). \tag{97}$$

Look at the way in which the quantum formalism restores the zeropoint field energy (for the resonant modes; the rest is noise and has been left aside). Undoubtedly, the new picture is very clean and efficient as a calculating tool, although physical transparency is not its best attribute.

---

[11] This is an example of what we have called (in *The Dice*) a *second level theory*. We start from a theory founded on classical laws and leave the complex system (field+matter) free to evolve by itself. We find that the interaction takes the system to a state that is most naturally described by means of quantum laws. So, we abandon the classical rules and take the new ones as the description of the final system. Now we cannot go back on our steps, since we arrived at a qualitatively different theory. Alternatively, one can say that *emergent properties* have arisen in the system that denote a qualitative transformation.

Let us add some very tentative comments suggested by the above results. The energy eigenvalue of an excited mode is $\hbar\omega\left(n+\frac{1}{2}\right)$; however, in the absence of external fields we should consider the excitation as a field in the immediate neighborhood of the atomic system. This is what seems to correspond to a virtual field, that is, a physical field that is there, but only seen by the system that generates it through its interaction with the field. The quantum formalism, involving field operators and field states, appears in this form as an efficient tool to deal with the field sensed by the atomic system, including its excitations (taken into account with the state vector), without presupposing the existence of an external field, which would be described by means of fixed (maybe random) *numerical* values of the amplitudes. This can be understood by observing that the differential equation describing the full field is

$$\frac{1}{c^2}\frac{\partial^2 \mathbf{A}}{\partial t^2} - \nabla^2 \mathbf{A} = \frac{1}{c}\mathbf{J}(\mathbf{x},t), \tag{98}$$

where $\mathbf{J}(\mathbf{x},t)$ is the current associated with the motion of the charged particles. Thus the field generated by the particles is a velocity field, that decreases rapidly with distance (as $1/R^2$, whereas the acceleration fields decrease as $1/R$) and may be considered essentially as a bound field, visible to the particle, but not to outsiders [55].

The picture that emerges is the following. The particle in motion generates a (bound) field in its neighbourhood, as follows from equation (98), giving rise to the radiation reaction force on the particle and to the corresponding loss of power according to Larmor's formula, $m\tau\overline{\left(\ddot{\mathbf{x}}^2\right)}$. [56] In equilibrium, this loss is compensated by the power absorbed from the field, proportional to $(e^2/m)\,\rho(\omega)$, were $\rho(\omega)$ represents the spectral density of the whole field felt by the particle at the frequency of the radiated field.

The quantum formalism is particularly apt to perform the trick. We can verify this qualitatively by means of a simple calculation, as follows. Let us consider the circular orbits of the electron of the H atom; assuming that modes of the frequency ω corresponding to the orbital motion have been excited to the nth level, we express the average powers radiated and absorbed at this frequency as

$$\overline{P_{\text{rad}}} \simeq \frac{2e^2}{3c^3}\omega^4\overline{r^2}, \tag{99}$$

$$\overline{P_{\text{abs}}} = \frac{e^2}{m}\rho(\omega) = \frac{4e^2\hbar}{3mc^3}\omega^3(n+\tfrac{1}{2}). \tag{100}$$

The equilibrium condition $\overline{P_{\text{rad}}} = \overline{P_{\text{abs}}}$ leads after minor simplifications to

$$\tfrac{1}{2}m\omega\overline{r^2} = \hbar(n+\tfrac{1}{2}). \tag{101}$$

We recognize here the (approximate) old quantization rule for the angular momentum (or for an action variable). The result suggests that we see in each quantization rule the expression of a balance condition. Most important is to note that the same mechanism that supports the stationarity of the ground state ($n = 0$), supports that of the excited states. For the atom, the field is there; since we do not perceive it, we call it virtual. It is very interesting to note that despite the purely 'mechanical' appearance of the quantization rules, they express an electromagnetic condition.

The above estimate allows us to understand also why the quantization rules, expressing a condition for electric equilibrium, nevertheless do not contain the electric charge explicitly, but Planck's constant. The point is that, using the Heisenberg equation of motion $m\ddot{x} = f(x)$, where $f(x)$ represents the external force acting on the particle,[12] one can recast equation (99) in the form

$$\overline{P_{\text{rad}}} \simeq \frac{2e^2}{3c^3}\overline{\ddot{r}^2} = \frac{2e^2}{3m^2c^3}\overline{f^2(r)}, \tag{102}$$

so that the balance condition gives (defining an 'average' spectral density $\overline{\rho(\omega)}$ that appropriately takes into account the contribution of all relevant frequencies)

$$\overline{f^2(r)} = \tfrac{3}{2}mc^3\overline{\rho(\omega)}. \tag{103}$$

The electric charge has cancelled out from this (exact) 'quantization' condition, which at the same time depends explicitly on $\hbar$ through the spectral density of the field.

III.C On the Spin of the Electron

The previous results suggest that in quantum theory the background noise is a nuisance to be swept away as soon as possible. However, in addition to quantization itself (both of matter and field) there seems to be at least one other (very) important quantum element that is much more than a mere radiative correction and that owes its existence to the field fluctuations, namely the electron spin.

We recall that in the calculation of the average stochastic field amplitudes $\mathbf{a}_{\alpha\beta}(\mathbf{t})$, the two polarizations were considered, see equation (39). However, in the long wavelength approximation the polarizations can be taken separately; one can actually approximate the field from the very beginning by taking $e^{i\mathbf{k}\cdot\mathbf{x}} \to 1$, and the spaces $\mathbf{x}$ and $\mathbf{k}$ become then totally independent. This means that the direction Oz in $\mathbf{k}$-space can be fixed arbitrarily in any given direction $\hat{\mathbf{n}}$; in what follows we take $\hat{\mathbf{k}}_z = \hat{\mathbf{n}}$. Now we perform the same averaging that led to equation (39), except we

---

[12] Of course, the force $f(x)$ may depend on the electric charge, as happens in the atomic case. But from the point of view of the present discussion this would be a mere accident. In such case, the e that enters into the force should be taken as a parameter and be distinguished from the e that couples the particle to the background field.

express the polarization vectors in terms of circular polarization around the axis $\hat{k}_z$ and do *not* perform the summation over polarizations. The result of such a procedure is that instead of one amplitude (averaged over all directions) for each frequency, we get a pair (one for each circular polarization), which we denote with $\mathbf{a}_{\alpha\beta}^{(\sigma)}$, where σ takes the values $\pm$ (or ± 1). Similarly, the field itself contains two parts, so that

$$\mathbf{a}_{\alpha\beta} = \mathbf{a}_{\alpha\beta}^{(+)} + \mathbf{a}_{\alpha\beta}^{(-)}, \tag{104}$$

$$\mathbf{A} = \mathbf{A}^{(+)} + \mathbf{A}^{(-)}. \tag{105}$$

For the rest, the theory runs as before, so that for a *linear* problem we may write, e.g., (for simplicity we omit the indices that refer to the frequencies)

$$x(t) = \sum_\sigma \sum_\omega \tilde{x}_\omega^{(\sigma)} a_\omega^{(\sigma)} + \text{noise}. \tag{106}$$

For non linear or multiply periodic systems there are products of all orders involving both polarizations; however, it is not too complicated to see that amplitudes corresponding to different polarizations are uncorrelated. For example, one gets for the correlation function of the components of the zeropoint electric field

$$\left\langle E_i^{(\sigma)}(t) E_j^{(\sigma')}(t') \right\rangle = \delta_{\sigma\sigma'} \frac{\hbar}{6\pi^2 c^3} \int_0^\infty d\omega \omega^3 \times$$

$$\times \left[\delta_{ij} \cos\omega(t-t') - \sigma\eta\varepsilon_{ijk}\hat{n}_k \sin\omega(t-t') \right]. \tag{107}$$

We have assigned the values ± 1 to the polarization index, and $\eta = 3/4$. This shows that the two subensembles of different polarizations are uncorrelated. Since the amplitudes $a_{\alpha\beta i}^{(+)}$ and $a_{\alpha\beta j}^{(-)}$ are uncorrelated, their products behave as if they belonged to different frequencies. Thus, using the principles previously developed, we can write quite generally

$$x^n(t) = \sum \left[\widetilde{(x^n)}_\omega^{(+)} a_\omega^{(+)}(t) + \widetilde{(x^n)}_\omega^{(-)} a_\omega^{(-)}(t) \right] + \text{noise}. \tag{108}$$

This means that the theory applies separately for each value of the index; for instance, we have two Schrödinger equations instead of one (plus their complex conjugates, of course). In other words, the state function (scalar) has transformed into a state spinor with two independent components.

The physical picture behind these considerations is quite simple. The stochastic motions impressed by the background field on the particle produce not only a random linear momentum and a random component of energy, but also random rotations around an arbitrarily oriented axis. Of course, such rotations average to zero under normal conditions, but there should be conditions under which they may manifest themselves. Here we are forcing them to evince by the artificial separation of the total field into two subensembles with different polarizations. Since in the real world both subensembles coexist, such angular motions remain unseen. However, we may expect them to appear under appropriate external influences, such as the addition of a magnetic field that gives a physical sense to the assumed directed axial symmetry.

To investigate whether this mechanism can be considered at the root of the spin of the electron we calculate the angular momentum $\mathbf{J} = \mathbf{r} \times \mathbf{p}$. For the average angular momentum of an isotropic harmonic oscillator in its ground state one gets[13]

$$\mathbf{S}^{(\sigma)} \equiv \left\langle \mathbf{J}^{(\sigma)} \right\rangle = \sigma \eta \frac{\hbar}{2} \hat{\mathbf{n}}. \tag{109}$$

This result merits at least a brief comment. First of all the dependence on $\hat{\mathbf{n}}$ is quite natural, since it is a direct result of the assumed spherical symmetry of the zeropoint field ($\hat{\mathbf{n}}$ is *any* direction in space). However, it predicts space quantization for the spin: in every direction of space that we freely choose to measure the spin projection, we would get the same (absolute) value $\eta \frac{\hbar}{2}$ (pity that the η is here). This of course will never happen with a (single) rotating ball (although it will happen *on the average* with an ensemble of rotating balls and a uniform distribution of the axis of rotation).

If we now add an external constant uniform magnetic field oriented along $\hat{\mathbf{n}}$, and repeat the calculations, we get

$$\left\langle J^{(\sigma)} \right\rangle = \frac{\omega_L}{2\omega_s} \hbar + S_n^{(\sigma)}. \tag{110}$$

In the first term on the right hand side we recognize the mean angular momentum around the magnetic field $\left\langle L_n^{\text{orb}} \right\rangle = (\omega_L / 2\omega_s) \hbar$, with $\omega_L = |e| B / 2mc$ the Larmor frequency, $\omega_s^2 = \omega_0^2 + \omega_L^2$, and ω_0 the oscillator frequency. These results strongly suggest to take $\mathbf{S}^{(\sigma)}$ as the spin of the electron, the most significant weakness of the theory being the factor $\eta = 3/4$. Similarly, a calculation of the energy to be assigned to each separate polarization gives to lowest order

$$E^{(\sigma)} = \tfrac{1}{2} \left[\tfrac{3}{2} \hbar \omega_0 + \omega_L \left(\left\langle L_n^{\text{orb}} \right\rangle + 2 S_n^{(\sigma)} \right) \right]. \tag{111}$$

The factor 2 in front of the spin term is there to compensate for the global factor $\tfrac{1}{2}$ that enters because we are calculating over just one of the subensembles (denoted by σ); the orbital momentum contributes to the energy an equal amount with each subensemble (thus the factor $\tfrac{1}{2}$), whereas the spin gives either its full contribution (the factor $\tfrac{1}{2} 2 = 1$), or nothing. This appears to give a new meaning to the characteristic gyromagnetic ratio of 2 for the spin of the electron.

---

[13] It is much easier to work with the oscillator than with the free particle. For the detailed calculations see [57,58]. Since for these calculations only the correlations between pairs of field amplitudes are needed, the fact they were made within the old theory does not impair the value of the results.

In the transition to the usual quantum formalism, the dynamical variables are transformed into operators, and this same change should be made with the spin. Equation (109) gives the hint to perform the transition, which is rather direct. Since σ can take only the values ± 1, the corresponding Hilbert space is two dimensional and the operator associated with $S_n = S_3$ becomes proportional to the Pauli matrix $\hat{\sigma}_3$. The remaining three linearly independent operators on this space can be selected as the remaining two Pauli matrices and the unit 2×2 matrix. Note the very indirect and abstract description of spin to which this procedure leads: of the physical (average) spin angular momentum variable $S_n^{(\sigma)}$ we only take the fact that it is a two-valued function (with a given magnitude) having the statistical properties of an angular momentum; that is all. Of course all these formal steps obscure the physical picture of what is happening behind the mathematics when the analysis is performed entirely within quantum theory, which affords no means to enrich the description with pictures such as the one we have just disclosed.

An interesting question that arises with the proposed model refers to the average radius of the random rotations associated with a component of the spin. A related question could be that of the spectrum of such rotations. We are visualizing the spin as a random rotation (or vibrations) around the trajectory (itself random) followed by the particle; this image recalls that of the zitterbewegung of the electron in Dirac's theory. There seems to be a fundamental difference between the two jitters, since in the present case we must assume that the random rotations are realized with all possible frequencies, whereas the zitterbewegung is a vibration of characteristic frequency of about twice the Compton frequency $\omega_C = mc^2/\hbar$. However, a more detailed analysis shows that the difference is more apparent than real. To see this we must go back as far as the theory of the radiation reaction force.

A useful starting point is the following formula for the self force on the (classical) radiating particle due to radiation:

$$\mathbf{F}_{\text{self}} = -\frac{4e^2}{3\pi c^3} \int_0^\infty d\omega\, \omega^2 \int_0^t dt'\, K(\omega s)\dot{\mathbf{x}}(t')\cos\omega(t-t'), \tag{112}$$

where

$$K(\omega s) = -3\left(\frac{\cos\omega s}{\omega^2 s^2} - \frac{\sin\omega s}{\omega^3 s^3}\right), \qquad s = \frac{|\mathbf{x}-\mathbf{x}'|}{c}. \tag{113}$$

The dependence of this expression on the trajectory of the particle is too complicated to allow for an explicit evaluation, even in the simplest cases. To simplify it we observe that the factor $K(\omega s)$ plays the role of a structure factor or cutoff function. For low frequencies such that $\omega s \ll 1$, $K(\omega s)$ has a long plateau of value $K(0) = 1$, whereas for high frequencies it oscillates very fast, with an amplitude that decreases as ω^{-2}. Usually this function is oversimplified by putting $K = 1$, which takes us to the Abraham-Lorentz equation (with an (infinite) mass renormalization) after performing two integrations by parts to get rid of the factor ω^2 in the integrand,

$$\mathbf{F}_{\text{self}} = -\frac{4e^2}{3c^3}\left(\tfrac{1}{2}\dddot{\mathbf{x}}(t) - \ddot{\mathbf{x}}(t)\frac{1}{\pi}\int_0^\infty d\omega\right) = m\tau\,\dddot{\mathbf{x}}(t) - \delta m\ddot{\mathbf{x}}(t), \tag{114}$$

where

$$\tau = \frac{2e^2}{3mc^3}, \quad \delta m = \frac{4e^2}{3\pi c^3}\int_0^\infty d\omega. \tag{115}$$

As is well known this approximate equation violates causality (for instance, it predicts preacceleration effects); hence to recover causality we look for a finer description [10,54]. This can be achieved by observing that the variable s in (113) never attains negative values, so that any average of it is a positive number; we are therefore entitled to introduce an *effective time* s_0 (corresponding to an *effective radius* $a = cs_0$) and to replace the function $K(\omega s)$ by its value at s_0: $K(\omega s) \Rightarrow K(\omega s_0) \equiv K_0(\omega)$. Then for any $t > s_0$ the self-force can be approximated by

$$\mathbf{F}_{\text{self}} = -\frac{4e^2}{3\pi c^3}\int_0^\infty d\omega\,\omega^2 K_0(\omega)\int_0^t dt'\dot{\mathbf{x}}(t')\cos\omega(t-t'), \tag{116}$$

with

$$K_0(\omega) = \frac{3\sin\omega s_0}{\omega^3 s_0^3} - \frac{3\cos\omega s_0}{\omega^2 s_0^2}. \tag{117}$$

Alternatively one can write

$$\mathbf{F}_{\text{self}} = -\int_0^t \mathbf{G}_0(t-t')\dot{\mathbf{x}}(t')dt', \quad \mathbf{G}_0(t) = \frac{2m\tau}{\pi}\int_0^\infty d\omega\,\omega^2 K_0(\omega)\cos\omega(t). \tag{118}$$

Equation (118) shows that the theory is explicitly causal because the self-force depends on the past trajectory only. Performing the integrations one obtains

$$\mathbf{F}_{\text{self}} = -\frac{3m\tau}{s_0^3}[\mathbf{x}(t) - \mathbf{x}(t-s_0) - s_0\dot{\mathbf{x}}(t-s_0)] \tag{119}$$

for $t > s_0$. This explicitly causal expression for the self-force is quite unusual. Contact with the familiar results can be made by performing a Taylor series expansion of $\mathbf{x}(t-s_0)$ and $\dot{\mathbf{x}}(t-s_0)$ around $\mathbf{x}(t)$, whence

$$\mathbf{F}_{\text{self}} = -\frac{3m\tau}{s_0}\ddot{\mathbf{x}} + m\tau\dddot{\mathbf{x}} + \ldots \tag{120}$$

This is equivalent to the Abraham-Lorentz result with a cutoff at $\omega_c = (3\pi/2s_0)$. When the force given by (119) acts on an otherwise (classical) free particle, the equation of motion admits a solution of the form $\text{const}\times e^{\lambda t}$, $\lambda = -\sigma + i\omega'$. After minor algebra one is left with the condition

$$\mu^2 = -k\left(1 - e^{-\mu} - \mu e^{-\mu}\right), \quad k = \frac{3\tau}{s_0}, \quad \mu = s_0\lambda. \tag{121}$$

A numerical calculation shows that there exists a solution for $0.15 \lesssim k < 1$ for $\mu \simeq -2 + 4i$, where the value 4 for the frequency is almost insensitive to k within the given range, but the real part is more sensitive; we have taken an average value. Consequently, one can consider that the particle has acquired an effective structure of size $cs_0 \gtrsim 3\tau c$ and performs violent oscillations with a frequency of the order of

$$\omega' = \frac{2k}{\alpha}\omega_C \sim \frac{\omega_C}{\alpha}. \tag{122}$$

This result suggests that radiation reaction engenders oscillations of very high frequency in the system.[14] At the same time, the electron acquires an effective radius of the order of cs_0, or a Compton wavelength for $\omega' \sim \omega_C$. One can see a (nonrelativistic) correspondence with the zitterbewegung predicted by the Dirac equation, even if the disparity in the order of magnitude is important. A well known feature in QED is that a point particle in interaction with the vacuum acquires an effective extension of the order of Compton's wavelength λ_C, a result that we may adopt here, on the grounds that SED leads to QED, as has been shown.

The picture that emerges can be stated simply as follows. The vibrations of (very) high frequency endow the otherwise pointlike particle with an effective extension for all of its properties that stretch over times greater than ω_C^{-1} (which amounts practically to all observations, at least at present). To this effective structure corresponds a cutoff of order ω_C. This dressing of the particle by the field not merely involves an increment of its internal energy, but it also generates the spin, when the vibrations are appropriately expressed (in terms of the circular polarizations).[15] It should also be noted that this coarse graining over a volume of radius of order λ_C and times of order ω_C^{-1} is perfectly in agreement with usual quantum theory, since for these quantities the Heisenberg inequalities are in full force, as follows from $(\lambda_C)(m\lambda_C\omega_C) = \hbar$.

The qualitative correspondence just revealed between the high frequency oscillations of the radiating electron and the zitterbewegung finds additional support in the following argument. Integrating the equation of motion for the free SED particle (we neglect radiation damping, which may be added to the electric field, and work in the long wavelength approximation)

$$m\ddot{x} = eE(t) = -\frac{e}{c}\dot{A}(t) \tag{123}$$

---

[14] The value obtained for the frequency of oscillation is too high by a factor of the order of α^{-1} (ω_C is the Compton frequency $mc^2/\hbar$). One might speculate that this poor value is due to the fact that the only intrinsic time available to construct it is $\tau = 2e^2/3mc^3 = \frac{2}{3}\alpha\omega_C^{-1}$, whereas in a relativistic context the quantity ω_C enters quite naturally.

[15] The importance of high frequency vibrations in quantum mechanics, even for non relativistic particles, has been emphasized over and over again by G. Cavalleri and by A. Rueda [59,60].

we get

$$m\dot{x} = k - \frac{e}{c}A(t), \qquad (124)$$

$$x = x_1 + \frac{k}{m}t - \frac{e}{c}\int_{-t_-}^{t} A(t')dt'. \qquad (125)$$

We have assumed that the interaction has been connected in the far past, at a time $-t_1$, $t_1 > 0$. The constant k is the canonical momentum of the particle. Equation (124) shows that the velocity of the particle and its (canonical) momentum have been 'divorced'. In classical physics (and also in quantum mechanics) we set $A(t) = 0$ and get $m\dot{x} = k$, so that conservation of canonical momentum k and of velocity $\dot{x}$ are essentially equivalent. But in SED, owing to the presence of the background field, whereas the canonical momentum k of the free particle is conserved, its velocity is a fluctuating quantity,

$$v = \dot{x} = \frac{k}{m} - \frac{e}{mc}A(t). \qquad (126)$$

The two quantities $k = m\dot{x} + (e/c)A(t)$ and $m\dot{x}$ are therefore different things, with differing properties. This is just at the basis of the zitterbewegung in Dirac's theory for the free electron, where to the operator $\hat{p}$ corresponds an integral of motion, while the velocity operator is given by a Dirac matrix, $\hat{\dot{x}}_i = c\hat{\alpha}_i$ and is not a conserved quantity. As is well known, this is explained in quantum theory by demonstrating that the velocity carries a contribution due to the conserved momentum plus another one due to the very fast motions associated with the zitterbewegung. Thus, despite its non relativistic formulation, non relativistic SED incorporates already an element suggestive of the relativistic zitterbewegung. Just as in the relativistic theory the zitterbewegung appears automatically with the incorporation of the spin, in SED the spin is incorporated by taking explicitly into account the random motion of the electron [61,62]. The interpretation of the zitterbewegung as the term that makes the difference between k and $m\dot{x}$ is the same in both theories; however, there is considerable disparity on the *meaning* of such difference. Here it is due to the presence of the background field; in Dirac's theory, the usual view is that it is produced by the interference between positive and negative energy solutions.

Let us develop the foregoing argument a bit further. In Dirac's theory for the free electron one gets

$$mc\hat{\alpha}_i = p_i + \text{zitterbewegung terms},$$

whereas in SED one has (with a small change in the notation)

$$mv = m\dot{x} = p - \frac{e}{c}A(t).$$

So we see that the representation of the velocity in terms of matrices in the relativistic theory simultaneously introduces various elements into the description:

a) The spin through the expansion of the Hilbert space, or, equivalently, the introduction of an internal degree of freedom (more precisely, two degrees of freedom, associated with the spin and with the sign of the energy, that amounts to raising the Hilbert space dimension by a factor 2×2; of course, the double sign associated with the energy is characteristic of any relativistic treatment);

b) the zitterbewegung through the fact that the velocity is given by a matrix (even for eigenstates of $\hat{H}$ and $\hat{p}$), and therefore refers to a stochastic variable; in the previous non relativistic treatment this contribution is (poorly) represented by the term $\sim A(t)$, which plays the role denoted by "zitterbewegung terms" in the relativistic expression.

All three elements are intimately related in this description.

REFERENCES

1. Bell, J.S., *Speakable and unspeakable in quantum mechanics* (Cambridge U. P., Cambridge), 1987.
2. Wick, D., *The Infamous Boundary. Seven Decades of Heresy in Quantum Physics* (Copernicus, New York), 1995.
3. Baggott, J., *The Meaning of Quantum Theory* (Oxford U.P., Oxford), 1992.
4. Einstein, A., Letter to Besso in *Albert Einstein, Correspondence avec Michele Besso 1903-1955*, P. Speziale, ed. (Hermann, Paris 1972), 1951.
5. Scully, M. O., and Sargent, M., *Physics Today*, March, p. 38 (1972).
6. de Broglie, L., "Les ondes de la mécanique ondulatoire", *Decouverte* No. 3432, April, p. 39-47 (1971).
7. Milonni, P. W., James, D. F. V., and Fearn, H., *Phys. Rev.* **52**, 1525 (1995).
8. Santos, E., in *New Developments on Fundamental Problems in Quantum Physics*, Ferrero, M., and van der Merwe, A., Eds. (Kluwer, Dordrecht), 1997.
9. Gilbert, B. C., and Sules, S., *Found. Phys.* **26**, 1401 (1996).
10. de la Peña, L., and Cetto, A. M., *The Quantum Dice. An introduction to Stochastic Electrodynamics* (Kluwer Acad. Publ., Dordrecht), 1996.
11. Marshall, T. W., and Santos, E., in *Problems in Quantum Physics: Gdańsk 87*, Kostro, L., Posiewnik, A., Pykacz, J., and Żukowski, M., Eds. (World Scientific, Singapore), 1988.
12. Marshall, T. W., and Santos, E., *Phys. Rev. A* **39**, 6271 (1989).
13. Casado, A., Fernández-Rueda, A., Marshall, T., Risco-Delgado, R., and Santos, E., *Phys. Rev. A* **55**, 3879 (1997).
14. Casado, A., Marshall, T., and Santos, E., *J. Opt. Soc. Am. B* **14**, 494 (1997).
15. Marshall, T. W., *Proc. Royal Soc. A* **276**, 475 (1963).
16. Boyer, T.H., *Phys. Rev. D* **11**, 790 (1975).
17. Power, E. A., *Introductory Quantum Electrodynamics* (Elsevier, New York), 1965.

18. Boyer, T. H., *Annals of Phys.* (NY) **56**, 474 (1970).
19. Spruch, L., and Kelsey, E. J., *Phys. Rev. A* **18**, 845 (1978).
20. Weinberg, S., *Rev. Mod. Phys.* **61**, 1 (1989).
21. Milonni, P. W., Ackerhalt, J. R. and Smith, W. A., *Phys. Rev. Lett.* **31**, 958 (1973).
22. Senitzky, I. R., *Phys. Rev. Lett.* **31**, 955 (1973).
23. Dalibard, J., Dupont-Roc, J., and Cohen-Tannoudji, C., *J. Phys.* (Paris) **43**, 1617 (1982).
24. Milonni, P. W., *The Quantum Vacuum. An Introduction to Quantum Electrodynamics* (Academic Press, San Diego), 1994.
25. Feynman, R. P., and Hibbs, A. R., *Quantum mechanics and path integrals* (McGraw-Hill, New York), 1965.
26. Marshall, T.W., *Proc. Camb. Phil. Soc.* **61**, 537 (1965).
27. Boyer, T. H., *Phys. Rev.* **182**, 1374 (1969).
28. Boyer, T. H., *Phys. Rev. D* **21**, 2137 (1980).
29. Santos, E., *An. Real Soc. Esp. Fís. Quím.*, **LXIV**, 317 (1968).
30. Santos, E., *Nuovo Cim. B* **22**, 201 (1974).
31. Goedecke, G. H., *Found. Phys.* **13**, 1101 (1983).
32. de la Peña, L., in *Stochastic Processes Applied to Physics and other Related Fields*, Gómez, B., Moore, S. M., Rodríguez Vargas, A.M., and Rueda, A., Eds. Proceedings of the ELAF 1982 (World Scientific, Singapore), 1983.
33. Cole, D. C., *Phys. Rev. A* **42**, 7006 (1990).
34. Papoulis, A., *Probability, Random Variables, and Stochastic Processes* (McGraw-Hill, Boston), 1965, 1991.
35. Marshall, T. W., *Nuovo Cim.* **38**, 206 (1965).
36. Boyer, T. H., *Phys. Rev.* **174**, 1631 (1968).
37. Boyer, T. H., *Phys. Rev.* **174**, 1764 (1968).
38. Boyer, T. H., *Phys. Rev. A* **11**, 1650 (1975).
39. Cetto, A. M., and de la Peña, L., *Phys. Rev. A* **37**, 1952, 1960 (1988).
40. França, H., Marshall, T. W., and Santos, E., *Phys. Rev. A* **45**, 6436 (1992).
41. Hillery, M., O'Connell, R. F., Scully, M. O., and Wigner, E. P., *Phys. Rep.* **106** No. 3, 121 (1984).
42. Boyer, T. H., *Phys. Rev. A* **21**, 1246 (1980).
43. Barranco, A. V., and França, H., *Found. Phys. Lett.* **5**, 25 (1992).
44. Boyer, T. H., *Phys. Rev. D* **13**, 2832 (1976).
45. Boyer, T. H., *Phys. Rev. A* **18**, 1228 (1978).
46. Boyer, T. H., in *Foundations of Radiation Theory and Quantum Electrodynamics*, Barut, A. O., Ed. (Plenum, London), 1980.
47. Claverie, P., Diner, S., and Israel, J., *Chem.* **19**, 54 (1980).
48. Marshall, T. W., and Claverie, P., *J. Math. Phys.* **21**, 1819 (1980).
49. Claverie, P., and Soto, F., *J. Math. Phys.* **23**, 753 (1982).
50. Pesquera, L., and Claverie, P., *J. Math. Phys.* **23**, 1315 (1982).
51. van Vleck, J. H., and Huber, D. L., *Rev. Mod. Phys.* **49**, 939 (1977).
52. de la Peña, L., and Cetto, A. M., *Found. Phys.* **24**, 753, 917 (1994).
53. de la Peña, L., and Cetto, A. M., *Found. Phys.* **25**, 573 (1995).
54. de la Peña, L., and Cetto, A. M., in *Proceedings Oviedo Symposium on Fundamental*

Problems in Quantum Physics, Ferrero, M., and van der Merwe, A., Eds. (Kluwer, Amsterdam), 1995.
55. Compagno, G., Passante, R., and Persico, F., *Atom-Field Interactions and Dressed Atoms* (Cambridge U. P., Cambridge), 1995.
56. Cohen-Tannoudji, C., Dupont-Roc, J., and Grynberg, G., *Photons and Atoms. Introduction to Quantum Electrodynamics* (Wiley, New York), 1989.
57. Jáuregui, A., and de la Peña, L., *Phys. Lett. A* **86**, 280 (1981).
58. de la Peña, L., and Jáuregui, A., *Found. Phys.* **12**, 441 (1982).
59. Cavalleri, G., *Lett. Nuovo Cim.* **43**, 285 (1985).
60. Rueda, A., *Found. Phys. Lett.* **6**, 75 (1993).
61. Feshbach, H., and Villars, F., *Rev. Mod. Phys.* **30**, 25 (1958).
62. Hestenes, D., *Found. Phys.* **15**, 63 (1985).

Conventional Quantum Mechanics Without Wave Function and Density Matrix

Vladimir I. Man'ko

P.N. Lebedev Physical Institute, Leninskii Pr. 53, Moscow 117924, Russia
e-mail: manko@na.infn.it
manko@lebedev.ru

Abstract. The tomographic invertable map of the Wigner function onto the positive probability distribution function is studied. Alternatives to the Schrödinger evolution equation and to the energy level equation written for the positive probability distribution are discussed. Instead of the transition probability amplitude (Feynman path integral) a transition probability is introduced. A new formulation of the conventional quantum mechanics (without wave function and density matrix) based on the "probability representation" of quantum states is given. An equation for the propagator in the new formulation of quantum mechanics is derived. Some paradoxes of quantum mechanics are reconsidered.

I INTRODUCTION

During more than 70 years of existence of quantum mechanics, there was a dream to reduce misterious and intuitively very unusual notions of this theory to the well-known and intuitively acceptable classical notions. There was a common prejudice that it is impossible to describe the notion of quantum state in the framework of the conventional quantum mechanics in terms of the probability density and one is obliged to use either complex wave function or density matrix in different representations. Fortunately, it turns out that it is possible to associate with a quantum state the usual probability density and the use of the wave function or density matrix is not mandatory in the conventional quanrum mechanics. This course of lectures is devoted to the new formulation of quantum mechanics.

Quantum mechanics is based on the concept of a complex wave function which satisfies the Schrödinger equation [1]. Several attempts to give classical-like interpretations of the wave function were done in [2-4] (see also [5]). These attempts [2-4] and related constructions of quasidistribution functions in the phase space of the system [6-9] give the idea that for quantum mechanics it is impossible to describe the state of the quantum system in terms of measurable positive probability analogously to the case of classical statistical mechanics, where the state of the system is described by the positive probability distribution due to the presence of classical fluctuations.

Nevertheless, it was shown recently [10–15] that in the framework of the symplectic tomography scheme [16,17], which generalizes the optical tomography scheme [18,19], it is possible to introduce a classical-like description of a quantum state using the measurable positive probability (instead of the complex probability amplitude).

This result was obtained because, in addition to considering a measured physical observable in a fixed reference frame in the phase space of the quantum system, different reference frames in the phase space were considered. In the spirit of methodology, it is close to special relativity theory, where to get unusual effects due to motion with high velocities, different reference frames connected by Lorentz transform must be used. In the quantum case, the extra parameters distinguishing different reference frames replace the information coded by a phase of the wave function. This approach can be considered as introducing a new representation in quantum mechanics which can be called the "probability representation" [20,21].

The description of the quantum state in terms of positive probability was obtained not only for continuous observables like the position [14,16,17], but also for pure quantum observables like spin [22,23] (see also [24]).

A classical formulation of quantum evolution was suggested, and for the marginal distribution a new quantum evolution equation was found [10] which is an alternative to the time-dependent Schrödinger equation. This equation gives the classical-like description of quantum evolution in terms of a normalized positive distribution containing complete information on the state of the system. Examples of free motion and some excited states of the harmonic oscillator were also considered [10,11]. The evolution of even and odd coherent states [25] of a particle in a Paul trap was investigated [26] in the framework of the classical-like description [10,11]. The even and odd coherent states of a trapped ion were discussed in recent papers [27,28]. Experimentally these states were realized in [29,30]. A review of the metod of integrals of motion and its application to oscillator's models used in the Paul trap problem is given in Ref. [31].

The aim of the course of lectures is to discuss, following [13–15,32], the notion of a quantum state in the new formulation of quantum mechanics. We review the classical-like description of transition probabilities between stationary states (energy levels) of quantum systems and obtain analogs of the orthogonality and the completeness relations. We show that, if the evolution equation describing the dynamics of a quantum system is determined by the imaginary part of the system's potential energy considered as a function of a complex coordinate, the energy states of the system are determined by the real part of the potential energy. The energy levels of the harmonic oscillator are rederived in the framework of classical-like alternatives to Schrödinger evolution and stationary equations. A new type of eigenvalue problem is formulated for the positive and normalized marginal distributions.

II CLASSICAL STATISTICAL MECHANICS AND TOMOGRAPHY MAP

In quantum mechanics, the tomography methods of measuring quantum states [16,18,19] gave the possibility to introduce new approach to the notion of quantum states [10,11,13]. It turned out that the tomography methods can be used in classical statistical mechanics [32,33]. Following [32] we start from introducing the positive probability distribution function for a state of a classical system. In the course of lectures, we will show that both classical and quantum states are described by the same probability distribution function (called the marginal distribution function).

Main expressions for marginal distributions in the optical tomography method [18,19] as well as in the symplectic tomography method [16] are based on a theorem which connects the characteristic function with the probability distribution function. This connection is valid for quantum states described by a density matrix [34]. It is obvious that the same connection also exists for classical systems in the framework of classical statistical mechanics. One can prove that the Fourier transform of the characteristic function (calculated by means of a classical probability distribution) is a positive distribution function. We illustrate this statement by an example of a one-dimensional system.

"States" in classical statistics are described by the function $f(q, p)$, which is the probability distribution function in the phase space, i.e.,

$$f(q, p) \leq 0, \qquad \int f(q, p)\, dp = P(q), \qquad \int f(q, p)\, dq = \tilde{P}(p),$$

with $P(q)$ and $\tilde{P}(p)$ probability distributions for position and momentum, respectively, (*marginals*).

Let the nonnegative function $f(q, p)$ be a distribution function of the classical system in the phase space. The coordinates $-\infty < q < \infty$ and $-\infty < p < \infty$ are the position and momentum of the system, respectively. The function $f(q, p)$ is taken to be normalized

$$\int f(q, p)\, dq\, dp = 1. \tag{1}$$

We consider an observable $X(q, p)$ which is a function on the phase space of the system under study. For the case of classical statistical mechanics, the characteristic function for the observable $X(q, p)$

$$\chi(k) = \langle e^{ikX} \rangle \tag{2}$$

is given by the relation

$$\chi(k) = \int e^{ikX(q,p)} f(q, p)\, dq\, dp. \tag{3}$$

The Fourier transform of the characteristic function

$$w(X) = \frac{1}{2\pi} \int \chi(k) e^{-ikX} dk \qquad (4)$$

is a real nonnegative function which is normalized

$$\int w(X) \, dX = 1. \qquad (5)$$

In fact, due to Fourier representation of Dirac delta-function, one has

$$w(X) = \int f(q, p) \, \delta(X(q, p) - X) \, dq \, dp. \qquad (6)$$

The distribution function is nonnegative and the delta-function is also nonnegative. So we integrate the product of two nonnegative functions over the phase space. The result of the integration $w(X)$ is obviously a nonnegative function.

Let us now check the normalization of the function $w(X)$. We have

$$\int w(X) \, dX = \int f(q, p) \, \delta(X(q, p) - X) \, dq \, dp \, dX. \qquad (7)$$

In view of the definition of delta-function, one has

$$\int \delta(X(q, p) - X) \, dX = 1. \qquad (8)$$

This means that

$$\int w(X) \, dX = \int f(q, p) \, dq \, dp. \qquad (9)$$

Since the distribution function $f(q, p)$ satisfies the normalization condition 1), we have shown that the Fourier transform of the characteristic function $w(X)$ given by 4) is normalized too, i.e., it satisfies the normalization condition 5). As a classical analog of the quantum symplectic-tomography observable introduced in Ref. [16] we consider the classical observable which is a linear function on the phase space of the system,

$$X(q, p) = \mu q + \nu p, \qquad (10)$$

where the real parameters μ and ν are interpreted as the parameters of symplectic transform of the position and momentum of the system under study. (We discuss only one variable —the position $X(q, p)$— and do not take into account the conjugate momentum.)

The variable $X(q, p)$ can be considered from two equivalent points of view. It can be interpreted as a canonically transformed position which is a linear combination of position and momentum in a fixed reference frame in the phase space of the system. Another equivalent interpretation of the variable $X(q, p)$ given by Eq. 10) is that it is a position of the system measured in the rotated and scaled reference frame in the classical phase space of the system.

We use the second interpretation, according to which the real parameters μ and ν determine the reference frame in the phase space of the system in which the position is measured. For the position in the transformed reference frame 10), we get from Eq. 3) the distribution function (the tomography map)

$$w(X, \mu, \nu) = \frac{1}{2\pi} \int e^{-ik(X-\mu q-\nu p)} f(q, p) \, dq \, dp \, dk. \tag{11}$$

Another form for the probability distribution is given by Eq. 6)

$$w(X, \mu, \nu) = \int f(q, p) \, \delta(\mu q + \nu p - X) \, dq \, dp. \tag{12}$$

One can see that the marginal distribution is a homogenious function, i.e.,

$$w(\lambda X, \lambda\mu, \lambda\nu) = |\lambda|^{-1} w(X, \mu, \nu). \tag{13}$$

We introduced the notation $w(X, \mu, \nu)$ for the probability distribution of the position of the classical system in the transformed reference frame in the phase space to point out the dependence of the distribution on the parameters μ and ν determining the reference frame. Due to the dependence of the distribution $w(X, \mu, \nu)$ on these parameters, we call the distribution a marginal distribution function.

The partial case of the canonical transform 10) is a rotation in the phase space

$$X = q \cos\varphi + p \sin\varphi. \tag{14}$$

This means that we choose the parameters of the symplectic transform

$$\mu = \cos\varphi, \qquad \nu = \sin\varphi. \tag{15}$$

By introducing the notation for the marginal distribution of the rotated position

$$w(X, \varphi) = w(X, \mu = \cos\varphi, \nu = \sin\varphi), \tag{16}$$

we get, in view of 12),

$$w(X, \varphi) = \int f(q, p) \, \delta(q \cos\varphi + p \sin\varphi - X) \, dq \, dp. \tag{17}$$

Using Eq. 11) we have another representation for the marginal distribution of the rotated position, namely,

$$w(X, \varphi) = \frac{1}{2\pi} \int e^{-ik(X-q\cos\varphi-p\sin\varphi)} f(q, p) \, dq \, dp \, dk. \tag{18}$$

Introducing the transformed position and momentum

$$Q = q\cos\varphi + p\sin\varphi; \qquad P = -q\sin\varphi + p\cos\varphi, \qquad (19)$$

in view of the invariance of the volume in the phase space

$$dq\,dp = dQ\,dP, \qquad (20)$$

we get, using the Fourier representation of delta-function,

$$w(X, \varphi) = \int \delta(X - Q)\, f(Q\cos\varphi - P\sin\varphi,\, Q\sin\varphi + P\cos\varphi)\, dQ\,dP, \qquad (21)$$

or

$$w(X, \varphi) = \int f(X\cos\varphi - P\sin\varphi,\, X\sin\varphi + P\cos\varphi)\, dP. \qquad (22)$$

Formula 22) is mathematically identical to Eq. 12) of Ref. [19] where the marginal distribution for the homodyne observable was considered. But in 22), the positive classical distribution $f(q, p)$ in the phase space is used instead of the Wigner function $W(q, p)$ elaborated in Eq. 12) of Ref. [19]. It is worth noting that the form of the expression for the marginal distribution $w(X, \varphi)$ is invariant. The only difference between the quantum and classical statistics in the context of the expression for the marginal distribution $w(X, \varphi)$ is in the difference between the classical distribution in the phase space and the Wigner function. The Wigner function $W(q, p)$ can take negative values. The classical distribution function $f(q, p)$ takes only nonnegative values. Nevertheless, the result of integration in both cases gives the nonnegative marginal distribution $w(X, \varphi)$.

Formula 11) has the inverse

$$f(q, p) = \frac{1}{4\pi^2} \int w(X, \mu, \nu) \exp[-i(\mu q + \nu p - X)]\, dX\,d\mu\,d\nu. \qquad (23)$$

In classical statistical mechanics, the admissible marginal distributions in formula 23) always satisfy the condition that the result of convolution $f(q, p)$ is a nonnegative function.

We have shown that instead of the distribution function $f(q, p)$ the state of the classical system in the framework of classical statistical mechanics can be determined by the marginal distribution function $w(X, \mu, \nu)$, in complete analogy with the quantum case where the symplectic tomography procedure is used [16]. Since the map

$$f(q, p) \Longrightarrow w(X, \mu, \nu)$$

is invertable, the information contained in the distribution function $f(q, p)$ is equivalent to the information contained in the marginal distribution $w(X, \mu, \nu)$. For $\mu = \cos\varphi$ and $\nu = \sin\varphi$, we have an analog of the optical tomography procedure developed for the quantum case in [19]. We have to invert formula 22). The inverse is given by the Radon transform (see, for example, Eq. 13) in [19] and also [35]). For example, if one introduces the distribution function in the form

$$f(q, p) = \delta(q - x_0)\delta(p - p_0), \tag{24}$$

the marginal distribution takes the form

$$w(X, \mu, \nu) = \delta(X - \mu x_0 - \nu p_0) \tag{25}$$

and

$$w(X, \varphi) = \delta(X - x_0 \cos\varphi - p_0 \sin\varphi). \tag{26}$$

For classical statistical mechanics, the tomography maps discussed connect the positive distributions, and in this context our understanding of the notion of the classical state for systems with fluctuations is unchanged.

The evolution equation for the classical distribution function for a particle with mass $m = 1$ and potential $U(q)$,

$$\frac{\partial f(q, p, t)}{\partial t} + p\frac{\partial f(q, p, t)}{\partial q} - \frac{\partial U(q)}{\partial q}\frac{\partial f(q, p, t)}{\partial p} = 0 \tag{27}$$

can be rewritten in terms of the marginal distribution $w(X, \mu, \nu, t)$

$$\dot{w} - \mu\frac{\partial}{\partial \nu}w - \frac{\partial U}{\partial q}(\tilde{q})\left[\nu\frac{\partial}{\partial X}w\right] = 0, \tag{28}$$

where the argument of the function $\partial U/\partial q$ is replaced by the operator

$$\tilde{q} = -\left(\frac{\partial}{\partial X}\right)^{-1}\frac{\partial}{\partial \mu}. \tag{29}$$

For the harmonic oscillator with frequency $\omega = 1$, the potential energy term $U(q) = q^2/2$ gives in Eq. 28) the following evolution equation

$$\dot{w} - \mu\frac{\partial w}{\partial \nu} + \nu\frac{\partial w}{\partial \mu} = 0. \tag{30}$$

We used the equality

$$\frac{1}{2}\frac{\partial q^2}{\partial q}(\tilde{q}) = -\left(\frac{\partial}{\partial X}\right)^{-1}\frac{\partial}{\partial \mu} \tag{31}$$

and the property

$$\left(\frac{\partial}{\partial X}\right)^{-1}\frac{\partial}{\partial \mu}\nu\frac{\partial}{\partial X} = \nu\frac{\partial}{\partial \mu}. \tag{32}$$

For the mean value of position in classical statistics, we have

$$\langle q \rangle = \int f(q, p) \, q \, dq \, dp = i \int w(X, \mu, \nu) \, e^{iX} \, \delta'(\mu) \, \delta(\nu) \, dX \, d\mu \, d\nu. \tag{33}$$

One can see that in classical statistical mechanics there exists a function associated to the position, and by means of this function one can calculate the mean value of position using the marginal distribution $w(X, \mu, \nu)$.

In classical statistical mechanics, one can introduce the propagator $\Pi_{cl}(X, \mu, \nu, X', \mu', \nu', t_2, t_1)$ that connects the two marginal distributions given for times t_1 and t_2 $(t_2 > t_1)$

$$w(X, \mu, \nu, t_2) = \int \Pi_{cl}(X, \mu, \nu, X', \mu', \nu', t_2, t_1) \, w(X', \mu', \nu', t_1) \, dX' \, d\mu' \, d\nu'. \tag{34}$$

The propagator satisfies the following equation

$$\frac{\partial \Pi_{cl}}{\partial t_2} - \mu \frac{\partial}{\partial \nu} \Pi_{cl} - \frac{\partial U(q)}{\partial q} (\hat{q}) \, \nu \frac{\partial}{\partial X} \Pi_{cl}$$
$$= \delta(t_2 - t_1) \, \delta(X - X') \, \delta(\mu - \mu') \, \delta(\nu - \nu'), \tag{35}$$

which follows from the evolution equation 28).

Any integral of motion $I(q, p, t)$ in classical statistical mechanics satisfies the equation

$$\frac{dI}{dt} = \frac{\partial I}{\partial t} + \frac{\partial I}{\partial q} \frac{\partial H}{\partial p} - \frac{\partial I}{\partial p} \frac{\partial H}{\partial q} = 0, \tag{36}$$

where $H(q, p, t)$ is the Hamiltonian of the classical system.

Equation 36) coincides for

$$H = \frac{p^2}{2} + U(q)$$

with the equation for the classical distribution function 27),

$$\frac{\partial I(q, p, t)}{\partial t} + p \frac{\partial I(q, p, t)}{\partial q} - \frac{\partial U(q)}{\partial q} \frac{\partial I(q, p, t)}{\partial p} = 0. \tag{37}$$

This follows from the fact that the distribution function itself is the integral of motion.

If one introduces the map 11) for the integrals of motion

$$\mathcal{I}(X, \mu, \nu, t) = \frac{1}{2\pi} \int e^{-ik(X - \mu q - \nu p)} \, I(q, p, t) \, dq \, dp \, dk, \tag{38}$$

the integral of motion $\mathcal{I}(X, \mu, \nu, t)$ satisfies Eq. 28) in which one has to make the replacement $w \to \mathcal{I}$.

In classical statistical mechanics, the distribution function $f(q, p, t)$ is a function of the integrals of motion, and the propagator that determines the evolution of the distribution function has the form

$$P(q, p, q', p', t) = \delta(q' - q_0(q, p, t))\, \delta(p' - p_0(q, p, t)), \tag{39}$$

where $q_0(q, p, t)$ and $p_0(q, p, t)$ are integrals of motion which have the following property:

$$q_0(q, p, 0) = q, \qquad p_0(q, p, 0) = p. \tag{40}$$

Using Eq. 39) one can find the propagator for the marginal distribution function. For example, the initial distribution 24) takes the form

$$f_0(q, p, t) = \delta(p - p_0)\, \delta(q - tp - x_0) \tag{41}$$

and the initial marginal distribution 25) reads

$$w_0(X, \mu, \nu, t) = \delta(X - \mu t p_0 - \mu x_0 - \nu p_0). \tag{42}$$

In quantum and classical statistical mechanics, the forms of the propagators determining the evolution of the marginal distributions $w(X, \mu, \nu, t)$ are identical for linear systems like an oscillator or free motion.

III   NEW NOTION OF QUANTUM STATE

We consider now a new approach to the notion of quantum state. It was shown [16] that for the generic linear combination of quadratures which is a measurable observable ($\hbar = 1$)

$$\widehat{X} = \mu \hat{q} + \nu \hat{p}, \tag{43}$$

where $\hat{q}$ and $\hat{p}$ are the position and momentum, respectively; the marginal distribution $w(X, \mu, \nu)$ (normalized with respect to the variable X), depending on the two extra real parameters μ and ν, is related to the state of the quantum system expressed in terms of its Wigner function $W(q, p)$ as follows:

$$w(X, \mu, \nu) = \int \exp[-ik(X - \mu q - \nu p)]\, W(q, p)\, \frac{dk\, dq\, dp}{(2\pi)^2}. \tag{44}$$

We use the same notation as in the classical case. If one has a pure state with the wave function $\Psi(y)$, the marginal distribution has the form found in Ref. [36]

$$w(X, \mu, \nu) = \frac{1}{2\pi|\nu|} \left| \int \Psi(y) \exp\left(\frac{i\mu y^2}{2\nu} - \frac{iyX}{\nu}\right) dy \right|^2. \tag{45}$$

The physical meaning of the parameters μ and ν is that they describe an ensemble of rotated and scaled reference frames in which the position X is measured. For $\mu = \cos\varphi$ and $\nu = \sin\varphi$, the marginal distribution 44) is the distribution for the homodyne-output variable used in optical tomography [19]. Formula 44) can be inverted and the Wigner function of the state can be expressed in terms of the marginal distribution [16]:

$$W(q, p) = \frac{1}{2\pi} \int w(X, \mu, \nu) \exp\left[-i(\mu q + \nu p - X)\right] d\mu \, d\nu \, dX. \qquad (46)$$

Since the Wigner function determines completely the quantum state of a system and, on the other hand, this function itself is completely determined by the marginal distribution, one can understand the notion of the quantum state in terms of the classical marginal distribution for squeezed and rotated quadrature.

So, we say that the quantum state is given if the position probability distribution $w(X, \mu, \nu)$ in an ensemble of rotated and squeezed reference frames in the classical phase space is given.

It is worth noting, that the information contained in the marginal distribution $w(X, \mu, \nu)$ is overcomplete. To determine the quantum state completely, it is sufficient to give the function for arguments with the constraints ($\mu^2 + \nu^2 = 1$) which corresponds to the optical tomography scheme [19,37], i.e., $\mu = \cos\varphi$ and the rotation angle φ labels the reference frame in the classical phase space.

So, we formulate also the notion of quantum states as follows:

We say that the quantum state is given if the position probability distribution $w(X, \varphi)$ in an ensemble of rotated reference frames in the classical phase space is given.

Since the quantum state is defined by the position distribution, one could associate an entropy with the state using the standard relation known in classical probability theory, i.e., the entropy $S(\mu, \nu)$ is given by the formula

$$S(\mu, \nu) = -\int dX \, w(X, \mu, \nu) \ln\left[w(X, \mu, \nu)\right]. \qquad (47)$$

If we use the distribution $w(X, \varphi)$, the entropy $S(\varphi)$ depends only on the rotation angle.

The discription of quantum states by the probability function gives the possibility to formulate quantum mechanics without using the wave function or density matrix. These ingredients of the quantum theory can be considered as objects which are not mandatory ones since they are not directly measurable. The marginal probability distribution function $w(X, \mu, \nu)$, which can be measured directly, replaces the wave function in the new formulation of quantum mechanics. Since the quantum mechanics formalism is reduced to the formalism of classical probability theory, well-known results of the probability theory can be used to get new results in quantum theory (including quantum computing, teleportation, and quantum cryptography).

One can also use the introduced formulation of the notion of quantum states to describe situations in which the states are either close or essentially different. We say that two states are close if their distributions are close, i.e., all the highest momenta of the distributions differ very slightly. We also say that two states are substantially different if their distributions differ substantially, i.e., there are highest momenta for the two distributions with large corresponding differences. The notion of distance in quantum mechanics using the tomography map was discussed in [38].

IV QUANTUM EVOLUTION AND ENERGY LEVELS

As was shown in [10], for systems with the Hamiltonian

$$H = \frac{\hat{p}^2}{2} + V(\hat{q}), \tag{48}$$

the marginal distribution satisfies the quantum time-evolution equation, being the integral equation determined by the imaginary part of the potential energy considered as a function of a complex coordinate. The evolution equation reads

$$\dot{w} - \mu \frac{\partial}{\partial \nu} w - i \left[V \left(-\frac{1}{\partial/\partial X} \frac{\partial}{\partial \mu} - i \frac{\nu}{2} \frac{\partial}{\partial X} \right) \right. \\ \left. - V \left(-\frac{1}{\partial/\partial X} \frac{\partial}{\partial \mu} + i \frac{\nu}{2} \frac{\partial}{\partial X} \right) \right] w = 0. \tag{49}$$

This equation is alternative to the Schrödinger equation

$$i\dot{\Psi} = H\Psi \tag{50}$$

and it can be obtained from the equation for density matrix

$$\dot{\rho} + i[H, \rho] = 0, \tag{51}$$

in view of the following formulas:

$$q W(q, p) \longrightarrow -\left(\frac{\partial}{\partial X}\right)^{-1} \frac{\partial}{\partial \mu} w(X, \mu, \nu),$$

$$\frac{\partial}{\partial q} W(q, p) \longrightarrow \mu \frac{\partial}{\partial X} w(X, \mu, \nu),$$

$$p W(q, p) \longrightarrow -\left(\frac{\partial}{\partial X}\right)^{-1} \frac{\partial}{\partial \nu} w(X, \mu, \nu), \tag{52}$$

$$\frac{\partial}{\partial p} W(q, p) \longrightarrow \nu \frac{\partial}{\partial X} w(X, \mu, \nu)$$

and

$$\frac{\partial}{\partial X}\rho(X, X') \longrightarrow \left(\frac{1}{2}\frac{\partial}{\partial q} + ip\right) W(q, p),$$

$$\frac{\partial}{\partial X'}\rho(X, X') \longrightarrow \left(\frac{1}{2}\frac{\partial}{\partial q} - ip\right) W(q, p),$$

$$X\rho(X, X') \longrightarrow \left(q + \frac{i}{2}\frac{\partial}{\partial p}\right) W(q, p),$$

$$X'\rho(X, X') \longrightarrow \left(q - \frac{i}{2}\frac{\partial}{\partial p}\right) W(q, p). \tag{53}$$

Equation 49) can be considered as a Fokker–Planck-like equation of classical probability theory. The measurable position is a cyclic variable for the evolution equation.

In order to compare the classical and quantum evolution equations, let us rewrite the quantum evolution equation 49) in the form of a series,

$$\dot{w} - \mu\frac{\partial w}{\partial \nu} + 2\sum_{n=0}^{\infty} \frac{V^{2n+1}(\hat{q})}{(2n+1)!} \left(\frac{\nu}{2}\frac{\partial}{\partial X}\right)^{2n+1} (-1)^{n+1} w = 0. \tag{54}$$

Here

$$V^{2n+1}(\hat{q}) = \frac{\partial^{2n+1} V}{\partial q^{2n+1}} (q = \hat{q}), \tag{55}$$

where the operator $\hat{q}$ is given by Eq. 29).

Equation 55) can also be presented in the form

$$\dot{w} - \mu\frac{\partial w}{\partial \nu} - \frac{\partial V}{\partial q}(\hat{q}) \nu \frac{\partial}{\partial X} w + 2\sum_{n=1}^{\infty} \frac{V^{2n+1}(\hat{q})}{(2n+1)!} \left(\frac{\nu}{2}\frac{\partial}{\partial X}\right)^{2n+1} (-1)^{n+1} w = 0. \tag{56}$$

The three first terms give the $\hbar \to 0$ classical Boltzman equation. It is important that both classical and quantum evolution equations are written for the same function $w(X, \mu, \nu)$.

Let us rewrite, following [13], the Schrödinger equation for the stationary state density matrix ρ_E of the quantum system with Hamiltonian 48)

$$H\rho_E = \rho_E H = E\rho_E \tag{57}$$

in terms of the time-independent marginal distribution $w_E(X, \mu, \nu)$ of the squeezed and rotated quadrature introduced in [16]. We have

$$\frac{1}{2}\left(\frac{\partial}{\partial X}\right)^{-2} \frac{\partial^2}{\partial \nu^2} w_E - \frac{1}{8}\mu^2 \frac{\partial^2}{\partial X^2} w_E$$

$$+ \mathrm{Re}\, V\left[\frac{i}{2}\nu\frac{\partial}{\partial X} - \left(\frac{\partial}{\partial X}\right)^{-1}\frac{\partial}{\partial \mu}\right] w_E = E\, w_E. \tag{58}$$

The positive marginal distribution (eigendistribution) satisfies this eigenvalue equation and also the equation

$$-\mu \frac{\partial}{\partial \nu} w_E = 2 \operatorname{Im} V \left[\frac{i}{2} \nu \frac{\partial}{\partial X} - \left(\frac{\partial}{\partial X} \right)^{-1} \frac{\partial}{\partial \mu} \right] w_E. \qquad (59)$$

Equation 59) follows from the evolution equation 49) for the marginal distribution of the quantum system (see Ref. [10]), if the marginal distribution does not depend on time. Thus, the normalized marginal distributions of stationary states of quantum systems satisfy the system of two equations 58) and 59).

We consider an example of the quantum harmonic oscillator since it is one of the most important quantum systems. For this case, using the Hamiltonian

$$H = \frac{\hat{p}^2}{2} + \frac{\hat{q}^2}{2}, \qquad (60)$$

we reduce Eq. 49) to the following one (in view of Ref. [10]):

$$\dot{w} - \mu \frac{\partial}{\partial \nu} w + \nu \frac{\partial}{\partial \mu} w = 0. \qquad (61)$$

The marginal distribution of the oscillator's ground state is

$$w_0^{(\mathrm{os})}(X, \mu, \nu) = \frac{1}{\sqrt{\pi (\mu^2 + \nu^2)}} \exp\left[-\frac{X^2}{\mu^2 + \nu^2} \right]. \qquad (62)$$

The marginal distribution must be consistent with the uncertainty relation, i.e.,

$$\left[\int w(X, 1, 0) X^2 \, dX - \left\{ \int w(X, 1, 0) X \, dX \right\}^2 \right]$$
$$\times \left[\int w(X, 0, 1) X^2 \, dX - \left\{ \int w(X, 0, 1) X \, dX \right\}^2 \right] \geq \frac{1}{4}. \qquad (63)$$

V PROPAGATOR

In Ref. [11], the classical transition-probability density from an initial position X' measured at time $t = 0$ in the reference frame in the classical phase space labeled by the parameters μ'; ν' to the position X measured at time t in the reference frame in the classical phase space labeled by the parameters μ; ν was introduced. This classical transition-probability density is the propagator for the evolution equation 49) for the marginal distribution and the propagator is the kernel of the integral relation

$$w(X, \mu, \nu, t) = \int \Pi(X, \mu, \nu, X', \mu', \nu', t) \, w(X', \mu', \nu', 0) \, dX' \, d\mu' \, d\nu'. \qquad (64)$$

The classical propagator has a specific feature, it takes into account that the transition probability is considered in different references frames in the phase space. In view of this fact, parameters of reference frames μ and ν are present in the evolution equation. Due to this, the equation for the propagator slightly differs from the Smoluchowski–Chapman–Kolmogorov equation elaborated in the classical probability theory.

The classical propagator can be related to a quantum propagator (the Green function) for the density matrix $\rho(X, X', t)$ in the coordinate representation. For a pure state with the wave function $\Psi(X, t)$, we have

$$\rho(X, X', t) = \Psi(X, t)\Psi^*(X', t). \tag{65}$$

The Green function of the Schrödinger equation $G(X, X', t)$ connects the wave functions at the initial time moment $t = 0$ and at time t

$$\Psi(X, t) = \int G(X, X', t)\Psi(X')\,dX'. \tag{66}$$

We have for the density matrix 65), in view of relation 66), the following expression:

$$\rho(X, X', t) = \int K(X, X', Y, Y', t)\rho(Y, Y', t=0)\,dY\,dY', \tag{67}$$

where the propagator $K(X, X', Y, Y', t)$ is expressed in terms of the Green function (for unitary evolution)

$$K(X, X', Y, Y', t) = G(X, Y, t)G^*(X', Y', t). \tag{68}$$

Since the relation of the density matrix to the marginal distribution is known for any time t (given by 65) and 66)), it is possible to obtain

$$K(X, X', Z, Z', t) = \frac{1}{(2\pi)^2}\int \frac{1}{|\nu'|}\exp\left\{i\left(Y - \mu\frac{X+X'}{2}\right)\right.$$
$$\left. - i\frac{Z-Z'}{\nu'}Y' + i\frac{Z^2 - Z'^2}{2\nu'}\mu'\right\}$$
$$\times \Pi(Y, \mu, X - X', Y', \mu', \nu', t)\,d\mu\,d\mu'\,dY\,dY'\,d\nu'. \tag{69}$$

Thus, given the classical propagator for the classical marginal distribution, the propagator for the density matrix is also given. Formula 69) can be converted.

Deriving formula 69) we used the relations

$$W(q, p) = \frac{1}{2\pi}\int w(X, \mu, \nu)\exp[-i(\mu q + \nu p - X)]\,d\mu\,d\nu\,dX,$$

$$\rho(X, X') = \frac{1}{2\pi}\int w(Y, \mu, X - X')\exp\left[i\left(Y - \mu\frac{X+X'}{2}\right)\right]d\mu\,dY,$$

and

$$w(X, \mu, \nu) = \frac{1}{2\pi|\nu|} \int \rho(Z, Z')$$
$$\times \exp\left[-i\frac{Z-Z'}{\nu}\left(X - \mu\frac{Z+Z'}{2}\right)\right] dZ\, dZ'. \tag{70}$$

The last formulas give some relationships between the marginal distribution $w(X, \mu, \nu)$, the Wigner function, and the density matrix in the coordinate representation.

In Ref. [6], the Wigner function was introduced in terms of the density matrix

$$W(q, p) = \int \rho\left(q + \frac{u}{2}, q - \frac{u}{2}\right) e^{-ipu}\, du, \tag{71}$$

which can be rewritten as

$$W(q, p) = \int \rho(Z, Z')\, \delta\left(Z - q - \frac{u}{2}\right) \delta\left(Z' - q + \frac{u}{2}\right) e^{-ipu}\, du\, dZ\, dZ',$$

or

$$W(q, p) = 2 \int \rho(Z, Z')\, e^{-2ip(Z-q)}\, \delta(Z' + Z - 2q)\, dZ\, dZ'. \tag{72}$$

Comparing formulas 72) and 70) one can conclude that the Wigner quasidistribution function $W(q, p)$ and the classical probability distribution $w(X, \mu, \nu)$, the latter being a positive and normalized function, are obtained using similar integral transforms of the density matrix.

The difference between the two functions is determined by the difference in the kernels of the integral transforms. In the case of the Wigner transform, the kernel reads

$$K_W(Z, Z', q, p) = 2\, e^{-2ip(Z-q)}\, \delta(Z' + Z - 2q). \tag{73}$$

In the case of the symplectic tomography transform suggested in Ref. [16], the kernel reads

$$K_M(Z, Z', X, \mu, \nu) = \frac{1}{2\pi|\nu|} \exp\left[-i\frac{Z-Z'}{\nu}\left(X - \mu - \frac{Z+Z'}{2}\right)\right]. \tag{74}$$

Due to the difference of the kernels, the Wigner function takes negative values and the marginal probability distribution is nonnegative function.

If one writes the classical propagator as a function of the initial time moment t_1 and the final time moment t_2 (i.e., $t_1 \neq 0$), relation 64) can be rewritten as

$$w(X, \mu, \nu, t_2) = \int \Pi(X, \mu, \nu, X', \mu', \nu', t_2, t_1)\, w(X', \mu', \nu', t_1)\, dX'\, d\mu'\, d\nu'. \tag{75}$$

From the physical meaning of the classical propagator, the nonlinear integral relation follows

$$\Pi\left(X, \mu, \nu, X', \mu', \nu', t_2, t_1\right) = \int \Pi\left(X, \mu, \nu, X'', \mu'', \nu'', t_2, t'\right)$$
$$\times \Pi\left(X'', \mu'', \nu'', X', \mu', \nu', t', t_1\right) dX'' d\mu'' d\nu''. \qquad (76)$$

This relation means that if the system is initially located at the point X' at time t_1 in the reference frame in the phase space labeled by the parameters μ'; ν', the probability for the system to arrive at the point X in the reference frame in the phase space labeled by the parameters μ; ν at time t_2 is equal to the probabilities to arrive at an intermediate point X'' in the reference frame in the phase space labeled by the parameters μ''; ν'' at time t' integrated over all the intermediate positions and all the intermediate reference frames.

The above integral equation 76) is an analog of the Smoluchowski–Chapman–Kolmogorov relation which in the approach introduced in Ref. [11] is generalized to the case of families of conditional probabilities if different reference frames in the phase space (parameters μ and ν) are taken into account. Also the propagator satisfies the differential equation (see Ref. [11])

$$\frac{\partial \Pi}{\partial t_2} - \mu \frac{\partial}{\partial \nu} \Pi - i \left[V\left(-\frac{1}{\partial/\partial X}\frac{\partial}{\partial \mu} - i\frac{\nu}{2}\frac{\partial}{\partial X}\right)\right.$$
$$\left. - V\left(-\frac{1}{\partial/\partial X}\frac{\partial}{\partial \mu} + i\frac{\nu}{2}\frac{\partial}{\partial X}\right)\right] \Pi$$
$$= \delta(t_2 - t_1)\delta(X - X')\delta(\mu - \mu')\delta(\nu - \nu'). \qquad (77)$$

The classical propagator satisfies the initial condition

$$\Pi(X, \mu, \nu, X', \mu', \nu', t, t) = \delta(X - X')\delta(\mu - \mu')\delta(\nu - \nu'). \qquad (78)$$

The relation that could be used to express the classical propagator in terms of the functional integral can be also written

$$\Pi(X, \mu, \nu, X', \mu', \nu', t_\mathrm{f}, t_\mathrm{in})$$
$$= \int \prod_{k=1}^{N-1} \{\Pi(X_{k+1}, \mu_{k+1}, \nu_{k+1}, X_k, \mu_k, \nu_k, t_{k+1}, t_k) dX_k d\mu_k d\nu_k\}, \qquad (79)$$

where the time interval $t_\mathrm{f} - t_\mathrm{in} = N\tau$; $t_k = t_\mathrm{in} + k\tau$; $k = 1, 2, \ldots, N$. Taking in relation 79) the limit $\tau \to 0$; $N \to \infty$, one obtains the expression for the classical propagator in terms of the functional integral.

VI QUANTUM TRANSITION PROBABILITIES

We express quantum-transition probabilities in terms of an overlap integral of classical-like marginal distributions describing the initial and final quantum states.

If the initial pure state of a quantum system is described by the marginal distribution

$$w_{\text{in}} = w_1(X, \mu, \nu) \tag{80}$$

and the final state of the quantum system is described by the marginal distribution

$$w_{\text{f}} = w_2(X, \mu, \nu), \tag{81}$$

the probability of the quantum transition P_{12} ($1 \Longrightarrow 2$) can be obtained using the known expression for the probability in terms of an overlap integral of the Wigner functions $W_1(q, p)$ and $W_2(q, p)$ of the initial and final states (see, for example, Ref. [39])

$$P_{12} = \frac{1}{2\pi} \int W_1(q, p) W_2(q, p) \, dq \, dp. \tag{82}$$

For the transition probability, one can obtain, in view of relation 82), the expression in terms of the marginal distributions

$$P_{12} = \int w_1(X, \mu, \nu) w_2(Y, -\mu, -\nu) \exp[i(X+Y)] \frac{d\mu \, d\nu \, dX \, dY}{2\pi}. \tag{83}$$

As follows from relation 83), any pure normalized quantum state is described by the marginal distribution $w_{\text{p}}(X, \mu, \nu)$, which satisfies the additional condition

$$\int w_{\text{p}}(X, \mu, \nu) w_{\text{p}}(Y, -\mu, -\nu) \exp[i(X+Y)] \frac{d\mu \, d\nu \, dX \, dY}{2\pi} = 1. \tag{84}$$

The complex wave functions, which belong to different energy levels of a quantum system, are orthogonal. This orthogonality condition is expressed in terms of the classical marginal distribution as the relation

$$\int w_n(X, \mu, \nu) w_m(Y, -\mu, -\nu) \exp[i(X+Y)] \frac{d\mu \, d\nu \, dX \, dY}{2\pi} = \delta_{mn}, \tag{85}$$

where the labels m, n correspond to the energy levels E_m, E_n. The pure states $|n\rangle$ satisfy the completeness relation

$$\sum_n |n\rangle\langle n| = \hat{1}. \tag{86}$$

This relation can be rewritten as the condition for the marginal distributions of the pure states with the energies E_n

$$\sum_n w_n(X, \mu, \nu) = w^{(\text{wn})}(X, \mu, \nu), \tag{87}$$

where the distribution $w^{(\text{wn})}$ describes the white noise,

$$w^{(\text{wn})}(X, \mu, \nu) = \int \frac{dx \, dy \, dk}{2\pi} \exp\left[ik(\mu x + \nu y) - ikx - ik^2 \frac{\mu\nu}{2}\right]. \tag{88}$$

Thus, the classical marginal distributions describing the energy levels of quantum systems are positive solutions to the system of equations 58) and 59) and these solutions satisfy the orthogonality condition 85) and the analog of the completeness relation 87). The distributions form an interesting mathematical set that differs substantially from the usual Hilbert space of states described by the normalized complex wave functions. Of course, the structure of the Hilbert space can be traced using the map, which connects the states expressed in terms of the density matrix and the states expressed in terms of the marginal distribution functions.

VII PROPAGATOR FOR SYSTEMS WITH QUADRATIC HAMILTONIANS

As an example, we consider the system with the quadratic Hermitian Hamiltonian

$$H = \frac{1}{2}(Q\,B\,Q) + C\,Q, \qquad (89)$$

where one has the vector-operator $Q = (p,\,q)$. The symmetric 2×2 matrix B and real 2-vector C depend on time. The system has linear integrals of motion (see Ref. [39,40]):

$$I(t) = \Lambda(t)\,Q + \Delta(t). \qquad (90)$$

Here the real symplectic 2×2 matrix $\Lambda(t)$ and the real vector $\Delta(t)$ satisfy the equations

$$\dot{\Lambda} = i\,\Lambda\,B\,\sigma_y, \qquad \dot{\Delta} = i\,\Lambda\,\sigma_y\,C, \qquad (91)$$

and the initial conditions

$$\Lambda(0) = 1; \qquad \Delta(0) = 0. \qquad (92)$$

As follows from relation 44) and from the property of the Wigner function (see Ref. [39]), the classical propagator is

$$\Pi(X,\,\mu,\,\nu,\,X',\,\mu',\,\nu',\,t) = \delta(X - X' + \mathcal{N}\,\Lambda^{-1}\Delta)\,\delta(\mathcal{N}' - \mathcal{N}\,\Lambda^{-1}), \qquad (93)$$

where the vectors $\mathcal{N}$ and $\mathcal{N}'$ are

$$\mathcal{N} = (\nu,\,\mu),$$
$$\mathcal{N}' = (\nu',\,\mu').$$

For the quadratic systems without linear terms $(C = 0)$, the classical propagator is

$$\Pi(X,\,\mu,\,\nu,\,X',\,\mu',\,\nu',\,t) = \delta(X - X')\,\delta(\mathcal{N}' - \mathcal{N}\,\Lambda^{-1}). \qquad (94)$$

Thus, if one knows the linear integrals of motion, i.e., the matrix $\Lambda(t)$ and the vector $\Delta(t)$, one knows the classical propagator.

For free motion with the Hamiltonian

$$H = \frac{p^2}{2}, \tag{95}$$

there are two linear invariants found in Ref. [39,40]

$$p_0(t) = p, \qquad q_0(t) = q - pt. \tag{96}$$

This means that $\Delta(t) = 0$, and the symplectic 2×2 matrix reads

$$\Lambda(t) = \begin{pmatrix} 1 & 0 \\ -t & 1 \end{pmatrix}. \tag{97}$$

Thus, we have

$$\mathcal{N}\Lambda^{-1}(t) = (\nu + \mu t, \mu). \tag{98}$$

Consequently, the classical propagator of free motion has the form

$$\Pi^{(f)}(X, \mu, \nu, X', \mu', \nu', t) = \delta(X - X')\,\delta(\nu' - \nu - \mu t)\,\delta(\mu - \mu'). \tag{99}$$

For the harmonic oscillator with the Hamiltonian 60), linear invariants are known (see Ref. [39,40]), and the matrix $\Lambda(t)$ is

$$\Lambda(t) = \begin{pmatrix} \cos t & \sin t \\ -\sin t & \cos t \end{pmatrix}. \tag{100}$$

This means that for the harmonic oscillator

$$\mathcal{N}\Lambda^{-1}(t) = (\nu \cos t - \mu \sin t,\ \nu \sin t + \mu \cos t). \tag{101}$$

Consequently, the classical propagator of the harmonic oscillator is

$$\Pi^{(\text{os})}(X, \mu, \nu, X', \mu', \nu')$$
$$= \delta(X - X')\,\delta(\nu' - \nu \cos t + \mu \sin t)\,\delta(\mu' - \nu \sin t - \mu \cos t). \tag{102}$$

VIII ENERGY LEVELS OF THE HARMONIC OSCILLATOR

The marginal distribution of the coherent state of the harmonic oscillator has the form obtained in Ref. [11]:

$$w_\alpha(X, \mu, \nu) = [\pi(\mu^2 + \nu^2)]^{-1/2} \exp\left[-|\alpha|^2 - \frac{X^2}{\mu^2 + \nu^2} + \frac{\alpha^2(\nu + i\mu)^2}{2(\mu^2 + \nu^2)}\right.$$
$$\left. + \frac{\alpha^{*2}(\nu - i\mu)^2}{2(\mu^2 + \nu^2)} - \frac{i\sqrt{2}\alpha X(\nu + i\mu)}{\mu^2 + \nu^2} + \frac{i\sqrt{2}\alpha^* X(\nu - i\mu)}{\mu^2 + \nu^2}\right]. \quad (103)$$

The eigendistribution function for the energy level of the harmonic oscillator satisfies the eigenvalue equation

$$\left\{\frac{1}{2}\left[\left(\frac{\partial}{\partial\nu}\right)^2 + \left(\frac{\partial}{\partial\mu}\right)^2\right]\left(\frac{\partial}{\partial X}\right)^{-2} - \frac{1}{8}(\mu^2 + \nu^2)\left(\frac{\partial}{\partial X}\right)^2\right\} w_E(X, \mu, \nu)$$
$$= E\, w_E(X, \mu, \nu). \quad (104)$$

This equation can be rewritten for the Fourier component of the marginal distribution

$$\tilde{w}_E(k, \mu, \nu) = \frac{1}{2\pi} \int w_E(X, \mu, \nu) \exp(-ikX)\, dX \quad (105)$$

in the form

$$\left\{-\frac{1}{2k^2}\left[\left(\frac{\partial}{\partial\nu}\right)^2 + \left(\frac{\partial}{\partial\mu}\right)^2\right] + \frac{1}{8}k^2(\mu^2 + \nu^2)\right\} \tilde{w}_E(k, \mu, \nu) = E\, \tilde{w}_E(k, \mu, \nu). \quad (106)$$

Since the marginal distribution of the stationary state of the harmonic oscillator must satisfy the stationarity condition found in Ref. [12],

$$\left(\mu\frac{\partial}{\partial\nu} - \nu\frac{\partial}{\partial\mu}\right) w_E(X, \mu, \nu) = 0, \quad (107)$$

Eq. 106) is equivalent to the equation for axially symmetric wave functions of a two-mode harmonic oscillator with mass $m = k^2$, frequency $\omega = 1/2$, and angular momentum $M = 0$. The wave function corresponding to zero angular momentum is expressed in terms of the Laguerre polynomials

$$\tilde{w}_n(k, \mu, \nu) = \frac{1}{2\pi} \exp\left[-\frac{k^2(\mu^2 + \nu^2)}{4}\right] L_n\left(\frac{k^2\mu^2 + k^2\nu^2}{2}\right). \quad (108)$$

The main quantum number n of the one-dimensional harmonic oscillator under discussion is equal to the integer radial quantum number n_r of the artificial two-mode oscillator

$$n = n_r, \qquad n_r = 0, 1, 2, \ldots. \quad (109)$$

The energy level of the artificial symmetric two-mode oscillator labeled by the radial quantum number n_r and the angular momentum M as

$$E_{n_r, M} = \omega\left(|M| + 1 + 2n_r\right) \tag{110}$$

for $\omega = 1/2$, $M = 0$, $n_r = n$ gives exactly the spectrum of the one-dimensional oscillator

$$E_n = E_{n_r, M} = n + \frac{1}{2}, \qquad n = 0, 1, 2, \ldots \tag{111}$$

To find the marginal distribution, we have to calculate

$$w_n(X, \mu, \nu) = \frac{1}{2\pi} \int \exp\left[-\frac{k^2(\mu^2 + \nu^2)}{4} + ikX\right] L_n\left(\frac{k^2\mu^2 + k^2\nu^2}{2}\right) dk. \tag{112}$$

In view of the integral

$$\frac{1}{2\pi} \int_{-\infty}^{\infty} \exp\left(-\frac{k^2}{4} + ikX\right) L_n\left(\frac{k^2}{2}\right) dk$$
$$= \pi^{-1/2} \, 2^{-n} (n!)^{-1} \exp(-X^2) H_n^2(X), \tag{113}$$

one obtains the marginal distribution

$$w_n(X, \mu, \nu) = [\pi(\mu^2 + \nu^2)]^{-1/2} \, 2^{-n} (n!)^{-1}$$
$$\times \exp\left(-\frac{X^2}{\mu^2 + \nu^2}\right) H_n^2\left(\frac{X}{\sqrt{\mu^2 + \nu^2}}\right), \tag{114}$$

It is worth noting that the normalization condition for the marginal distribution $w_n(X, \mu, \nu)$ implies the condition for the Fourier component

$$\int \widetilde{w}_n(k, \mu, \nu) \exp(ikX) \, dk \, dX = 2\pi \widetilde{w}_n(k = 0, \mu, \nu) = 1. \tag{115}$$

We take solutions 108) without using the normalization condition in terms of the variables μ and ν of the artificial two-mode oscillator, but using the normalization condition of the marginal distribution in terms of the variable X and the corresponding property of its Fourier component.

IX TOMOGRAPHY OF SPIN STATES

Let us introduce the probability distribution for the spin projection in a given direction considered in a rotated reference frame. For arbitrary values of spin, let the spin state have the density matrix

$$\rho_{mm'}^{(j)} = \langle jm \mid \hat{\rho}^{(j)} \mid jm' \rangle, \qquad m = -j, -j+1, \ldots, j-1, j, \tag{116}$$

where

$$\hat{j}_3 \mid jm\rangle = m \mid jm\rangle,$$

$$\hat{j}^2 \mid jm\rangle = j(j+1) \mid jm\rangle, \qquad (117)$$

and

$$\hat{\rho}^{(j)} = \sum_{m=-j}^{j} \sum_{m'=-j}^{j} \rho_{mm'}^{(j)} \mid jm\rangle\langle jm' \mid . \qquad (118)$$

The operator $\rho^{(j)}$ is the density operator of the state under discussion. The diagonal elements of the density matrix determine the positive probability distribution

$$\rho_{mm}^{(j)} = w_0(m), \qquad (119)$$

which is normalized,

$$\sum_{m=-j}^{j} w_0(m) = 1. \qquad (120)$$

In Refs. [41–43], a general group construction of tomographic schemes was discussed, and this scheme was also used for spin tomography in Refs. [22,23]. The idea is to consider the diagonal elements of the density matrix in another reference frame. The density matrix in another reference frame reads

$$\rho_{m_1 m_2}^{(j)} = \left(\mathcal{D}\rho\mathcal{D}^\dagger\right)_{m_1 m_2}. \qquad (121)$$

Here the unitary rotation transform $\mathcal{D}$ depends on the Euler angles α, β, γ and, by definition, the diagonal matrix elements of the density matrix yield the positive normalized probability distribution. For the diagonal elements of the density matrix 121),

$$m_1 = m_2.$$

We introduce new notation and rewrite equality 121) for $m_1 = m_2$ in the form

$$\tilde{w}(m_1, \alpha, \beta, \gamma) = \sum_{m_1'=-j}^{j} \sum_{m_2'=-j}^{j} D_{m_1 m_1'}^{(j)}(\alpha, \beta, \gamma) \, \rho_{m_1' m_2'}^{(j)} \, D_{m_1 m_2'}^{(j)*}(\alpha, \beta, \gamma). \qquad (122)$$

Here the matrix elements $D_{m_1 m_1'}^{(j)}(\alpha, \beta, \gamma)$ (the Wigner function) are the matrix elements of the rotation-group representation

$$D_{m'm}^{(j)}(\alpha, \beta, \gamma) = e^{im'\gamma} d_{m'm}^{(j)}(\beta) \, e^{im\alpha}, \qquad (123)$$

where

$$d^{(j)}_{m'm}(\beta) = \left[\frac{(j+m')!(j-m')!}{(j+m)!(j-m)!}\right]^{1/2} \left(\cos\frac{\beta}{2}\right)^{m'+m} \left(\sin\frac{\beta}{2}\right)^{m'-m}$$
$$\times P^{(m'-m,m'+m)}_{j-m'}(\cos\beta), \qquad (124)$$

and $P^{(a,b)}_n(x)$ is the Jacobi polynomial.

Since
$$D^{(j)*}_{m'm}(\alpha,\beta,\gamma) = (-1)^{m'-m} D^{(j)}_{-m'-m}(\alpha,\beta,\gamma), \qquad (125)$$

the marginal distribution depends only on two angles, α and β.

Thus, let us denote
$$w(m_1,\alpha,\beta) = \tilde{w}(m_1,\alpha,\beta,\gamma), \qquad (126)$$

which satisfies
$$\sum_{m_1=-j}^{j} w(m_1,\alpha,\beta) = 1. \qquad (127)$$

For a spin-1/2 state with spin projection $+1/2$ and wave function
$$\psi_{+1/2} = \begin{pmatrix} 1 \\ 0 \end{pmatrix},$$

or with density matrix
$$\rho_+ = \begin{pmatrix} 1 & 0 \\ 0 & 0 \end{pmatrix},$$

the marginal distribution is equal to
$$w\left(\frac{1}{2},\alpha,\beta\right) = \cos^2\frac{\beta}{2} \quad \text{for} \quad m_1 = +\frac{1}{2}, \qquad (128)$$

and, correspondingly,
$$w\left(-\frac{1}{2},\alpha,\beta\right) = \sin^2\frac{\beta}{2} \quad \text{for} \quad m_1 = -\frac{1}{2}. \qquad (129)$$

In Ref. [23], by using the properties of the Wigner function and the Clebsch–Gordan coefficients, formula 122) was inverted and the density matrix was expressed in terms of the marginal distribution

$$(-1)^{m'_2} \sum_{j_3=0}^{2j} \sum_{m_3=-j_3}^{j_3} (2j_3+1)^2 \sum_{m_1=-j}^{j} \int (-1)^{m_1} w(m_1,\alpha,\beta) D^{(j_3)}_{0m_3}(\alpha,\beta,\gamma)$$
$$\times \begin{pmatrix} j & j & j_3 \\ m_1 & -m_1 & 0 \end{pmatrix} \begin{pmatrix} j & j & j_3 \\ m'_1 & -m'_2 & m_3 \end{pmatrix} \frac{d\omega}{8\pi^2} = \rho^{(j)}_{m'_1 m'_2}. \qquad (130)$$

Here $m, m' = -j, -j+1, \ldots, j$ and one integrates over rotation parameters α, β, γ.
To derive formula 130), we used the known property of the Wigner function:

$$\int D^{(j_1)}_{m'_1 m_1}(\omega) D^{(j_2)}_{m'_2 m_2}(\omega) D^{(j_3)}_{m'_3 m_3}(\omega) \frac{d\omega}{8\pi^2} = \begin{pmatrix} j_1 & j_2 & j_3 \\ m'_1 & m'_2 & m'_3 \end{pmatrix}$$
$$\times \begin{pmatrix} j_1 & j_2 & j_3 \\ m_1 & m_2 & m_3 \end{pmatrix}, \quad (131)$$

where

$$\int d\omega = \int_0^{2\pi} d\alpha \int_0^{\pi} \sin\beta \, d\beta \int_0^{2\pi} d\gamma, \quad (132)$$

along with the orthogonality property of $3j$-symbols:

$$(2j+1) \sum_{m_1=-j_1}^{j_1} \sum_{m_2=-j_2}^{j_2} \begin{pmatrix} j_1 & j_2 & j \\ m_1 & m_2 & -m \end{pmatrix} \begin{pmatrix} j_1 & j_2 & j' \\ m_1 & m_2 & -m' \end{pmatrix} = \delta_{jj'} \delta_{mm'},$$

$$\sum_{j=|j_1-j_2|}^{j_1+j_2} \sum_{m=-j}^{j} (2j+1) \begin{pmatrix} j_1 & j_2 & j \\ m_1 & m_2 & -m \end{pmatrix} \begin{pmatrix} j_1 & j_2 & j \\ m'_1 & m'_2 & -m \end{pmatrix} = \delta_{m_1 m'_1} \delta_{m_2 m'_2}.$$

Formula 130), being the inverse of 122), is an analog of the Radon transform for spin states. *Given a measurable marginal distribution for arbitrary spin, one can reconstruct the state density matrix by means of this relation.*

The results obtained enable one to measure the spin state by measuring a spin projection on a given axis. One obtains the experimental probability-distribution function which depends on two angles determining the axis. Using the relationship between the probability distribution and the state density matrix, one reconstructs all the information about the quantum spin state. This means that the probability distribution can be used instead of complex spinors and density matrices for the spin-state description since it contains complete (even overcomplete) information on the state.

X QUANTUM MEASUREMENTS AND COLLAPSE OF WAVE FUNCTION

We will review the discussion of quantum measurements done in Refs. [10,11,14]. It is known (see, for example, Refs. [44,45]) that quantum mechanics is problematic in the sense that it is incomplete and needs the notion of a classical device measuring quantum observables as an important ingredient of the theory. Due to this, one accepts that there exist two worlds: the classical one and the quantum one. In the classical world, the measurements of classical observables are produced by classical devices. In the framework of standard theory, in the quantum world the measurements of quantum observables are produced by classical devices, too.

Due to this, the theory of quantum measurements is considered as something very specifically different from classical measurements.

It is psycologically accepted that to understand the physical meaning of a measurement in the classical world is much easier than to understand the physical meaning of analogous measurement in the quantum world.

As was pointed out in Refs. [10,11], all the roots of the difficulties of quantum measurements are present in classical measurements, as well. Using the relations of the quantum states in the standard representation and in the classical one (described by classical distributions), one can conclude that complete information on a quantum state is obtained from purely classical measurements of the position of a particle made by classical devices in each reference frame of an ensemble of classical reference frames, which are scaled and rotated in the classical phase space.

These measurements do not need any quantum language if we know how to produce, in the classical world (using the notion of classical position and momentum), reference frames in the classical phase space differing from each other by rotation and scaling of the axis of the reference frame and how to measure only the position of the particle from the viewpoint of these different reference frames. So, knowing how to obtain the classical marginal distribution function $w(X, \mu, \nu)$ which depends on the parameters μ; ν, labeling each reference frame in the classical phase space, we reconstruct the quantum density operator.

Thus, we avoid the paradox of the quantum world which requires for its explanation measurements by a classical apparatus accepted in the framework of standard treatment of quantum mechanics. But the difficulties of the quantum approach are present, since we need to understand better the procedure of measurement in a rotated reference frame in the phase space of the classical system. The problem of wave function collapse [44,45] reduces to the problem of a reduction of the probability distribution which occurs as soon as we "pick" a classical value of the classical random observable in the classical framework of [10,11]. This means that we "solved" the paradox of the wave function collapse reducing it to the problem of standard measurement of a classical random variable used in the probability theory.

The approach developed in [10,11] enables one to transform such an unpleasant problem of standard quantum mechanics as the need of a classical device and the reduction of wave packets into the standard problem of classical measurements of classical random variables. In fact, this means that the problem of classical measurements is as difficult as the problem of quantum measurements. An important analogy with methodology of special relativity arises: It turns out that it is necessary to introduce a consideration of events in the set of moving reference frames in space–time in order to explain relativistic effects, and it is necessary to introduce a consideration of events in the set of rotated and scaled reference frames in the phase space in order to explain the nonrelativistic quantum mechanics in terms of only classical concepts of classical fluctuation theory. But these reference frames are the reference frames in the phase space (not in space–time). Possibly, a combination of these two approaches can be generalized to give a classical description of relativistic quantum mechanics.

One can conclude that the stationary states of quantum systems (for example, of a harmonic oscillator) can be obtained using classical-like alternative equations to the Schrödinger equations. A new type of eigenvalue problems for real positive marginal distributions is formulated. The analogs of orthogonality and completeness relations for the wave functions are formulated in terms of conditions for the marginal distributions as well as the transition probabilities among the energy levels. The criterion for determining the pure states of the quantum system is given in terms of the classical marginal distribution.

Thus, using the marginal distribution one can formulate the standard quantum mechanics without the complex wave function and density matrix. But the position distributions in an ensemble of classical reference frames in the phase space play an important role.

It should be pointed out that in the standard formulation of quantum mechanics there exist different representations such as the coordinate representation, the momentum representation, etc. The counterpart of this variety in the classical formulation is related to different tomography schemes like optical tomography [19,37], symplectic tomography [11,16], and photon number tomography [41,46,47]. The photon number tomography uses the marginal distribution of a discrete variable, which corresponds to a number representation. Just as different representations in the standard formulation of quantum mechanics are related by some transformations in the Hilbert space of states, the marginal distributions of different tomography schemes can be transformed into each other. This transformation consists of two steps. First, one makes a map of the marginal distribution (in one of the tomography schemes) onto the Wigner function and then one makes another map (of the different tomography scheme) of this Wigner function onto the corresponding marginal distribution.

The construction introduced in spirit is similar to the Moyal approach [48] which considers quantum mechanics as a statistical theory. But in Ref. [48], the quantum state was described by a quasidistribution function in the phase space that is identical to the Wigner function. Thus, the "negative probabilities" to find the system in some domain in the phase space is an unavoidable feature of the Moyal approach. The density matrix was introduced in Ref. [49,50]. In the framework of the new formulation of quantum mechanics, the density matrix is not mandatory to be used.

In the introduced formulation of quantum mechanics, only positive probabilities of the measurable position in an ensemble of reference frames in the phase space of the system is used. It is remarkable that in the positive probability representation the states in quantum mechanics are described identically with the states in classical statistical mechanics if one uses the positive marginal distribution $w(X, \mu, \nu)$ (though the sets of the distributions in the classical and quantum cases are different).

The difference between classical statistical mechanics and quantum mechanics in the formulation introduced appears in the dynamics of the marginal distributions, since in quantum mechanics the evolution equation for the positive probability distribution has a different form from that in the classical case. It is remarkable that the relations of the propagators (conditional probabilities) in the phase space representation and in the probability representation are described by the same formula both in classical statistical mechanics and in quantum mechanics.

For linear systems (oscillators), the propagators in classical stastistical mechanics and in quantum mechanics coincide. The difference for these systems in the quantum and classical cases is due to the fact that not all positive probability distributions $w(X, \mu, \nu)$ are realized for classical systems. Also not all marginal distributions are admissible in the quantum case, but only which satisfy uncertainty relations.

We have demonstrated that spin states and states of a trapped ion [51] can be described by measurable positive probability distributions. This implies that quantum-mechanical systems can be considered in the framework of the same formalism of probability theory as classical statistical systems. Thus, the known results of classical probability theory can be applied to the study of quantum states. For example, the central limit theorem can be used for describing multimode systems. The approach developed can be elaborated for solving many problems of quantum optics [31,52,53] and quantum computing. It is important to study the Schrödinger uncertainty relation [54] in the framework of the new approach. Linear integrals of motion for quadratic systems [55,56] are useful to obtain the propagator of the new evolution equation for the marginal distribution [57]. The new approach can be also applied to study nonlinear coherent states [58–60].

We have shown that quantum mechanics can be formulated without wave function and density matrix using the tomographic probability representation. The general approach to the tomographic map and relations among different tomography schemes are discussed in Refs. [61,62]. The tomography of spin states for two particles is described in Ref. [63]. A review of the new approach to quantum mechanics is given in Ref. [64]. The generalization of the metod to the case of the field theory one can find in Ref. [65]. One can conclude that the problem of formulation of quantum theory using only probabilities both for continuous and discrete observables has the solution in the framework of the tomographic probability representation of quantum mechanics.

ACKNOWLEDGMENTS

The author thanks the Organizers of the XXXI Latin-American School of Physics for invitation to give a course of lectures and El Colegio Nacional for kind hospitality.

This study has been partially supported by the Russian Foundation for Basic Research under Project No. 17222.

REFERENCES

1. Schrödinger, E., *Ann. d. Physik* (Leipzig) **79**, 489 (1926).
2. De Broglie, L., *Compt. rend.* **183**, 447 (1926); **184**, 273 (1927); **185**, 380 (1927).
3. Madelung, E., *Zeits. f. Physik* **40**, 332 (1926).
4. Bohm, D., *Phys. Rev.* **85**, 166; 180 (1952).
5. Terletskii, Ya. A., and Gusev, A. A., (Eds.), *Problems of Causality in Quantum Mechanics* (Inostrannaya Literatura, Moscow, 1959) [in Russian].
6. Wigner, E., *Phys. Rev.* **40**, 749 (1932).
7. Husimi, K., *Proc. Phys. Math. Soc. Jpn.* **23**, 264 (1940).
8. Glauber, R. J., *Phys. Rev. Lett.* **10**, 84 (1963).
9. Sudarshan, E. C. G., *Phys. Rev. Lett.* **10**, 277 (1963).
10. Mancini, S., Man'ko, V. I., and Tombesi, P., *Phys. Lett. A* **213**, 1 (1996).
11. Mancini, S., Man'ko, V. I., and Tombesi, P., *Found. Phys.* **27**, 801 (1997).
12. Man'ko, V. I., "Energy levels of a harmonic oscillator in the classical-like formulation of quantum mechanics," in Dremin, I. M., and Semikhatov, A. M., (Eds.), *Proceedings of the Second International A. D. Sakharov Conference on Physics*, Moscow, May 1996 (World Scientific, Singapore, 1997), p. 486; "Optical symplectic tomography and classical probability instead of wave function in quantum mechanics," in Doebner, H.-D., Scherer, W., and Schultz, C., (Eds.), *GROUP21. Physical Applications and Mathematical Aspects of Geometry, Groups, and Algebras*, Goslar, Germany, June–July 1996 (World Scientific, Singapore, 1997), Vol. 2, p. 764; "Transition probabilities between energy levels in the framework of the classical approach," invited lecture presented at the Inaguration Conference of APCTP, Seoul, June 1996 (to appear in the Proceedings of the Conference, World Scientific, Singapore, 1998); "Classical description of quantum states and tomography," in Han, D., Janszky, J., Kim, Y. S., and Man'ko, V. I., (Eds.), *Fifth International Conference on Squeezed States and Uncertainty Relations*, Balatonfüred, Hungary, May 1997 (NASA Conference Publication, Goddard Space Flight Center, Greenbelt, Maryland, 1998), Vol. NASA/CP-1998-206855, p. 523.
13. Man'ko, V. I., *J. Russ. Laser Research* (Plenum Press) **17**, 579 (1996).
14. Man'ko, V. I., "Quantum mechanics and classical probability theory," in Gruber, B., and Ramek, M., (Eds.), *Symmetries in Science IX* (Plenum Press, New York, 1997), p. 215.
15. Man'ko, Olga, "Tomography of spin states and classical formulation of quantum mechanics," in Gruber, B., and Ramek, M., (Eds.), *Symmetries in Science X* (Plenum Press, New York, 1998, to appear).
16. Mancini, S., Man'ko, V. I., and Tombesi, P., *Quantum Semiclass. Opt.* **7**, 615 (1995).
17. D'Ariano, G. M., Mancini, S., Man'ko, V. I., and Tombesi, P., *Quantum Semiclass. Opt.* **8**, 1017 (1996).
18. Bertrand, J., and Bertrand, P., *Found. Phys.* **17**, 397 (1987).
19. Vogel, K., and Risken, H., *Phys. Rev. A* **40**, 2847 (1989).
20. Man'ko, V. I., and Safonov, S. S., *Teor. Mat. Fiz.* **112** 1172 (1997).
21. Man'ko, V. I., and Safonov, S. S., *Phys. Atom. Nucl.* **61**, 585 (1998).
22. Dodonov, V. V., and Man'ko, V. I., *Phys. Lett. A* **229**, 335 (1997).

23. Man'ko, V. I., and Man'ko, O. V., *JETP* **85**, 430 (1997).
24. Leonhardt, U., *Phys. Rev. A* **53**, 2998 (1996).
25. Dodonov, V. V., Malkin, I. A., and Man'ko, V. I., *Physica* **72**, 597 (1974).
26. Man'ko, O. V., "Symplectic tomography of nonclassical states of a trapped ion," *Preprint* IC/96/39 (ICTP, Trieste, 1996); Los Alamos Report No. quant-ph/9604018; *J. Russ. Laser Research* (Plenum Press) **17**, 439 (1996); "Symplectic tomography of Schrödinger cat states of a trapped ion," in Ferrero, M., and van der Merwe, A., (Eds.), *Proceedings of the Second International Symposium on Fundamental Problems in Quantum Mechanics*, Oviedo, Spain, July 1996 (Kluwer Academic Press,1997), p. 225; "Symplectic tomography of nonlinear Schrödinger cats of a trapped ion," in Han, D.,Janszky, J., Kim, Y. S., and Man'ko, V. I., (Eds.), *Fifth International Conference on Squeezed States and Uncertainty Relations*, Balatonfüred, Hungary, May 1997 (NASA Conference Publication, Goddard Space Flight Center, Greenbelt, Maryland, 1998), Vol. NASA/CP-1998-206855, p. 309.
27. de Matos Filho, R. L., and Vogel, W., *Phys. Rev. Lett.* **76**, 608 (1996).
28. Nieto, M. M., *Phys. Lett. A* **219**, 180 (1996).
29. Meekhof, D. M., Monroe, G., King, B. E., Itano, W. M., and Wineland, D., *Phys. Rev. Lett.* **76**, 1796 (1996).
30. Haroche, S., *Nuovo Cimento B* **110**, 545 (1995).
31. Man'ko, V. I., "Introduction to quantum optics," in: Castanõs, O., Lopez-Peña, R., Hirsh, J. G., and Wolf, K.-B., (Eds.), *Latin-American School of Physics, XXX ELAF, Group Theory and Its Applications*, Mexico, July–August 1995, *AIP Conference Proceedings* **365** (AIP, New York, 1996), p. 337.
32. Man'ko, Olga, and Man'ko, V. I., *J. Russ. Laser Research* (Plenum Press) **18**, 407 (1997).
33. Fedele, R., and Man'ko, V. I., *Phys. Rev. E* **58**, 992 (1998).
34. Cahill, K. E., and Glauber, R. J., *Phys. Rev.* **177**, 1882 (1969).
35. Wünsche, A., *J. Mod. Opt.* **44**, 2293 (1997).
36. Man'ko, V. I., and Mendes, R. V., "Noncommutative time–frequency tomography of analytic sugnals," LANL Physics/9712022 Data Analysis, Statistics, and Probability, *IEEE Signal Processing* (submitted, 1998).
37. Smithey, D. T., Beck, M., Raymer, M. G., and Faridani, A., *Phys. Rev. Lett.* **70**, 1244 (1993).
38. Dodonov, V. V., Man'ko, O. V., Man'ko, V. I., and Wünsche, A., "Energy-sensitive and 'classical-like' distances between quantum states," *Phys. Scr.* (to appear , 1998).
39. Dodonov, V. V., and Man'ko, V. I., *Invariants and Evolution of Nonstationary Quantum Systems, Proceedings of the Lebedev Physical Institute* (Nova Science, New York, 1989), Vol. 183.
40. Malkin, I. A., and Man'ko, V. I., *Dynamical Symmetries and Coherent States of Quantum Systems* (Nauka, Moscow, 1979) [in Russian].
41. Mancini, S., Man'ko, V. I., and Tombesi, P., *Europhys. Lett.* **37**, 79 (1997).
42. Mancini, S., Man'ko, V. I., and Tombesi, P., *J. Mod. Opt.* **44**, 2281 (1997).
43. Mancini, S., Man'ko, V. I., and Tombesi, P., *Quantum Semiclass. Opt.* **9**, 987 (1997).
44. Wheeler, J. A., and Zurek, W. H., (Eds.), *Quantum Theory and Measurement* (Princeton University Press, 1983).

45. Bell, J. S., *Speakable and Unspeaskable in Quantum Mechanics* (Cambridge University Press, 1987).
46. Wallentowitz, S., and Vogel, W., *Phys. Rev. A* **53**, 4528 (1996).
47. Banaszek, K., and Wodkiewicz, K., *Phys. Rev. Lett.* **76**, 4344 (1996).
48. Moyal, J. E., *Proc. Cambridge Philos. Soc.* **45**, 99 (1949).
49. Landau, L. D., *Z. Physik*, **45**, 430 (1927).
50. von Neumann, J., *Mathematische Grundlagen der Quantenmechanik* (Springer, Berlin, 1932).
51. Man'ko, O. V., *Phys. Lett. A*, **228**, 29 (1997); *SPIE* , (1997).
52. Man'ko, V. I., and Wünsche, A., *Quantum Semiclass. Opt.* **9**, 381 (1997).
53. Schrade, G., Man'ko, V. I., Schleich, W. P., and Glauber, R. J., *Quantum Semiclass. Opt.* **7**, 307 (1995).
54. Schrödinger, E., *Sitzungsber. Preuss. Acad. Wiss.* **24**, 296 (1930).
55. Malkin, I. A., and Man'ko, V. I., *Phys. Lett. A* **32**, 243 (1970).
56. Malkin, I. A., Man'ko, V. I., and Trifonov, D. A., *Phys. Rev. D* **2**, 1371 (1970).
57. Man'ko, V. I., Rosa, L., and Vitale. P., *Phys. Rev A* **58**, 3291 (1998).
58. Man'ko, V. I., Marmo, G., Sudarshan, E. C. G., and Zaccaria, F., "f–Oscillators," in Atakishiyev, N. M., Seligman, T. H., and Wolf, K.-B., (Eds.), *Proceedings of the Fourth Wigner Sysmposium*, Guadalajara, Mexico, July 1995 (World Scientific, Singapore, 1996), p. 421.
59. Man'ko, V. I., Marmo, G., Sudarshan, E. C. G., and Zaccaria, F., *Phys. Scr.* **55**, 528 (1997).
60. de Matos Filho, R. L., and Vogel, W., *Phys. Rev. A* **54**, 4560 (1996).
61. Mancini, S., Man'ko, V. I., and Tombesi, P., *J. Mod. Opt.* **44**, 2281 (1997).
62. Mancini, S., Man'ko, V. I., and Tombesi, P., *Quantum Semiclass. Opt.* **9**, 987 (1997).
63. Andreev, V. A., and Man'ko, V. I., *JETP* (1998).
64. Andreev, V. A., Man'ko, O. V., Man'ko, V. I., and Safonov, S. S., *J. Russ. Laser Research* (Plenum Press) **19**, No. 4 (1998).
65. Man'ko, V. I., Rosa, L., and Vitale, P., *Phys. Lett.* B (1998, to appear).

A Tutorial on Quantum Distribution Functions for Spin-$\frac{1}{2}$ Systems and Einstein-Podolsky-Rosen Correlations*

Marlan O. Scully and Hwang Lee

Institute for Quantum Studies and Department of Physics, Texas A & M University, College Station, Texas 77843-4242, U. S. A.
Max-Planck-Institut für Quantenoptik, 85748 Garching, Germany

and

Eduardo Gómez and Roberto Ortega-Martínez

Centro de Instrumentos, Universidad Nacional Autónoma de México, AP 70-186 Cd. Universitaria, 04510 D.F., México

Abstract. In earlier paper entitled *How to make quantum mechanics look like a hidden-variable theory and vice versa*, Phys. Rev. D **28**, 2477 (1983) [1], we developed a quantum distribution function analysis for spin-$\frac{1}{2}$ systems, which had many of the characteristics of a hidden variable theory but was rigorously quantum mechanical. In the same paper, we also presented a hidden variable theory which was modeled after the Wigner like quantum distribution analysis and which agreed with the predictions of Einstein-Podolsky-Rosen (EPR) experiments but was non-local. These considerations are useful in understanding various subtle aspects of locality and reality in EPR type experiments. This has not been widely appreciated by students of philosophy and of the foundations of quantum mechanics. This is at least in part no doubt because the mathematical underpinnings are of a somewhat different nature than those usually associated with quantum mechanics. That is, the elements of quantum distribution theory are not the usual material of quantum mechanics course. The main goal of the present notes is therefore to provide a tutorial on quantum distribution theory as applied to spin one half particles. To that end, the original Phys. Rev. paper is included, as are our extensions of that paper. The technical variations, which are naturally included in an abbreviated form in the Phys. Rev. paper, are here presented as a series of worked problems.

---

*) Dedicated to Prof. Marcos Moshinsky who has given us many insights into quantum mechanics.

I INTRODUCTION

Quantum Mechanics, like so much in life, is in one way easy another way hard. It is easy to get used to the basic facts of quantum life. Complementarity (e.g. wave-particle duality) is one such fact. The Schrödinger equation and the wonderful agreement between its predictions and experiments in atomic and condensed matter physics is another. It is easy to see that quantum mechanics works and will, in one form or another, always be with us.

On the other hand, it is hard to know exactly how we should think about some of the deeper interpretational issues such as the Einstein-Podolsky-Rosen (EPR) problem, Bell's theorem, and the problems of locality and reality in quantum mechanics. Such is the focus of the present school and of these notes. In particular, we would like to make the case for a local interpretation of the EPR correlations by appealing to quantum distribution theory. These considerations derive from earlier work of Belinfante [2] in which he was trying to cook up a hidden variable theory which would give insight into Bell's inequalities. He was not able to come up with a hidden variable theory which actually reproduced EPR two-particle correlations, and concluded that his approach, involving the angle of orientation of a spin-$\frac{1}{2}$ system relative to some quantization axis, did not have a basis in quantum mechanics. In his words, "the angle α does not even exist in quantum mechanics". It was shown in the enclosed 1983 Physical Review paper that it is possible to use distribution theory to extend Belinfante's ideas to provide a hidden-variable theory which is in complete agreement with the two particle correlation functions along the lines that he wanted to see.

At the same time, it was discovered that a formulation of the EPR problem in terms of Wigner-like quantum distribution theory [3–5], was interesting and cast quantum mechanics in a form making it look like a hidden variable theory. The basic point being that the hidden variable theory turned out to be non-local. On the other hand, the corresponding quantum mechanical distribution turned out to involve local physics but negative "probabilities" and thus the reality of various "physical" quantities has to be reassessed.

These considerations have been enjoyed by many of us, but have not been widely appreciated by students of the foundations of quantum mechanics, partly because the mathematical underpinnings are of a different type than those usually taught in quantum mechanics courses. That is, the elements of quantum distribution theory as we apply it so frequently in, for example quantum optics, are not part what we teach in a two semester course in quantum mechanics. It is, therefore, the purpose of these notes to provide the technical details (in problem form) associated with the 1983 paper and related extensions.

Thus main goal of the present lecture notes is to provide a tutorial on the quantum distribution theory for spin-$\frac{1}{2}$ systems and by so doing to show that quantum mechanics is in fact closer to a hidden variable theory than has been generally appreciated.

In the next two sections we sketch the results of the 1983 paper. And in Sec. IV we extend these considerations in such a way as to make quantum mechanics look nonlocal. But we argue that it is still best understood as local physics involving long range correlations and information transfer. The 1983 Physical Review paper is reprinted[1] in Section V. Detailed technical and computational aspects of this work are presented in section VI.

II QUANTUM DISTRIBUTION FUNCTIONS FOR SPIN-$\frac{1}{2}$ SYSTEMS

The measurement of one member of a EPR spin singlet can be thought of as preparing the state of the other particle which may be a long distance away as per Fig. 1. Thus, we may write the passage probability of particle 1 through its Stern-Gerlach Apparatus (SGA) given that we have prepared the state of particle 1 by the (distant) measurement on particle 2 as

$$\mathcal{P}(\theta_1, \uparrow_2) = |\langle \theta_1 | \Psi_1(\uparrow_2) \rangle|^2, \tag{1a}$$

$$|\Psi_1(\uparrow_2)\rangle = \langle \uparrow_2 | \Psi_{1,2} \rangle. \tag{1b}$$

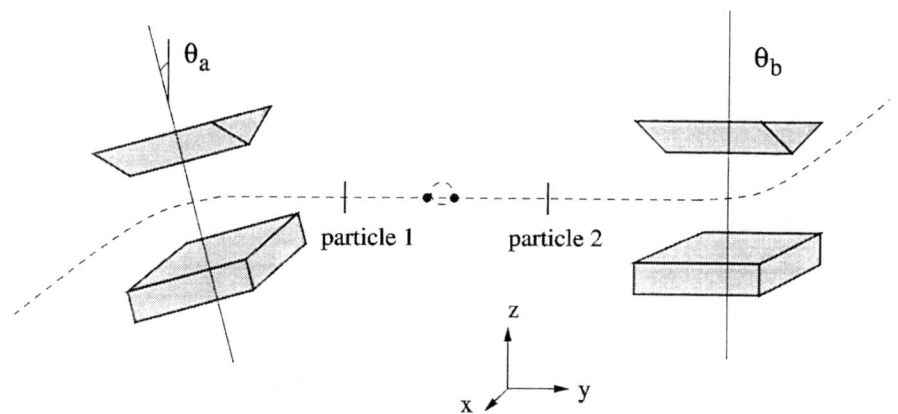

FIGURE 1. Schematic of EPR gedanken experiment. A spin-zero system is split such that the two spin-$\frac{1}{2}$ particles proceed in the opposite directions.

It is often argued that this "long-distant" preparation demonstrates the nonlocal nature of quantum mechanics.

---

[1]) With permission from AIP.

But what if we are in a moving frame such that particle 1 goes through its SGA before particle 2 as viewed in the moving frame? Then we should write

$$\mathcal{P}(\theta_1, \uparrow_2) = |\langle \uparrow_2 | \Psi_2(\theta_1) \rangle|^2, \tag{2a}$$

where

$$|\Psi_2(\theta_1)\rangle = \langle \theta_1 | \Psi_{1,2} \rangle, \tag{2b}$$

since particle 1 is now providing the state preparation for particle 2.

Hence we see that the nonlocal state preparation is not unique and it is better to write the joint passage probability in the symmetric form

$$\mathcal{P}_{1,2}(\theta, \uparrow) = \langle \Psi_{1,2} | \hat{\pi}_\theta^{(1)} \hat{\pi}_\uparrow^{(2)} | \Psi_{1,2} \rangle, \tag{3a}$$

where the projection operators for the 1st and 2nd particle are

$$\hat{\pi}_\theta^{(1)} = |\theta_1\rangle\langle\theta_1|, \tag{3b}$$

and

$$\hat{\pi}_\uparrow^{(2)} = |\uparrow_2\rangle\langle\uparrow_2|. \tag{3c}$$

Clearly the expressions (1), (2), and (3) for $\mathcal{P}_{1,2}(\theta, \uparrow)$ are identical but the physical interpretations can be argued somewhat differently. Forms (1) and (2) are often argued as evidence for the nonlocal character of quantum mechanics. But form (3a) points our thinking in a different direction. Equation (3) is telling us something quantum optical types have bred into their bones: focus on the experiments, it is the coincidence count rates, which describe the physics. Problems 1-5 are based on these considerations.

Assuming the second SGA is also tipped, the joint probability that particle 1 is passed through a SGA oriented at an angle θ_a to the vertical direction (z) and that particle 2 is passed through a SGA oriented at an angle θ_b to the vertical is given by

$$\mathcal{P}(\theta_a, \theta_b) = \langle \Psi | \hat{\pi}_a^{(1)} \hat{\pi}_b^{(2)} | \Psi \rangle, \tag{4}$$

where the projection operator $\hat{\pi}_a = |\theta_a\rangle\langle\theta_a|$ given that $|\theta_a\rangle = e^{-i\sigma_y \theta_a/2} |\uparrow\rangle$ (see Prob. 1-4). For the spin singlet

$$|\Psi\rangle = \frac{1}{\sqrt{2}}[|\uparrow, \downarrow\rangle - |\downarrow, \uparrow\rangle], \tag{5}$$

we find (see Prob. 5)

$$\mathcal{P}(\theta_a, \theta_b) = \frac{1}{4}[1 - \cos(\theta_a - \theta_b)]. \tag{6}$$

Let us consider one particle prepared in $|\uparrow\rangle$ state. The probability of its passing through a SGA tipped by an angle θ_a is given as (see Fig. 2)

$$\mathcal{P}_\uparrow(\theta_a) = \frac{1}{2}(1 + \cos\theta_a). \tag{7}$$

Now in EPR situation if spin 2 goes down then we have prepared spin 1 in up state. Then from Eq. (6) $\mathcal{P}(\theta_a, \downarrow) = \frac{1}{4}[1 - \cos(\theta_a - \pi)]$ and if we use Bayes theorem

$$\mathcal{P}(\theta_a | \downarrow) = \mathcal{P}(\theta_a, \downarrow)/\mathcal{P}(\downarrow) = \frac{1}{2}(1 + \cos\theta_a), \tag{8}$$

since $\mathcal{P}(\downarrow) = \frac{1}{2}$. Thus EPR is just a way of preparing state $|\uparrow\rangle$.

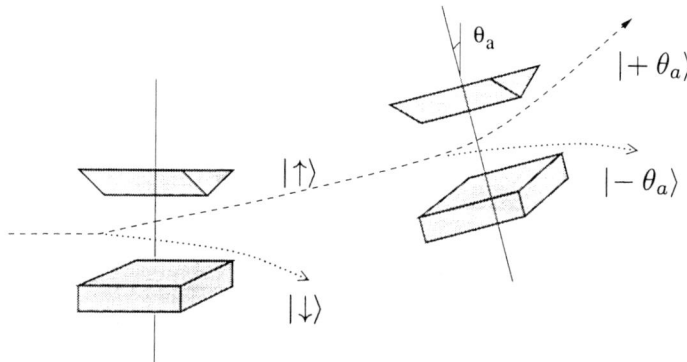

FIGURE 2. Simultaneous passage of a spin-$\frac{1}{2}$ particle through two SGA, which obeys the spin analogy of Malus's cosine law: $\mathcal{P}_\uparrow(\theta_a) = \frac{1}{2}[1 + \cos\theta_a]$.

We proceed to the issue of locality and quantum reality by appealing to the formulation indicated in Table 1, ordinary quantum mechanics involves a density matrix specifying the initial state of our spin singlet, which then interacts with the two SGA devices. The specification of the SGA orientation is contained in the projection operators, i.e., $\hat{\pi}_a$ and $\hat{\pi}_b$.

In another completely equivalent formulation of the problem we make use of phase space distribution theory like (but not necessarily identical to) the Wigner distribution function, applied now to spin-$\frac{1}{2}$ particles. For a spin-$\frac{1}{2}$ system, as shown in PRD (Ref. [1])-Sec. III, the Wigner-like distribution function $P(m_x, m_y, m_z)$ can be introduced by replacing σ_x, σ_y, σ_z with c-numbers m_x, m_y, and m_z, respectively, such that for any operator $\hat{Q}(\sigma_x, \sigma_y, \sigma_z)$ (see Probs. 6-11)

$$\langle \hat{Q}(\sigma_x, \sigma_y, \sigma_z)\rangle = \int dm_x\, dm_y\, dm_z\, Q(m_x, m_y, m_z)\, P(m_x, m_y, m_z) \tag{9}$$

where $P(m_x, m_y, m_z) = \text{Tr}\{\delta(\sigma_x - m_x)\delta(\sigma_y - m_y)\delta(\sigma_z - m_z)\,\rho\}$.

The operators $\hat{\pi}_a$ and $\hat{\pi}_b$ then become c-number functions and the Wigner quasiprobability distribution represents the initial EPR state. It should be noted that whereas the Wigner density itself is everywhere positive, the individual $\hat{\pi}$ operators in this representation can be negative. For example, when m_x and m_z are represented in polar coordinates, then the projection operator $\hat{\pi}_a(\alpha)$ takes the form (see PRD-(3.12) and Prob. 2)

$$\pi_a(\alpha) = \frac{1}{2}\Big[1 + m\cos(\theta_a - \alpha)\Big], \tag{10}$$

where $m = \sqrt{2}$ and $m_x = m\sin\alpha$, $m_z = m\cos\alpha$.

The "density matrix" for a spin-up particle going with Eq. (10) is then given by PRD-(3.10a) or Prob. 8,

$$P_{\phi=0}(\alpha) = \frac{1}{2}\Big[\delta(\alpha - \frac{\pi}{4}) + \delta(\alpha + \frac{\pi}{4})\Big]\delta(m - \sqrt{2}). \tag{11}$$

Hence from PRD-(3.14a) or Prob. 10, for the state corresponding to an arbitrary angle ϕ,

$$\langle\hat{\pi}_a\rangle = \int d\alpha\, \pi_a(\alpha) P_\phi(\alpha) = \frac{1}{2}[1 + \cos(\theta_a - \phi)], \tag{12}$$

which is the expected (positive) result.

More interesting is the two-particle EPR correlation function given by

$$\int d\alpha d\beta\, \pi_a(\alpha)\pi_b(\beta) P_\phi(\alpha,\beta), \tag{13}$$

where $P_\phi(\alpha,\beta)$ for a spin-singlet state is shown in Prob. 13 to be (see also PRD-(4.4) and PRD-Fig. 4)

$$P_\phi(\alpha,\beta) = \delta(\alpha - \beta - \pi)\frac{1}{2}\Big\{\frac{1}{2}[\delta(\alpha - \frac{\pi}{4}) + \delta(\alpha + \frac{\pi}{4})] \\ + \frac{1}{2}[\delta(\alpha - \frac{3\pi}{4}) + \delta(\alpha + \frac{3\pi}{4})]\Big\}. \tag{14}$$

for $\phi = 0$. The joint probability as per Eq. (13) calculated by this quantum-distribution-function analysis gives the same as in Eq. (3) (see Prob. 15). It is, however, noted that when we define an angle α such that $m_x = m\sin\alpha$ and $m_z = m\cos\alpha$, $\pi_a(\alpha)$, the c-number representation of the operator $\hat{\pi}_a$, is not positive definite (see Prob. 2).

III HIDDEN VARIABLE TREATMENT USING SPIN POLARIZATION AS A HIDDEN VARIABLE

The idea in hidden variable theories involves the assumption that the two spins will have some properties which will have correlated values. These values will help the spin decide, upon reaching a SGA, whether or not it should pass through that apparatus.

As in PRD-Sec. II, we note that if a spin emerges from a SGA oriented at an angle α and then passes into a SGA tipped through an angle θ_a relative to vertical, then the likelihood that the particle will emerge from the second SGA is given by $\frac{1}{2}[1 + \cos(\theta_a - \alpha)]$. Thus we might say that a hidden-variable α determined whether the spin passed through the apparatus whose angle θ_a is determined by the experimenter and we define (as in PRD-(2.4)) the hidden-variable probability function

$$\tilde{\pi}_a(\alpha) \equiv \frac{1}{2}[1 + \cos(\theta_a - \alpha)], \qquad (15)$$

as giving the "simultaneous passage" through the α and θ_a oriented SGA. Note that this is like $\pi_a(\alpha)$ in Eq. (10) above.

Next we postulate a distribution (see PRD-(3.15a))

$$\tilde{P}_\phi(\alpha) = \delta(\alpha - \phi) \qquad (16)$$

and it gives

$$\int d\alpha \tilde{\pi}_a(\alpha) \tilde{P}_\phi(\alpha) = \frac{1}{2}[1 + \cos(\theta_a - \phi)]. \qquad (17)$$

And then, for EPR spin pairs we have

$$\tilde{P}_\phi(\alpha) = \delta(\alpha - \beta - \pi)\frac{1}{2}[\delta(\alpha - \phi) + \delta(\alpha - \phi - \pi)], \qquad (18)$$

and the joint probability can be calculated as

$$\tilde{\mathcal{P}}(\theta_a, \theta_b) = \int d\alpha d\beta \tilde{\pi}_a(\alpha) \tilde{\pi}_b(\beta) \tilde{P}_\phi(\alpha, \beta) = \frac{1}{4}[1 - \cos(\theta_a - \theta_b)], \qquad (19)$$

for $\phi = \theta_a$. While $\tilde{\pi}_a(\alpha)$ is positive definite and $\tilde{\mathcal{P}}(\theta_a, \theta_b)$ yields the same result as quantum mechanics (Probs. 14-16), this hidden-variable theory is nonlocal in the sense that $\tilde{P}_\phi(\alpha, \beta)$ is determined by the experimenter's choice of θ_a (see Probs. 17 and 18).

IV HOW TO FORCE QUANTUM MECHANICS INTO A FORM WHICH MAKES IT LOOK NONLOCAL

Having seen how our hidden-variable theory is in agreement with quantum mechanics but is nonlocal, we consider the distributions in which quantum mechanical operators are expressed as, for example,

$$\hat{\pi}_a = \int dx_a \, \delta(\hat{\pi}_a - x_a) \, x_a. \qquad (20)$$

This is inserted into the standard quantum mechanical expression and yields what we call the mixed representation (summarized in Table 1). We here find that the effective operators like quantities x_a and x_b are 0 or 1. The one particle distribution function for spin-up state (see Prob. 19) is given by

$$P_\uparrow(x_a) = \frac{1}{2}(1 + \cos\theta_a)\,\delta(x_a - 1) + \frac{1}{2}(1 - \cos\theta_a)\,\delta(x_a). \tag{21}$$

The probability of the particle "passing through" the SGA oriented at angle θ_a (i.e., $x_a = 1$) is $\cos^2(\theta_a/2)$ and that of "not passing through" ($x_a = 0$) is $\sin^2(\theta_a/2)$. Hence the Eq. (21) shows us another way to see the spin analogy of Malus law based on simple yes, no events [5].

If we consider the two-particle EPR correlation (as given in Prob. 20), the joint probability is written as

$$\langle \hat{\pi}_a \hat{\pi}_b \rangle = \int dx_a dx_b\, x_a\, x_b\, P_m(x_a, x_b), \tag{22}$$

where

$$\begin{aligned}P_m(x_a, x_b) = \frac{1}{4}[&(1 - \cos(\theta_a - \theta_b))\{\delta(x_a)\delta(x_b) + \delta(x_a - 1)\delta(x_b - 1)\} \\&+ (1 + \cos(\theta_a - \theta_b))\{\delta(x_a)\delta(x_b - 1) + \delta(x_a - 1)\delta(x_b)\}].\end{aligned} \tag{23}$$

The probability density $P_m(x_a, x_b)$ now depends upon the angle $\theta_a - \theta_b$ between the two SGA devices, and therefore looks nonlocal. That is, the appearance of $\vec{a}\cdot\vec{b}$ in $P_m(x_a, x_b)$ involves the angle of the two experimental apparatus, which are far removed (see Table 1).

Since $P_m(x_a) = \frac{1}{2}$, the conditional probability is then found as

$$\begin{aligned}P_m(x_b|x_a) &= P_m(x_a, x_b)/P(x_a) \\&= \frac{1}{2}[(1 - \cos(\theta_a - \theta_b))\{\delta(x_a)\delta(x_b) + \delta(x_a - 1)\delta(x_b - 1)\} \\&\quad + (1 + \cos(\theta_a - \theta_b))\{\delta(x_a)\delta(x_b - 1) + \delta(x_a - 1)\delta(x_b)\}],\end{aligned} \tag{24}$$

and it says that when particle 1 is passed through SGA 1 ($x_a = 1$), the probability that particle 2 being passed through SGA 2 ($x_b = 1$) is $\cos^2\{(\theta_a - \theta_b)/2\}$ and $\sin^2\{(\theta_a - \theta_b)/2\}$ for "not passing through" ($x_b = 0$).

One should not take this example as providing a strong case for quantum nonlocality. Instead it implies that if we choose to fold the experimental parameterization (θ_a and θ_b) into the state description, then we find that the resulting probability density is partly governed by the experimenter's choice of instrumental settings and partly by the preparation of our initial spin singlet. This is clearly counter to the theme of quantum mechanics in the standard or Wigner form. Never the less, this example is given because it shows that it is possible to write quantum mechanics in a quantum distribution form so as to make the case for quantum nonlocality if one is inclined to do so. We however, feel that this is beyond "quantum sense" and does not represent a faithful picture of the physics.

TABLE 1. Comparison between spin-correlation functions for various representations ($\vec{a}$ denotes $\hat{x}\sin\theta_a + \hat{z}\cos\theta_a$).

QM Standard	$\mathcal{P}_{ab} = \text{Tr}\{\hat{\pi}_a^{(1)}\, \hat{\pi}_b^{(2)}\, \rho\,\}$
	$\rho = |\Psi\rangle\langle\Psi|, \qquad |\Psi\rangle = \dfrac{1}{\sqrt{2}}[|\uparrow,\downarrow\rangle - |\downarrow,\uparrow\rangle]$
	$\hat{\pi}_a^{(1)} = \dfrac{1}{2}(1 + \vec{\sigma}^{(1)} \cdot \vec{a})$
QM Wigner	$\mathcal{P}_{ab} = \displaystyle\int d\vec{m}_1 d\vec{m}_2\, \pi_a(\vec{m}_1)\, \pi_b(\vec{m}_2)\, P_w(\vec{m}_1,\vec{m}_2)$
	$P_w(\vec{m}_1,\vec{m}_2) = \text{Tr}\{\delta(\sigma_{1x}-m_{1x})\delta(\sigma_{1z}-m_{1z})\delta(\sigma_{2x}-m_{2x})\delta(\sigma_{2z}-m_{2z})\rho\}$
	$\qquad = \dfrac{1}{4}[\delta(m_{1z}-1)\delta(m_{2z}+1) + \delta(m_{1z}+1)\delta(m_{2z}-1)]$
	$\qquad\quad \times [\delta(m_{1x}-1)\delta(m_{2x}+1) + \delta(m_{1x}+1)\delta(m_{2x}-1)]$
	$\pi_a(\vec{m}^{(1)}) = \dfrac{1}{2}(1 + \vec{m}^{(1)}\cdot\vec{a})$
HV$_3$	$\tilde{\mathcal{P}}_{ab} = \displaystyle\int d\alpha d\beta\, \tilde{\pi}_a(\alpha)\tilde{\pi}_b(\beta)\tilde{P}_\phi(\alpha,\beta)$
	$\tilde{P}_\phi(\alpha,\beta) = \delta(\alpha-\beta-\pi)\dfrac{1}{2}[\delta(\alpha-\phi) + \delta(\alpha-\phi-\pi)]$
	$\tilde{\pi}_a(\alpha) = \dfrac{1}{2}[1 + \cos(\theta_a - \alpha)]$
QM Mixed	$\mathcal{P}_{ab} = \displaystyle\int dx_a dx_b\, x_a\, x_b\, P_m(x_a, x_b)$
	$P_m(x_a, x_b) = \text{Tr}\{\delta(\hat{\pi}_a - x_a)\,\delta(\hat{\pi}_b - x_b)\,\rho\}$
	$\qquad = \dfrac{1}{4}(1 - \vec{a}\cdot\vec{b})[\delta(x_a)\delta(x_b) + \delta(x_a-1)\delta(x_b-1)]$
	$\qquad\quad + \dfrac{1}{4}(1 + \vec{a}\cdot\vec{b})[\delta(x_a)\delta(x_b-1) + \delta(x_a-1)\delta(x_b)]$
	$x_a = 1, 0$

V REPRINT OF PHYSICAL REVIEW D-1983

PHYSICAL REVIEW D   VOLUME 28, NUMBER 10   15 NOVEMBER 1983

How to make quantum mechanics look like a hidden-variable theory and vice versa

Marlan O. Scully

*Max-Planck-Institut für Quantenoptik, D-8046 Garching bei Munchen, West Germany
and Center for Advanced Studies, University of New Mexico, Albuquerque, New Mexico 87131*

(Received 9 June 1983)

The quantum theory of a singlet spin-$\frac{1}{2}$ system is developed in terms of angular variables using a quantum-distribution-function technique. These calculations demonstrate a much closer correspondence between quantum mechanics and certain hidden-variable theories than was previously appreciated. It is found that a new type of hidden-variable theory is suggested by the quantum-distribution-function treatment of the Einstein-Podolsky-Rosen-Bohm spin-spin correlation problem which is in agreement with quantum theory but is "nonlocal."

I. INTRODUCTION

A. Background and motivation

Conventional quantum mechanics is a superb calculational tool. It has successfully solved mysteries ranging from macroscopic superconductivity[11(a)] to the microscopic theory of the electron[11(b)] and has provided deeper insight into the nature of the vacuum[11(c)] on the one hand and the description of the nucleon[11(d)] on the other. Whole new fields[2(a)–2(f)] such as quantum optics and quantum electronics owe their very existence to this body of knowledge.

Nevertheless certain aspects, concerning the foundations and interpretation of the theory, are regarded by many[3] as incomplete, unsatisfactory, or contradictory. Thus, various alternative theories to quantum mechanics have been considered over the years, hidden-variable (HV) theories[4] being prominent among these.

In this latter HV context the inequalities of Bell[5] have been instrumental in rendering various philosophical arguments susceptible to experimental[6] test. Such experiments are important from at least two points of view: first they provide a sharp focus on the essential differences between quantum and HV theories and, of course, help us choose between theories. More important, perhaps, is the ability of such studies to assist us in gaining a better understanding and appreciation for what quantum theory is not. For example, in the current HV vernacular it is not a "local-realistic"[7] theory. It is in this latter context that the present author finds these studies most interesting.

It is thus not a desire to replace quantum theory, but rather a desire to better understand its foundations that stimulates this worker. With this motivation in mind let us consider the framework in which this paper fits.

The stimulus for the present study was provided by Belinfante's[4] chapter on "Some Examples of Hidden-Variables Theories of the Second Kind for 'Explaining' Polarization Correlations." In that chapter it is argued that certain "polarization hidden-variables" (α), as discussed in Sec. II of this paper, are interesting in a HV context, but do not exist in quantum theory. While it is true that conventional quantum mechanics involves "no such a thing as α," we show in Sec. III that a completely acceptable quantum theory can be built around this α variable via an associated quasiclassical quantum distribution function[8] $P(\alpha)$ [e.g., Eqs. (3.14)]. With this distribution function in hand, we are naturally motivated to consider a class of closely related HV distribution functions $\widetilde{P}(\alpha)$ [e.g., Eqs. (3.15)]. Following along these lines the present hidden-variables theory reproduces the quantum two-particle correlation.

To sum up, in this paper we show that

(1) various polarization "hidden variables" α have a proper quantum-mechanical interpretation which suggests a

(2) hidden-variable theory, in agreement[9,10] with quantum theory insofar as the current two-particle correlation experiments are concerned.

No consideration as to the "philosophical" pros and cons of the present HV calculational strategy is included here, although an essential difference between the epistemology of this work and quantum theory is developed in Sec. VI. Various extensions and further discussion of the ways in which our HV theory agrees and disagrees with quantum theory will be presented elsewhere.

B. A Pico review of the HV concept, EPRB, and orthodox quantum theory

Just as statistical mechanics provides a deeper "hidden-variable" description of thermodynamical phenomena, so, many people would like to see quantum mechanics supplanted by a deeper hidden-variable theory. According to this point of view the various parameters of interest (e.g., *all three* components of a particle's angular momentum) are knowable but just not known to us.[11] For example, a spin-$\frac{1}{2}$ system might be thought of as having a kind of hidden spin parameter ϕ as in Fig. 1, such that the measured spin projection is given by

$$S_z = \frac{\hbar}{2}[\theta(\pi - \phi) - \theta(\phi - \pi)] \ .$$

©1983 The American Physical Society

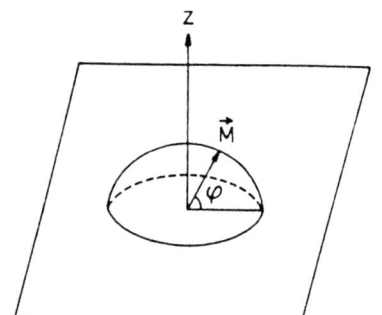

FIG. 1. Hidden-variable description of a spin-$\frac{1}{2}$ particle in which the hidden variable ϕ lying between 0 and π corresponds to spin-up while angle ϕ in the lower half plane corresponds to spin-down.

At the heart of most modern discussions of "hidden variables" in quantum mechanics lies a *Gedankenexperiment* of the Einstein-Podolsky-Rosen-Bohm[12] variety. As sketched in Fig. 2 the essence of this "experiment" involves the passage (or blockage) of particles 1 and 2 through appropriate Stern-Gerlach apparati (SGA).

In order to set the stage for our considerations, let us first review what orthodox quantum mechanics says about the problem of Fig. 2. To calculate the probability that particle 1 is passed by a SGA oriented at an angle ϕ to the vertical ($+z$) direction and that particle 2 is passed by a SGA oriented at an angle θ to the vertical, we must evaluate the correlation function

$$\mathcal{P}(\theta,\phi)=\langle\psi|\hat{\pi}_\theta^{(2)}\hat{\pi}_\phi^{(1)}|\psi\rangle. \tag{1.1}$$

In Eq. (1.1), the singlet state may be written as

$$|\psi\rangle=\frac{1}{\sqrt{2}}(|+_1-_2\rangle-|-_1+_2\rangle), \tag{1.2}$$

where, for example,

$$|+_i\rangle=|+\phi_i\rangle=e^{-i\phi_i\hat{\sigma}_y/2}|\uparrow_i\rangle, \quad i=1,2.$$

The projection operators are given by

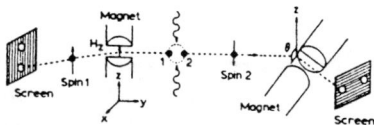

FIG. 2. Singlet-spin system, such as orthohydrogen, is split by external radiation field and the corresponding spin-$\frac{1}{2}$ particles (protons) proceed to opposite ends of laboratory where they are passed through SG apparati oriented along the z axis in the case of particle 1, and at an angle θ to the z axis in the case of spin 2.

$$\hat{\pi}_\phi^{(1)}=|\phi_1\rangle\langle\phi_1|$$
$$=e^{-i\phi\hat{\sigma}_y/2}|\uparrow_1\rangle\langle\uparrow_1|e^{i\hat{\sigma}_y\phi/2}$$
$$=\tfrac{1}{2}(1+\hat{\sigma}_z^{(1)}\cos\phi+\hat{\sigma}_x^{(1)}\sin\phi) \tag{1.3a}$$

and

$$\hat{\pi}_\theta^{(2)}=\tfrac{1}{2}(1+\hat{\sigma}_z^{(2)}\cos\theta+\hat{\sigma}_x^{(2)}\sin\theta). \tag{1.3b}$$

The Pauli spin operators appearing in Eqs. (1.3) are defined as

$$\hat{\sigma}_x=\begin{bmatrix}0 & 1\\ 1 & 0\end{bmatrix}, \quad \hat{\sigma}_y=\begin{bmatrix}0 & -i\\ i & 0\end{bmatrix},$$

and

$$\hat{\sigma}_z=\begin{bmatrix}0 & 1\\ 1 & 0\end{bmatrix}.$$

Inserting Eqs. (1.2), (1.3a), and (1.3b) into Eq. (1.1) we find the usual result

$$\mathcal{P}(\theta,\phi)=\tfrac{1}{4}[1-\cos(\theta-\phi)]. \tag{1.4}$$

II. A HIDDEN-VARIABLE TREATMENT OF SPIN CORRELATIONS FOLLOWING BELINFANTE[13]

A. Spin polarization as a possible hidden variable

The leading idea in hidden-variable theories involves the assumption that the two spins will have some properties which will have correlated values. These values will help the spin decide, upon reaching a SGA, whether or not it should pass through that apparatus.

Following the treatment of Belinfante, we require that in order to give such theories an air of possibility we want to invent them in such a way that at least in the simplest cases they yield the same results as quantum mechanics. For example, in an unpolarized beam only $\frac{1}{2}$ the particles should pass a given SGA. Further the probability of passing a second SGA placed behind (and at an angle θ) relative to the previous (vertical) SGA should be given by

$$\langle\uparrow|\pi_\theta|\uparrow\rangle, \tag{2.1}$$

which, using Eq. (1.3b), is

$$\tfrac{1}{2}(1+\cos\theta). \tag{2.2}$$

Or, more generally, if a spin emerges from a SGA oriented at an angle α and then passes into a SGA tipped through an angle θ relative to the vertical, then the likelihood that the particle will emerge from the second SGA is given by

$$\tfrac{1}{2}[1+\cos(\theta-\alpha)]. \tag{2.3}$$

Thus we might say that a "hidden variable" α determined whether the spin passed through the apparatus whose angle θ is determined by the experimenter.

With this in mind we define the hidden-variable probability function

$$\bar{\pi}_\theta(\alpha) \equiv \tfrac{1}{2}[1+\cos(\theta-\alpha)] \qquad (2.4)$$

as giving the "simultaneous passage" through the α and θ oriented SGA.

B. Correlation between singlet spins in hidden-variables theory

Consider the case where the two spins of our singlet system of Fig. 2 (having polarization angles α and β for spins 1 and 2) are correlated such that

$$I(\alpha,\beta)d\alpha\,d\beta \qquad (2.5)$$

is the probability that the spins carry polarizations α and β, while all other hidden variables can be randomly distributed. For maximum polarization correlation, we might then think that

$$I(\alpha,\beta)=\delta(\alpha-\beta-\pi)\frac{1}{2\pi} . \qquad (2.6)$$

Let us next ask: what is the probability of simultaneous passage of spins 1 and 2 through the double SG system of Fig. 1? That is, what is $\mathscr{P}(\theta,\phi)$ in hidden-variable theory? Following the above discussion and in view of Eqs. (2.4) and (2.5) we would expect

$$\mathscr{P}(\theta,\phi)=\int d\alpha\,d\beta\,I(\alpha,\beta)\bar{\pi}_\theta^{(2)}(\beta)\bar{\pi}_\phi^{(1)}(\alpha) , \qquad (2.7)$$

and from Eqs. (2.6) and (2.4) this implies

$$\mathscr{P}(\theta,\phi)=\int d\alpha\,d\beta\,\delta(\alpha-\beta-\pi)$$
$$\times \frac{1}{2\pi}\{\tfrac{1}{2}[1+\cos(\theta-\beta)]\tfrac{1}{2}[1+\cos(\phi-\alpha)]\} . \qquad (2.8)$$

Carrying out the simple integrations in Eq. (2.8) we find

$$\mathscr{P}(\theta,\phi)=\tfrac{1}{4}[1-\tfrac{1}{2}\cos(\theta-\phi)] . \qquad (2.9)$$

It is precisely the difference between the quantum correlation (1.4) and the hidden-variable result (2.9) which concerns us here. How should we react to this difference? From a pragmatic, calculational point of view quantum theory is a superb tool, hence Eq. (1.4) has never really been seriously doubted (although such results should be and have been verified experimentally).

What then should we think of the hidden-variable prediction Eq. (2.9)? It is not all that different from the quantum prediction. Might there not be a germ of "truth" hidden in Eq. (2.9) and, more to the point, the arguments leading to it? In this context we quote Belinfante[9]: "The polarization [spin] hidden-variable here introduced is, of course, a quantity which does not exist in quantum theory. ... in quantum theory no such a thing as α even exists."

In the next section we shall argue that a rigorous quantum mechanical treatment of the present problem can in fact be couched in terms of the angular variable α. In so doing we shall be led to reconsider the correlation function [Eq. (2.7)] and by a simple extension of Eq. (2.6) regain the quantum result [Eq. (1.4)] via a "hidden-variable" calculation.

III. QUANTUM DISTRIBUTION THEORY AND A CORRESPONDING HIDDEN-VARIABLE TREATMENT OF SPIN-$\tfrac{1}{2}$ SYSTEMS

A. Distribution functions

Some 50 years ago Eugene Wigner succeeded in developing a phase-space distribution[8] description of quantum mechanics in terms of what is now called the Wigner distribution $W(p,q)$. This distribution has many nice properties and has been fruitfully applied to a variety of problems in conventional quantum theory. Practicing quantum opticians have such notions bred into their bones at an early age. For example, the probability density function,[14] $P(E)$, for the quantized electromagnetic field on the one hand and the phase-space ("Bloch" type) description of two-level systems[15] on the other.

Proceeding toward an "α" description of our spin singlet problem, consider first the expectation value of an operator $\hat{Q}(\vec{\sigma})$ given in terms of the density matrix ρ:

$$\langle \hat{Q} \rangle = \text{Tr}[\rho(t)\hat{Q}(\sigma)] .$$

Introducing the operator δ function

$$\delta(\gamma-\hat{c}) \equiv \int \frac{d\xi}{2\pi}\exp[-i(\gamma-\hat{c})] , \qquad (3.1)$$

where $\hat{c}$ is an operator (e.g., $\hat{\sigma}_z$) and γ is the associated classical variable (e.g., m_z); we may then write $\hat{Q}(\vec{\sigma})$ as

$$\hat{Q}(\vec{\sigma}) = \int d^3m\,Q(m_x,m_y,m_z)\delta(m_x-\hat{\sigma}_x)$$
$$\times \delta(m_y-\hat{\sigma}_y)\delta(m_z-\hat{\sigma}_z) . \qquad (3.2)$$

Inserting Eq. (3.2) into Eq. (3.1) we then find the expectation value for $\hat{Q}$ to be given by

$$\langle \hat{Q} \rangle = \int d^3m\,P(\vec{m},t)Q(\vec{m}) , \qquad (3.3)$$

where

$$P(\vec{m},t) \equiv \text{Tr}[\rho(t)\delta(m_x-\hat{\sigma}_x)\delta(m_y-\hat{\sigma}_y)\delta(m_z-\hat{\sigma}_z)] . \qquad (3.4)$$

Recalling, however, from the discussion of Secs. I and II that we shall only be interested in operators Q involving $\hat{\sigma}_x$ and $\hat{\sigma}_z$, e.g., $\bar{\pi}_\theta$ of Eq. (1.3b) we may rigorously restrict our treatment to operator expansions of the form

$$\hat{Q}(\hat{\sigma}_x,\hat{\sigma}_z) = \int Q(m_x,m_z)\delta(m_x-\hat{\sigma}_x)\delta(m_z-\hat{\sigma}_z)$$
$$\times dm_x\,dm_z ,$$

and the associated quantum distribution function

$$P(m_x,m_z,t) = \text{Tr}[\rho(t)\delta(m_x-\hat{\sigma}_x)\delta(m_z-\hat{\sigma}_z)] . \qquad (3.5)$$

The distribution function Eq. (3.5) is the vehicle by which we shall realize a quantum treatment of the present spin-$\tfrac{1}{2}$ problem in terms of the angle α. In the next few paragraphs we treat the simple problem of a single "spin up" particle.

For the "spin up" case the density matrix is, of course,

$$\rho = |\uparrow\rangle\langle\uparrow| \quad (3.6)$$

and the associated distribution function is

$$P_\uparrow(m_x, m_z) = \text{Tr}[|\uparrow\rangle\langle\uparrow|\delta(m_x - \hat{\sigma}_x)\delta(m_z - \hat{\sigma}_z)]$$
$$= \langle\uparrow|\delta(m_x - \hat{\sigma}_x)\delta(m_z - \hat{\sigma}_z)|\uparrow\rangle, \quad (3.7)$$

which by Eq. (3.1) becomes

$$P_\uparrow(m_x, m_z) = \int \frac{d\xi}{2\pi} \int \frac{d\eta}{2\pi} \langle\uparrow|e^{i\xi\hat{\sigma}_x}e^{i\eta\hat{\sigma}_z}|\uparrow\rangle e^{-im_x\xi}e^{-im_z\eta}$$

$$= \int \frac{d\xi}{2\pi} \int \frac{d\eta}{2\pi} \cos\xi e^{i\eta} e^{-im_x\xi} e^{-im_z\eta}$$

$$= \tfrac{1}{2}[\delta(m_x + 1)\delta(m_z - 1) + \delta(m_x - 1)\delta(m_z - 1)]. \quad (3.8)$$

Consulting Fig. 3 we see that the quantum distribution function for the state $|\uparrow\rangle$ corresponds to equal admixtures of "probability" at $\pm\pi/4$.

Define the angle which m makes with the vertical ($+z$) direction as α, see Fig. 3, so that

$$m_x = m\sin\alpha \quad (3.9a)$$

and

$$m_z = m\cos\alpha. \quad (3.9b)$$

In terms of this angle α, it is clear that Eq. (3.8) now becomes

$$P_\uparrow(\alpha) = \tfrac{1}{2}\left[\delta\left(\alpha - \frac{\pi}{4}\right) + \delta\left(\alpha + \frac{\pi}{4}\right)\right]\delta(m - \sqrt{2}). \quad (3.10a)$$

In like manner it is easy to show that the probability density for "spin down" is given by

$$P_\downarrow(\alpha) = \tfrac{1}{2}\left[\delta\left(\alpha - \frac{3\pi}{4}\right) + \delta\left(\alpha + \frac{3\pi}{4}\right)\right]\delta(m - \sqrt{2}). \quad (3.10b)$$

Let us now rewrite the operator $\hat{\pi}_\theta$ of Eq. (1.3) in its associated c number representation, that is,

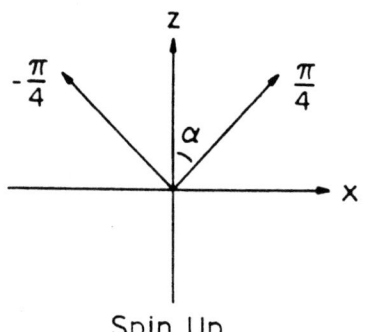

FIG. 3. Figure depicting spin-up particle in quantum distribution theory.

$$\pi_\theta = \tfrac{1}{2}(1 + m_z\cos\theta + m_x\sin\theta), \quad (3.11)$$

and, using Eq. (3.9), this becomes

$$\pi_\theta = \tfrac{1}{2}[1 + m\cos(\theta - \alpha)]. \quad (3.12)$$

We may check the vector length m by requiring that $\langle\uparrow|\hat{\pi}_0|\uparrow\rangle$ be unity so that from Eqs. (3.10) and Eq. (3.11b) we have

$$1 = \int d\alpha \, \tfrac{1}{2}\left[\delta\left(\alpha - \frac{\pi}{4}\right) + \delta\left(\alpha + \frac{\pi}{4}\right)\right]\tfrac{1}{2}(1 + m\cos\alpha)$$

$$= \tfrac{1}{2}\left[1 + m\cos\frac{\pi}{4}\right]. \quad (3.13)$$

This then fixes the value of m, as noted in (3.10), to be

$$m = \sqrt{2}.$$

In conclusion we note that the probability density calculated for states corresponding to an arbitrary angle ϕ are given by

$$P_{+\phi}(\alpha) = \tfrac{1}{2}\left[\delta\left(\alpha - \phi - \frac{\pi}{4}\right) + \delta\left(\alpha - \phi + \frac{\pi}{4}\right)\right] \quad (3.14a)$$

and

$$P_{-\phi}(\alpha) = \tfrac{1}{2}\left[\delta\left(\alpha - \phi - \frac{3\pi}{4}\right) + \delta\left(\alpha - \phi + \frac{3\pi}{4}\right)\right]. \quad (3.14b)$$

B. Connection with a revised hidden-variable scheme

Thus we see from the calculations of Sec. III A that the angle α "exists" in quantum theory. Furthermore we note that the probability/projection functions in hidden-variable and quantum theory, i.e., $\tilde{\pi}_\phi(\alpha)$ from Eq. (2.4) and $\pi_\phi(\alpha)$ as given by Eq. (3.11b) are quite similar. Indeed the quantum probability densities given by Eqs. (3.14a) and (3.14b) suggest that we introduce similar probability densities into our hidden-variable considerations as follows:

$$\tilde{P}_{+\phi}(\alpha) = \delta(\alpha - \phi) \quad (3.15a)$$

and

$$\tilde{P}_{-\phi}(\alpha) = \delta(\alpha - \phi - \pi). \quad (3.15b)$$

Belinfante, in his scholarly treatment of the subject, divides HV theories into two main groups:

1. Hidden-variable theories of the first kind (HV_1) in which the theory will make exactly the quantum predictions when the hidden variables are in "equilibrium." Deviations from quantum mechanics would occur within the context of HV_1 only for a nonequilibrium situation which is supposed to exist for very short times in any given experiment.

2. Hidden-variable theories of the second kind (HV_2) in which the theory disagrees with quantum mechanics even when the supposed hidden variables are in a steady state

TABLE I. Tabulation emphasizing close similarity between the single-particle descriptions in quantum mechanics and HV_3.

	QUANTUM	HV_3
PASSAGE "PROBABILITY"	$\pi_\theta(\alpha) = \frac{1}{2}[1 + m\cos(\theta-\alpha)]$ $m = \sqrt{2}$	$\tilde{\pi}_\theta(\alpha) = \frac{1}{2}[1 + \cos(\theta-\alpha)]$
DISTRIBUTION FOR SPIN UP	$P_\uparrow(\alpha) = \frac{1}{2}[\delta(\alpha-\frac{\pi}{4}) + \delta(\alpha+\frac{\pi}{4})]$ 	$\tilde{P}_\uparrow(\alpha) = \delta(\alpha)$ 
EXPECTATION OF PASSING $\theta = 0$ SGA FOR ↑ SPIN	$\mathcal{P}_\uparrow = \int P_\uparrow(\alpha)\pi_0(\alpha)d\alpha$ $= 1$	$\tilde{\mathcal{P}}_\uparrow = \int \tilde{P}_\uparrow(\alpha)\tilde{\pi}_0(\alpha)d\alpha$ $= 1$

or equilibrium condition. The current experimental studies, stimulated by the inequalities of Bell, support quantum theory in favor of HV_2.

In this section, as well as the next, we are lead to consider a hidden-variable theory which is a kind of mixture of HV_1 and HV_2, wherein correlation functions (associated with the above-mentioned type of experiments) as calculated from quantum theory and our HV theory are identical. In fact the present HV theory is a close cousin to and directly suggested by the quantum distribution treatment of the problem. In this sense there is a definite HV_1 flavor to our work although no mention of equilbrium (or nonequilibrium) hidden variables is made here. On the other hand, while this work is even closer to an HV_2 theory, it does not fall into this category either since to quote Belinfante: "... this (second) kind of hidden-variables theory may be expected to contradict quantum theory." To be sure, the HV theory here presented will be seen to deviate from quantum theory, but is not immediately contradictory in so far as the correlation functions of current interest are concerned.

For those reasons, and for want of a better term, we call the present HV theory a hidden-variable theory of the third kind (HV_3). A summary of the comparison between the quantum and hidden-variable descriptions for the present example is given in Table I.

IV. QUANTUM DISTRIBUTION AND HV_3 TREATMENT OF THE SINGLET SPIN PROBLEM

A. Quantum and hidden-variable correlation functions

Corresponding to the anticorrelated (singlet) state for a pair of spin-$\frac{1}{2}$ particles

$$|\psi\rangle = \frac{1}{\sqrt{2}}[|+_1-_2\rangle - |-_1+_2\rangle] \quad (4.1)$$

we have the associated probability distribution

$$P(\vec{m}_1,\vec{m}_2) = \text{Tr}[\rho\delta(m_x^{(1)}-\hat{\sigma}_x^{(1)})\delta(m_z^{(1)}-\hat{\sigma}_z^{(1)}) \\ \times \delta(m_x^{(2)}-\hat{\sigma}_x^{(2)})\delta(m_z^{(2)}-\hat{\sigma}_z^{(2)})] . \quad (4.2)$$

In Eq. (4.2) above,

$$\rho = |\psi\rangle\langle\psi| \text{ with } |\psi\rangle \text{ given by Eq. (4.1)}, \quad (4.3a)$$

$\hat{\sigma}_x^{(i)}$ and $\hat{\sigma}_z^{(i)}$ are the usual Pauli operators

$$\text{for the } i\text{th particle } (i=1,2), \quad (4.3b)$$

and $m_x^{(i)}$ and $m_z^{(i)}$ are the associated quasiclassical variables.

We must now carry out a calculation for $P(\vec{m}^{(1)},\vec{m}^{(2)})$ similar to the calculation of $P(\vec{m})$ given in Sec. III B; after some algebra, we find

$$P_\phi(\alpha,\beta)$$
$$= \delta(\alpha-\beta-\pi)\frac{1}{2}\left\{\frac{1}{2}\left[\delta\left(\alpha-\phi-\frac{\pi}{4}\right)+\delta\left(\alpha-\phi+\frac{\pi}{4}\right)\right]\right.$$
$$+\frac{1}{2}\left[\delta\left(\alpha-\phi-\frac{3\pi}{4}\right)\right.$$
$$\left.\left.+\delta\left(\alpha-\phi+\frac{3\pi}{4}\right)\right]\right\} \quad (4.4)$$

where α and β are the angular variables corresponding to particles 1 and 2. The physical meaning of the various terms in Eq. (4.4) is discussed in Fig. 4 for the case $\phi=0$.

We note immediately that the "spin anticorrelation" factor occurring in Eq. (4.4), i.e., the $\delta(\alpha-\beta-\pi)$ term, is

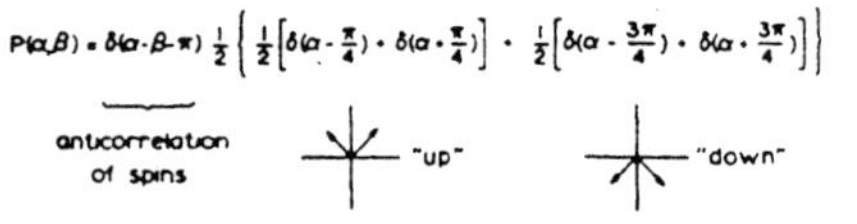

FIG. 4. Physical interpretation of the joint probability density for spin-spin correlation functions.

TABLE II. Table summarizing correspondence between two-particle spin-correlation functions in HV_2, quantum mechanics, and HV_3.

	$P(\alpha,\beta)$ TWO PARTICLE DISTRIBUTION	$\mathcal{P}(\theta,\phi)$ JOINT PROBABILITY
HV_2	$\delta(\alpha\cdot\beta\cdot\pi)\frac{1}{2\pi}$	$\frac{1}{4}(1-\frac{1}{2}\cos(\theta-\phi))$
QM	$\delta(\alpha\cdot\beta\cdot\pi)\frac{1}{2}\left\{\frac{1}{2}[\delta(\alpha\cdot\phi\cdot\frac{\pi}{4})+\delta(\alpha\cdot\phi\cdot\frac{\pi}{4})]+\frac{1}{2}[\delta(\alpha\cdot\phi\cdot\frac{3\pi}{4})+\delta(\alpha\cdot\phi\cdot\frac{3\pi}{4})]\right\}$	$\frac{1}{4}(1-\cos(\theta-\phi))$
HV_3	$\delta(\alpha\cdot\beta\cdot\pi)\frac{1}{2}\left\{\delta(\alpha\cdot\phi)+\delta(\alpha\cdot\phi\cdot\pi)\right\}$	$\frac{1}{4}(1-\cos(\theta-\phi))$

identical with that contained in Belinfante's $I(\alpha,\beta)$ given by Eq. (2.6). The difference, of course, is that the present result follows directly from quantum theory. Furthermore the curly bracketed factor corresponds to the sum of single-particle probability density functions for spin along $+\phi$ and $-\phi$.

In view of Eqs. (2.5), (4.4), (3.14), and (3.15), we are motivated to consider the "hidden-variable" probability function

$$\widetilde{P}_\phi(\alpha,\beta)=\delta(\alpha-\beta-\pi)\tfrac{1}{2}[\delta(\alpha-\phi)+\delta(\alpha-\phi-\pi)] \; . \tag{4.5}$$

The comparison between Eq. (4.5), Belinfante's equation (2.6), and the quantum-mechanical result [Eq. (4.4)] is summarized in Table II.

As we shall see in the next section $\widetilde{P}_\phi(\alpha,\beta)$ allows us to regain the quantum-mechanical result Eq. (1.4).

B. Joint count probabilities in HV_3

The joint count probabilities such as Eqs. (1.4) and (2.9) can now be easily calculated via the general expression

$$\widetilde{\mathcal{P}}(\theta,\phi)=\int \widetilde{P}_\phi(\alpha,\beta)\widetilde{\pi}_\theta(\beta)\widetilde{\pi}_\phi(\alpha)d\alpha\,d\beta \; . \tag{4.6}$$

The present hidden-variables "theory" involves inserting Eqs. (2.4) and (4.5) into Eq. (4.6) to obtain

$$\widetilde{\mathcal{P}}(\theta,\phi)=\int\delta(\alpha-\beta-\pi)\tfrac{1}{2}[\delta(\alpha-\phi)+\delta(\alpha-\phi-\pi)]\tfrac{1}{2}[1+\cos(\theta-\beta)]\tfrac{1}{2}[1+\cos(\phi-\alpha)]d\alpha\,d\beta \tag{4.7}$$

and, upon carrying out a couple of simple integrations, this yields

$$\widetilde{\mathcal{P}}(\theta,\phi)=\tfrac{1}{4}[1-\cos(\theta-\phi)] \; . \tag{4.8}$$

The present HV_3 result (4.8), agrees with the quantum-mechanical prediction [Eq. (1.4)]. The results of the various theories are summarized in Table II.

V. THE DIFFERENCE BETWEEN HV_3 AND QUANTUM MECHANICS

In this section we treat a somewhat subtle but important point. If one calculates $\mathcal{P}(\theta,\phi)$ via $P_0(\alpha,\beta)$ from Eq. (4.4) with $\phi=0$ we find

$$\mathcal{P}(\theta,\phi)=\int\delta(\alpha-\beta-\pi)\tfrac{1}{2}\left\{\tfrac{1}{2}\left[\delta\left(\alpha-\tfrac{\pi}{4}\right)+\delta\left(\alpha+\tfrac{\pi}{4}\right)\right]+\tfrac{1}{2}\left[\delta\left(\alpha-\tfrac{3\pi}{4}\right)+\delta\left(\alpha+\tfrac{3\pi}{4}\right)\right]\right\}$$

$$\times\tfrac{1}{2}[1+m\cos(\theta-\beta)]\tfrac{1}{2}[1+m\cos(\phi-\alpha)]d\alpha\,d\beta$$

$$=\tfrac{1}{4}[1-\cos(\theta-\phi)] \; , \tag{5.1}$$

which is the same as before, i.e., the same as that obtained when $P_\phi(\alpha,\beta)$ from Eq. (4.4) was used. This is as would be expected since use of either spherically symmetric state

$$|\psi\rangle=\frac{1}{\sqrt{2}}[|\uparrow_1\downarrow_2\rangle-|\downarrow_1\uparrow_2\rangle]\rightarrow P_0(\alpha,\beta) \tag{5.2a}$$

or

$$|\psi\rangle=\frac{1}{\sqrt{2}}[|+_1-_2\rangle-|-_1+_2\rangle]\rightarrow P_\phi(\alpha,\beta) \tag{5.2b}$$

should give the same results.

Note, however, that the operator associated quasiclassical expression $\pi_\phi(\alpha)$, corresponds to and suggests that the particular choice $P_\phi(\alpha,\beta)$ is to be preferred. That is, in any consistent treatment of quantum theory from an associated quasiclassical distribution point of view the counterparts to operators $\hat{Q}(\hat{c})$, i.e., $Q(\gamma)$ are chosen to correspond to the operator ordering, etc., associated with $P(\gamma)$.

However, if, within HV$_3$, we carry out the calculation of $\tilde{\mathcal{P}}(\theta,\phi)$, using $\tilde{P}_0(\alpha,\beta)$ as given in Fig. 4 we would have

$$\tilde{\mathcal{P}}(\theta,\phi) = \int \delta(\alpha-\beta-\pi)\tfrac{1}{2}[\delta(\alpha)+\delta(\alpha-\pi)]$$
$$\times \tfrac{1}{2}[1+\cos(\theta-\beta)]$$
$$\times \tfrac{1}{2}[1+\cos(\phi-\alpha)]d\alpha\,d\beta \quad (5.3)$$

which yields

$$\tilde{\mathcal{P}}(\theta,\phi) = \tfrac{1}{4}[1+\tfrac{1}{2}\cos(\theta-\phi)-\tfrac{1}{2}\cos(\theta+\phi)]. \quad (5.4)$$

This result is clearly in agreement with the quantum prediction only if $\phi=0$.

The physical content and interpretation of passage (projection) functions, e.g., $\pi_\phi(\alpha)$ and the conditional probability distribution $P_\phi(\alpha,\beta)$ in quantum theory and HV$_3$ differ, and the roles of state preparation and experimental parametrization enter into the theories in different ways.

In quantum theory state preparation involves choosing $|\psi\rangle$, i.e., $P_\phi(\alpha,\beta)$ and then the experiment in question tells us what operator, e.g., $\hat{\pi}_0$, we should calculate. However, if we change from $P_\phi(\alpha,\beta)$ to $P_0(\alpha,\beta)$ the calculation should remain unchanged since we are considering a spherically symmetric state.

In HV$_3$ on the other hand the choice of $\tilde{P}_\phi(\alpha,\beta)$, for the present spherically symmetric problem, must "match" the choice of $\pi_\phi(\alpha)$ as the calculations of Eqs. (5.3), (5.4), and (4.8) indicate. Both $\tilde{P}_\phi(\alpha,\beta)$ and $\tilde{\pi}_\phi(\alpha)$ involve state preparation and experimental specification. Thus while HV$_3$ reproduces the quantum spin-spin correlation function, it differs from quantum theory in this important aspect. As will be developed in more detail elsewhere this implies that HV$_3$ is a nonlocal theory.

VI. SUMMARY AND COMMENTS

By way of summary, the major points of this paper are as follows:

A. The variable α has a satisfactory and rigorous quantum-mechanical interpretation via quantum distribution theory.

B. The spin analogy of "Malus's cosine law," i.e.,

$$\tilde{\pi}_\phi(\alpha) = \tfrac{1}{2}[1+\cos(\theta-\alpha)] \quad (6.1)$$

is always positive and has a near transcription in quantum distribution theory, namely,

$$\pi_\phi(\alpha) = \tfrac{1}{2}[1+m\cos(\theta-\alpha)], \quad (6.2)$$

where $m=\sqrt{2}$, and $\pi_\phi(\alpha)$ is therefore not positive definite.

C. The quantum mechanical probability density for a spin-up particle is given by

$$P_0(\alpha) = \frac{1}{2}\left[\delta\left(\alpha-\frac{\pi}{4}\right)+\delta\left(\alpha+\frac{\pi}{4}\right)\right], \quad (6.3)$$

and the corresponding HV$_3$ quantity is

$$\tilde{P}_0(\alpha) = \delta(\alpha), \quad (6.4)$$

which is always positive.

D. The correlation $I(\alpha,\beta)$ implied by Ref. 1,

$$I(\alpha,\beta) \propto (\alpha-\beta-\pi), \quad (6.5)$$

is contained in the quantum treatment as indicated in Eq. (4.4).

E. The HV$_3$ probability density

$$\tilde{P}_\phi(\alpha,\beta) = \delta(\alpha-\beta-\pi)\tfrac{1}{2}[\delta(\alpha-\phi)+\delta(\alpha-\phi-\pi)] \quad (6.6)$$

taken together with Eq. (2.4) implies

$$\tilde{\mathcal{P}}(\theta,\phi) = \int \tilde{P}_\phi(\alpha,\beta)\tilde{\pi}_\phi(\alpha)\tilde{\pi}_\theta(\beta)d\alpha\,d\beta$$
$$= \tfrac{1}{4}[1-\cos(\theta-\phi)] \quad (6.7)$$

which reproduces the quantum joint probability distribution for the passage of the first spin through a SGA oriented at ϕ and the second spin through a corresponding apparatus tipped through an angle θ.

F. In quantum mechanics

$$\int P_\phi(\alpha,\beta)\pi_\phi(\alpha)\pi_\theta(\beta)d\alpha\,d\beta$$
$$= \int P_0(\alpha,\beta)\pi_\phi(\alpha)\pi_\theta(\beta)d\alpha\,d\beta \quad (6.8)$$

and is therefore a "local" theory. However in HV$_3$

$$\int \tilde{P}_\phi(\alpha,\beta)\tilde{\pi}_\phi(\alpha)\tilde{\pi}_\theta(\beta)d\alpha\,d\beta$$
$$\neq \int \tilde{P}_0(\alpha,\beta)\tilde{\pi}_\phi(\alpha)\tilde{\pi}_\theta(\beta)d\alpha\,d\beta \quad (6.9)$$

and is, in this sense, "nonlocal."

A central theme of this paper is the observation that quantum mechanics (in the α,β representation) is rather closer to our "HV intuition" than might have been expected. As mentioned earlier, other facets of this work will be presented elsewhere.[16]

ACKNOWLEDGMENTS

The author wishes to thank A. Aspect, I. Białynicki-Birula, A. Barut, J. Clauser, D. Leiter, P. Meystre, and S. Stenholm for their helpful suggestions concerning the manuscript. The author also wishes to express his appreciation to the institutions which supported and encouraged this work, in particular: The present work was stimulated by discussions with A. Barut and P. Meystre at the Max-Planck Institute for Quantum Optics, the calculation was then carried out while a lecturer at the 1982 International School of Physics (Islamabad, Pakistan) and extended at the University of New Mexico Center for Advanced Studies where this paper was written.

[1](a) J. Bardeen, L. Cooper, and J. Schrieffer, Phys. Rev. 108, 1175 (1957); (b) P. Dirac, *Quantum Mechanics* (Oxford University, London, 1935); (c) W. Lamb and R. Retherford, Phys. Rev. 72, 241 (1947); (d) see, for example, K. Huang, *Quarks, Leptons and Gauge Fields* (World Scientific, Singapore, 1982).

[2](a) *Quantum Optics and Electronics*, 1964 Les Houches Lectures, edited by C. DeWitt, A. Blandin, and C. Cohen-Tannoudji (Gordon and Breach, New York, 1965); (b) M. Lax, in *Brandeis University Summer Lectures*, edited by M. Chretien, S. Deser, and E. Gross (Gordon and Breach, New York, 1966); (c) H. Haken, *Handbuch Der Physik* (Springer, Berlin, 1920), Vol. 25/2C; (d) W. H. Louisell, *Quantum Statistical Properties of Radiation* (Wiley, New York, 1973); (e) R. Loudon, *The Quantum Theory of Light* (Oxford University, London, 1973); (f) M. Sargent, M. Scully, and W. Lamb, *Laser Physics* (Addison-Wesley, Reading, Mass., 1974).

[3]A. Einstein, B. Podolsky, and N. Rosen, Phys. Rev. 47, 777 (1935). The present spin-$\frac{1}{2}$ example is due to D. Bohm. In this regard B. Hiley quotes Dirac and Feynman [New Sci. 6 (1981)] as follows:

"Paul Dirac, in referring to the apparent non-local effects that arise in quantum mechanics, wrote, 'It is against the spirit of relativity, but it is the best we can do . . . We cannot be content with such a theory.'"

"Richard Feynman recently wrote, 'We have always had a great deal of difficulty in understanding the world view that quantum mechanics represents . . . It has not yet become obvious to me that there's no real problem.'"

We wish to thank J. Cresser for bringing to our attention the article by E. Jaynes, in *Foundations of Radiation—Theory and Quantum Electrodynamics*, edited by A. Barut (Plenum, New York, 1980), in which he says

"Quantum theory not only does not use—it does not even dare to mention—the notion of a 'real physical situation.' Defenders of the theory say that this notion is philosophically naive, a throwback to outmoded ways of thinking, and that recognition of this constitutes deep new wisdom about the nature of human knowledge. I say that it constitutes a violent irrationality, that somewhere in this theory the distinction between reality and our knowledge of reality has become lost, and the result has more the character of medieval necromancy than of science. It has been my hope that quantum optics, with its vast new technological capability, might be able to provide the experimental clue that will show us how to resolve these contradictions."

The present author's point of view on this subject is to be found in an article by M. Scully, R. Shea, and T. McCullen, Phys. Rep. 43C, 487 (1978).

[4]F. Belinfante, *A Survey of Hidden-Variable Theories* (Pergamon, New York, 1973); J. Clauser and A. Shimony, Rep. Prog. Phys. 41, 1881 (1978).

[5]J. Bell, Rev. Mod. Phys. 38, 447 (1966).

[6]A good summary of experimental tests up to 1978 is given by Clauser and Shimony in Ref. 4. The recent work of Aspect is especially interesting in this regard, see A. Aspect, P. H. Grangier and G. Roger, Phys. Rev. Lett. 49, 91 (1982).

[7]B. Hiley, New Sci. 6, 17 (1983).

[8]See, for example, the review article by M. Hillery, R. O'Connell, M. Scully, and E. Wigner (unpublished).

[9]F. Belinfante, in Ref. 4, p. 13

[10]F. Belinfante, in Ref. 4, p. 279.

[11]In this context see especially E. Wigner, Am. J. Phys. 38, 1005 (1970).

[12]See, for example, C. Cantrell and M. Scully, Phys. Rep. 43C, 499 (1978).

[13]The work of Belinfante referred to here actually involves photon polarization correlations rather than the spin correlations which we deal with. However, the two problems are holomorphic and we present here the arguments couched in spin-$\frac{1}{2}$ language. We emphasize however the intellectual content of the arguments is credited to Belinfante. In fact we have even tried to use his phraseology where appropriate in order to remain faithful to his logic.

[14]See especially R. Glauber, in Ref. 2(a) and L. Mandel and E. Wolf, Rev. Mod. Phys. 37, 231 (1965).

[15]For a nice overview of the subject, see E. Hahn in *NMR Grundlagen und Fortschritte* (Springer, Berlin), Vol. 13.

[16]A. Barut, P. Meystre, and M. Scully (unpublished).

V PROBLEMS

1. Show that $|\theta\rangle = e^{-i\frac{\theta}{2}\sigma_y}|\uparrow\rangle$.

Just as $|\Psi(t)\rangle = e^{-i\mathcal{H}t/\hbar}|\Psi(0)\rangle$ and $|x_0+x\rangle = e^{-ipx/\hbar}|x_0\rangle$ we have $|\theta\rangle = e^{-iJ_y\theta/\hbar}|\uparrow\rangle$ where $J_y = (\hbar/2)\sigma_y$. Thus $|\theta\rangle = e^{-i\sigma_y\theta/2}|\uparrow\rangle$. (see Fig. 3.)

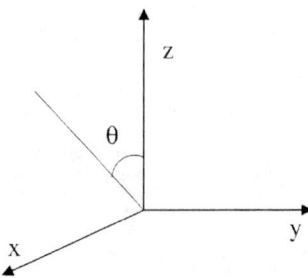

FIGURE 3. Rotation of the spin-up state by an angle θ.

2. Show that $|\theta\rangle\langle\theta| \equiv \hat{\pi}_\theta = \frac{1}{2}[1 + \sigma_z \cos\theta + \sigma_x \sin\theta]$.

$$e^{-i\sigma_y\theta/2} = 1 - i\sigma_y\left(\frac{\theta}{2}\right) - \frac{1}{2!}\sigma_y^2\left(\frac{\theta}{2}\right)^2 + \frac{i}{3!}\sigma_y^3\left(\frac{\theta}{2}\right)^3 + \cdots$$

$$= 1 - \frac{1}{2}\left(\frac{\theta}{2}\right)^2 + \cdots - i\sigma_y\left[\left(\frac{\theta}{2}\right) - \frac{1}{3!}\left(\frac{\theta}{2}\right)^3 + \cdots\right]$$

$$= \cos\frac{\theta}{2} - i\sigma_y\sin\frac{\theta}{2},$$

$$|\theta\rangle\langle\theta| = e^{-i\sigma_y\theta/2}|\uparrow\rangle\langle\uparrow|e^{i\sigma_y\theta/2}$$

$$= \left[\cos\frac{\theta}{2} - i\sigma_y\sin\frac{\theta}{2}\right]\begin{pmatrix}1 & 0\\ 0 & 0\end{pmatrix}\left[\cos\frac{\theta}{2} + i\sigma_y\sin\frac{\theta}{2}\right]$$

$$= \cos^2\frac{\theta}{2}\begin{pmatrix}1 & 0\\ 0 & 0\end{pmatrix} + i\left[\begin{pmatrix}0 & -i\\ 0 & 0\end{pmatrix} - \begin{pmatrix}0 & 0\\ i & 0\end{pmatrix}\right]\sin\frac{\theta}{2}\cos\frac{\theta}{2} + \sin^2\frac{\theta}{2}\begin{pmatrix}0 & 0\\ 0 & 1\end{pmatrix}$$

$$= \cos^2\frac{\theta}{2}\left(\frac{1+\sigma_z}{2}\right) + \sin\frac{\theta}{2}\cos\frac{\theta}{2}\sigma_x + \sin^2\frac{\theta}{2}\left(\frac{1-\sigma_z}{2}\right)$$

$$= \frac{1}{2}[1 + \sigma_z\cos\theta + \sigma_x\sin\theta] \equiv \hat{\pi}_\theta,$$

or using $\sigma_y\sigma_z = -\sigma_z\sigma_y$,

$$|\theta\rangle\langle\theta| = e^{-i\sigma_y\theta/2}\frac{1+\sigma_z}{2}e^{i\sigma_y\theta/2} = \frac{1}{2} + \frac{1}{2}e^{-i\sigma_y\theta/2}\sigma_z e^{i\sigma_y\theta/2}$$

$$= \frac{1}{2}[1 + e^{-i\sigma_y\theta/2}e^{-i\sigma_y\theta/2}\sigma_z] = \frac{1}{2}[1 + \{\cos(\sigma_y\theta) - i\sin(\sigma_y\theta)\}\sigma_z]$$

$$= \frac{1}{2}[1 + \{\cos\theta - i\sigma_y\sin\theta\}\sigma_z] = \frac{1}{2}[1 + \sigma_z\cos\theta + \sigma_x\sin\theta].$$

3. Show that the simultaneous passage probability for a spin 1/2 particle is given by $\mathcal{P}_\theta(\alpha) = \frac{1}{2}[1 + \cos(\theta - \alpha)]$.

$$\begin{aligned}
\mathcal{P}_\theta(\alpha) &= |\langle\theta|\alpha\rangle|^2 = |\langle\uparrow|e^{i(\frac{\theta}{2}-\frac{\alpha}{2})\sigma_y}|\uparrow\rangle|^2 \\
&= \langle\uparrow|e^{i(\frac{\theta}{2}-\frac{\alpha}{2})\sigma_y}|\uparrow\rangle\langle\uparrow|e^{-i(\frac{\theta}{2}-\frac{\alpha}{2})\sigma_y}|\uparrow\rangle \\
&= \frac{1}{2}\langle\uparrow|[1 + \cos(\theta - \alpha)\sigma_z + \sin(\theta - \alpha)\sigma_x]|\uparrow\rangle \\
&= \frac{1}{2}[1 + \cos(\theta - \alpha)] .
\end{aligned}$$

4. Show that the joint probability of two-spin state $|\Psi\rangle$ passing SGA at θ_a for particle 1 and θ_b for particle 2 is $\mathcal{P}(\theta_a, \theta_b) = \langle\Psi|\hat{\pi}^{(1)}_{\theta_a}\hat{\pi}^{(2)}_{\theta_b}|\Psi\rangle$.

$$\begin{aligned}
\mathcal{P}(\theta_a, \theta_b) &= |\langle\theta^{(1)}_a|\langle\theta^{(2)}_b|\Psi\rangle|^2 \\
&= \langle\Psi|\theta^{(1)}_a\rangle|\theta^{(2)}_b\rangle\langle\theta^{(1)}_a|\langle\theta^{(2)}_b|\Psi\rangle \\
&= \langle\Psi|\hat{\pi}^{(1)}_{\theta_a}\hat{\pi}^{(2)}_{\theta_b}|\Psi\rangle.
\end{aligned}$$

5. Find $\mathcal{P}(\theta_a, \theta_b)$ for spin singlet $|\Psi\rangle = \frac{1}{\sqrt{2}}[|\uparrow_1\downarrow_2\rangle - |\downarrow_1\uparrow_2\rangle]$.

$$\begin{aligned}
\mathcal{P}(\theta_a, \theta_b) &= \frac{1}{2}(\langle\uparrow_1\downarrow_2| - \langle\downarrow_1\uparrow_2|)\frac{1}{4}\left[1 + \sigma^{(1)}_z \cos\theta_a + \sigma^{(1)}_x \sin\theta_a\right] \\
&\quad \times \left[1 + \sigma^{(2)}_z \cos\theta_b + \sigma^{(2)}_x \sin\theta_b\right](|\uparrow_1\downarrow_2\rangle - |\downarrow_1\uparrow_2\rangle) \\
&= \frac{1}{8}\Big[(\langle\uparrow_1\downarrow_2| - \langle\downarrow_1\uparrow_2|) + (\langle\uparrow_1\downarrow_2| + \langle\downarrow_1\uparrow_2|)\cos\theta_a + (\langle\downarrow_1\downarrow_2| - \langle\uparrow_1\uparrow_2|)\sin\theta_a\Big] \\
&\quad \times \Big[(|\uparrow_1\downarrow_2\rangle - |\downarrow_1\uparrow_2\rangle) + (-|\uparrow_1\uparrow_2\rangle - |\downarrow_1\downarrow_2\rangle)\cos\theta_b + (|\uparrow_1\uparrow_2\rangle - |\downarrow_1\downarrow_2\rangle)\sin\theta_b\Big] \\
&= \frac{1}{4}\Big[1 - (\cos\theta_a \cos\theta_b + \sin\theta_a \sin\theta_b)\Big] \\
&= \frac{1}{4}[1 - \cos(\theta_a - \theta_b)].
\end{aligned}$$

6. Show that $\delta(\sigma_z - m_z)|\uparrow\rangle = \delta(m_z - 1)|\uparrow\rangle$ and $\delta(\sigma_z - m_z)|\downarrow\rangle = \delta(m_z + 1)|\downarrow\rangle$.

$$\begin{aligned}
\delta(\sigma_z - m_z)|\uparrow\rangle &= \int \frac{d\eta}{2\pi} e^{i\eta(\sigma_z - m_z)}|\uparrow\rangle \\
&= \int \frac{d\eta}{2\pi} e^{i\eta(1 - m_z)}|\uparrow\rangle \\
&= \delta(m_z - 1)|\uparrow\rangle,
\end{aligned}$$

and same for $|\downarrow\rangle$ but with $1 \to -1$ because $\sigma_z|\downarrow\rangle = -|\downarrow\rangle$.

7. Show that

$$\delta(\sigma_x - m_x) = \int \frac{d\xi}{2\pi} [\cos\xi + i\sigma_x \sin\xi] e^{-im_x\xi}.$$

$$\delta(\sigma_x - m_x) = \int \frac{d\xi}{2\pi} e^{i\xi(\sigma_x - m_x)}.$$

As in Prob. 2, $e^{i\xi\sigma_x} = \cos\xi + i\sigma_x \sin\xi$, and the result follows.

8. Find the distribution for spin up, $P_\uparrow(\alpha)$.

$$P_\uparrow(m_x, m_z) = \text{Tr}\{\delta(\sigma_x - m_x)\delta(\sigma_z - m_z)\,\rho\}$$

$$= \int \frac{d\xi\,d\eta}{2\pi\,2\pi} \langle\uparrow|e^{i\xi\sigma_x}e^{i\eta\sigma_z}|\uparrow\rangle e^{-im_x\xi}e^{-im_z\eta}$$

$$= \int \frac{d\xi\,d\eta}{2\pi\,2\pi} \langle\uparrow|(\cos\xi + i\sigma_x \sin\xi)\,e^{i\eta}|\uparrow\rangle e^{-im_x\xi}e^{-im_z\eta}$$

$$= \frac{1}{2}\int \frac{d\xi\,d\eta}{2\pi\,2\pi}\left[e^{i\xi(1-m_x)} + e^{-i\xi(1+m_x)}\right]e^{i\eta(1-m_z)}$$

$$= \frac{1}{2}\left[\delta(m_x - 1) + \delta(m_x + 1)\right]\delta(m_z - 1)$$

$$\Rightarrow\quad P_\uparrow(\alpha) = \frac{1}{2}\left[\delta(\alpha - \frac{\pi}{4}) + \delta(\alpha + \frac{\pi}{4})\right]\delta(m - \sqrt{2}). \quad \text{(see PRD-Fig. 3)}$$

9. Find the distribution function $P_\phi(m_x, m_z)$ for

$$|\Psi\rangle = \frac{1}{\sqrt{2}}[|+\phi, -\phi\rangle - |-\phi, +\phi\rangle],$$

where $|+\phi\rangle = e^{-i\sigma_y\phi/2}|\uparrow\rangle$ and $|-\phi\rangle = e^{-i\sigma_y(\phi+\pi)/2}|\uparrow\rangle$.

$$P_\phi(m_x, m_z) = \langle\phi_+|\,\delta(\sigma_x - m_x)\,\delta(\sigma_z - m_z)|\phi_+\rangle$$

$$= \int \frac{d\xi}{2\pi}\left[\cos\frac{\phi}{2}\langle\uparrow| + \sin\frac{\phi}{2}\langle\downarrow|\right][\cos\xi + i\sigma_x \sin\xi]\,\delta(\sigma_z - m_z)$$

$$\times \left[\cos\frac{\phi}{2}|\uparrow\rangle + \sin\frac{\phi}{2}|\downarrow\rangle\right]e^{-im_x\xi}$$

$$= \int \frac{d\xi}{2\pi}\left[\cos\frac{\phi}{2}\langle\uparrow| + \sin\frac{\phi}{2}\langle\downarrow|\right][\cos\xi + i\sigma_x \sin\xi]$$

$$\times \left[|\uparrow\rangle\cos\frac{\phi}{2}\delta(1-m_z) + |\downarrow\rangle\sin\frac{\phi}{2}\delta(1+m_z)\right]e^{-im_x\xi}$$

$$= \cos^2\frac{\phi}{2}\frac{1}{2}[\delta(m_x+1) + \delta(m_x-1)]\,\delta(m_z-1)$$

$$+ \sin^2\frac{\phi}{2}\frac{1}{2}[\delta(m_x+1) + \delta(m_x-1)]\,\delta(m_z+1)$$

$$+2\sin\frac{\phi}{2}\cos\frac{\phi}{2}\frac{1}{4}[\delta(m_x-1)-\delta(m_x+1)]\delta(m_z+1)$$
$$+2\sin\frac{\phi}{2}\cos\frac{\phi}{2}\frac{1}{4}[\delta(m_x-1)-\delta(m_x+1)]\delta(m_z-1)$$
$$=\frac{1}{2}\cos^2\frac{\phi}{2}\delta(m_z-1)[\delta(m_x+1)+\delta(m_x-1)]$$
$$+\frac{1}{2}\sin^2\frac{\phi}{2}\delta(m_z+1)[\delta(m_x+1)+\delta(m_x-1)]$$
$$+\frac{1}{4}\sin\phi\,\delta(m_x-1)[\delta(m_z+1)+\delta(m_z-1)]$$
$$-\frac{1}{4}\sin\phi\,\delta(m_x+1)[\delta(m_z+1)+\delta(m_z-1)].$$

10. Find $P_{\pm\phi}(\alpha)$ in $|+\phi\rangle, |-\phi\rangle$ basis (prove PRD-(3.14)).

First note that
$$P_\uparrow(\alpha)=\frac{1}{2}\left[\delta(\alpha-\frac{\pi}{4})+\delta(\alpha+\frac{\pi}{4})\right]\delta(m-\sqrt{2}).$$

Now in $|+\phi\rangle, |-\phi\rangle$ basis
$$P_{+\phi}(\alpha_\phi)=\frac{1}{2}[\delta(m_{\phi x}+1)+\delta(m_{\phi x}-1)]\delta(m_{\phi z}-1)$$
$$=\frac{1}{2}\left[\delta(\alpha_\phi-\frac{\pi}{4})+\delta(\alpha_\phi+\frac{\pi}{4})\right]\delta(m-\sqrt{2}).$$

Since $\alpha_\phi=\alpha-\phi$, we have (see Fig. 4)
$$P_{+\phi}(\alpha)=\frac{1}{2}\left[\delta(\alpha-\phi-\frac{\pi}{4})+\delta(\alpha-\phi+\frac{\pi}{4})\right]\delta(m-\sqrt{2}).$$

And since $\alpha_{-\phi}=\alpha-\phi-\pi$,
$$P_{-\phi}(\alpha)=\frac{1}{2}\left[\delta(\alpha-\phi-\frac{3\pi}{4})+\delta(\alpha-\phi+\frac{3\pi}{4})\right]\delta(m-\sqrt{2}).$$

11. Show that $\langle\sigma_z\rangle=\cos\phi$, $\langle\sigma_x\rangle=\sin\phi$ by using i)$|\Psi\rangle=|\phi\rangle$, ii)$P_\phi(m_x,m_z)$ from Prob. 9, and iii) $P_\phi(\alpha)$ from Prob. 10.

---

i) $$\langle\sigma_z\rangle=\langle\phi|\sigma_z|\phi\rangle$$
$$=\left[\cos\frac{\phi}{2}\langle\uparrow|+\sin\frac{\phi}{2}\langle\downarrow|\right]\sigma_z\left[\cos\frac{\phi}{2}|\uparrow\rangle+\sin\frac{\phi}{2}|\downarrow\rangle\right]$$
$$=\cos^2\frac{\phi}{2}-\sin^2\frac{\phi}{2}=\cos\phi,$$

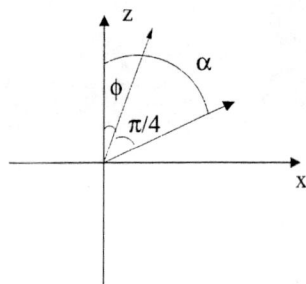

FIGURE 4. $\alpha_\phi = \alpha - \phi = \frac{\pi}{4}$.

$$\langle \sigma_x \rangle = \left[\cos \frac{\phi}{2} \langle \uparrow | + \sin \frac{\phi}{2} \langle \downarrow | \right] \sigma_x \left[\cos \frac{\phi}{2} | \uparrow \rangle + \sin \frac{\phi}{2} | \downarrow \rangle \right]$$
$$= 2 \sin \frac{\phi}{2} \cos \frac{\phi}{2} = \sin \phi \ .$$

ii) $\langle \sigma_z \rangle = \int dm_z dm_x m_z P_\phi$

$$= \int dm_x \left[\delta(m_x + 1) + \delta(m_x - 1) \right] \int dm_z m_z \delta(m_z - 1) \frac{1}{2} \cos^2 \frac{\phi}{2}$$
$$+ \int dm_x \left[\delta(m_x + 1) + \delta(m_x - 1) \right] \int dm_z m_z \delta(m_z + 1) \frac{1}{2} \sin^2 \frac{\phi}{2}$$
$$= \cos^2 \frac{\phi}{2} - \sin^2 \frac{\phi}{2} = \cos \phi \ ,$$

$\langle \sigma_x \rangle = \int dm_z dm_x m_x P_\phi$

$$= \int dm_z \left[\delta(m_z + 1) + \delta(m_z - 1) \right] \int dm_x m_x \delta(m_x - 1) \frac{1}{4} \sin \phi$$
$$- \int dm_z \left[\delta(m_z + 1) + \delta(m_z - 1) \right] \int dm_x m_x \delta(m_x + 1) \frac{1}{4} \sin \phi$$
$$= \sin \phi \ .$$

iii) $\langle \sigma_z \rangle = \int \frac{d\alpha}{2} \left[\delta(\alpha - \phi - \frac{\pi}{4}) + \delta(\alpha - \phi + \frac{\pi}{4}) \right] \delta(m - \sqrt{2}) m \cos \alpha$

$$= \frac{1}{2} \sqrt{2} \left[\cos(\phi + \frac{\pi}{4}) + \cos(\phi - \frac{\pi}{4}) \right]$$
$$= \frac{1}{2} \sqrt{2} \left[2 \cos \phi \cos \frac{\pi}{4} \right] = \cos \phi \ ,$$

$\langle \sigma_x \rangle = \int \frac{d\alpha}{2} \left[\delta(\alpha - \phi - \frac{\pi}{4}) + \delta(\alpha - \phi + \frac{\pi}{4}) \right] \delta(m - \sqrt{2}) m \sin \alpha$

$$= \frac{1}{2} \sqrt{2} \left[\sin(\phi + \frac{\pi}{4}) + \sin(\phi - \frac{\pi}{4}) \right]$$

$$= \frac{1}{2}\sqrt{2}\left[2\sin\phi\cos\frac{\pi}{4}\right] = \sin\phi .$$

12. Find the two-particle distribution $P_{\phi=0}(m_{1x}, m_{1z}, m_{2x}, m_{2z})$ for spin-singlet state.

$P_{\phi=0}(m_{1x}, m_{1z}, m_{2x}, m_{2z})$
$= \langle\Psi|\delta(\sigma_{1x} - m_{1x})\delta(\sigma_{1z} - m_{1z})\delta(\sigma_{2x} - m_{2x})\delta(\sigma_{2z} - m_{2z})|\Psi\rangle$
$= \frac{1}{2}(\langle\uparrow,\downarrow| - \langle\downarrow,\uparrow|)\delta(\sigma_{1x} - m_{1x})\delta(\sigma_{1z} - m_{1z})$
$\quad \times \delta(\sigma_{2x} - m_{2x})\delta(\sigma_{2z} - m_{2z})(|\uparrow,\downarrow\rangle - |\downarrow,\uparrow\rangle)$
$= \frac{1}{2}\Big[\langle\uparrow,\downarrow|\Delta|\uparrow,\downarrow\rangle - \langle\uparrow,\downarrow|\Delta|\downarrow,\uparrow\rangle - \langle\downarrow,\uparrow|\Delta|\uparrow,\downarrow\rangle + \langle\downarrow,\uparrow|\Delta|\downarrow,\uparrow\rangle\Big],$

where $\Delta = \delta(\sigma_{1x} - m_{1x})\delta(\sigma_{1z} - m_{1z})\delta(\sigma_{2x} - m_{2x})\delta(\sigma_{2z} - m_{2z})$.
Now as in Prob. 9 we obtain

$P_{\phi=0}(m_{1x}, m_{1z}, m_{2x}, m_{2z})$
$= \frac{1}{2}\Big[\frac{1}{2}\Big\{\delta(m_{1x} - 1) + \delta(m_{1x} + 1)\Big\}\delta(m_{1z} - 1)$
$\quad \times \frac{1}{2}\Big\{\delta(m_{2x} - 1) + \delta(m_{2x} + 1)\Big\}\delta(m_{2z} + 1)$
$\quad - \frac{1}{2}\Big\{\delta(m_{1x} - 1) - \delta(m_{1x} + 1)\Big\}\delta(m_{1z} + 1)$
$\quad \times \frac{1}{2}\Big\{\delta(m_{2x} - 1) - \delta(m_{2x} + 1)\Big\}\delta(m_{2z} - 1)$
$\quad - \frac{1}{2}\Big\{\delta(m_{1x} - 1) - \delta(m_{1x} + 1)\Big\}\delta(m_{1z} - 1)$
$\quad \times \frac{1}{2}\Big\{\delta(m_{2x} - 1) - \delta(m_{2x} + 1)\Big\}\delta(m_{2z} + 1)$
$\quad + \frac{1}{2}\Big\{\delta(m_{1x} - 1) + \delta(m_{1x} + 1)\Big\}\delta(m_{1z} + 1)$
$\quad \times \frac{1}{2}\Big\{\delta(m_{2x} - 1) + \delta(m_{2x} + 1)\Big\}\delta(m_{2z} - 1)\Big]$
$= \frac{1}{4}\Big[\delta(m_{1z} - 1)\delta(m_{2z} + 1) + \delta(m_{1z} + 1)\delta(m_{2z} - 1)\Big]$
$\quad \times \Big[\delta(m_{1x} - 1)\delta(m_{2x} + 1) + \delta(m_{1x} + 1)\delta(m_{2x} - 1)\Big] .$

13. Find $P_\phi(\alpha, \beta)$ for spin-singlet state (derive PRD-(4.4)).

From Prob. 12 we can write

$$P_{\phi=0}(\alpha,\beta) = \frac{1}{4}\Big[\delta(\alpha-\frac{\pi}{4})\delta(\beta+\frac{3\pi}{4}) + \delta(\alpha+\frac{\pi}{4})\delta(\beta-\frac{3\pi}{4})$$
$$+ \delta(\alpha-\frac{3\pi}{4})\delta(\beta+\frac{\pi}{4}) + \delta(\alpha+\frac{3\pi}{4})\delta(\beta-\frac{\pi}{4})\Big]$$
$$= \frac{1}{4}\delta(\alpha-\beta-\pi)\Big[\delta(\alpha-\frac{\pi}{4}) + \delta(\alpha+\frac{\pi}{4}) + \delta(\alpha-\frac{3\pi}{4}) + \delta(\alpha+\frac{3\pi}{4})\Big].$$

Then we immediately obtain

$$P_{\phi}(\alpha,\beta) = \frac{1}{4}\delta(\alpha-\beta-\pi)$$
$$\times \Big[\delta(\alpha-\phi-\frac{\pi}{4}) + \delta(\alpha-\phi+\frac{\pi}{4}) + \delta(\alpha-\phi-\frac{3\pi}{4}) + \delta(\alpha-\phi+\frac{3\pi}{4})\Big].$$

14. Use $P_{\phi=0}(\alpha,\beta)$ from Prob. 13, and $\pi^{(1)}_{\theta_a}(\alpha)$ and $\pi^{(2)}_{\theta_b}(\beta)$ as per PRD-(3.12) to show that the joint probability is given by

$$\mathcal{P}(\theta_a,\theta_b) = \frac{1}{4}[1-\cos(\theta_a-\theta_b)].$$

$$\pi^{(1)}_{\theta_a}(\alpha) = \frac{1}{2}[1 + m\cos(\alpha-\theta_a)] \quad (m=\sqrt{2}),$$

$$\mathcal{P}_{\uparrow}(\theta_a,\theta_b) = \langle \hat{\pi}^{(1)}_{\theta_a}(\alpha)\hat{\pi}^{(2)}_{\theta_b}(\beta)\rangle$$
$$= \int d\alpha d\beta P_{\phi=0}(\alpha,\beta)\pi^{(1)}_{\theta_a}(\alpha)\pi^{(2)}_{\theta_b}(\beta)$$
$$= \int d\alpha d\beta \frac{1}{4}\delta(\alpha-\beta-\pi)\Big[\delta(\alpha-\frac{\pi}{4}) + \delta(\alpha+\frac{\pi}{4}) + \delta(\alpha-\frac{3\pi}{4}) + \delta(\alpha+\frac{3\pi}{4})\Big]$$
$$\times \frac{1}{2}\Big[1+m\cos(\alpha-\theta_a)\Big]\frac{1}{2}\Big[1+m\cos(\beta-\theta_b)\Big]\delta(m-\sqrt{2})$$
$$= \frac{1}{16}\int d\alpha \Big[\delta(\alpha-\frac{\pi}{4}) + \delta(\alpha+\frac{\pi}{4}) + \delta(\alpha-\frac{3\pi}{4}) + \delta(\alpha+\frac{3\pi}{4})\Big]$$
$$\times \Big[1+\sqrt{2}\cos(\alpha-\theta_a)\Big]\Big[1-\sqrt{2}\cos(\alpha-\theta_b)\Big]$$
$$= \frac{1}{16}\Big[(1+\cos\theta_a+\sin\theta_a)(1-\cos\theta_b-\sin\theta_b)$$
$$+ (1+\cos\theta_a-\sin\theta_a)(1-\cos\theta_b+\sin\theta_b)$$
$$+ (1-\cos\theta_a+\sin\theta_a)(1+\cos\theta_b-\sin\theta_b)$$
$$+ (1-\cos\theta_a-\sin\theta_a)(1+\cos\theta_b+\sin\theta_b)\Big]$$
$$= \frac{1}{16}\Big[4 - 2(\cos\theta_a+\sin\theta_a)(\cos\theta_b+\sin\theta_b) - 2(\cos\theta_a-\sin\theta_a)(\cos\theta_b-\sin\theta_b)\Big]$$
$$= \frac{1}{4}\Big[1-(\cos\theta_a\cos\theta_b+\sin\theta_a\sin\theta_b)\Big]$$
$$= \frac{1}{4}\Big[1-\cos(\theta_a-\theta_b)\Big].$$

15 Find $\mathcal{P}(\theta_a, \theta_b)$ by using $P_{\phi=\theta_a}(\alpha, \beta)$ instead of $P_{\phi=0}(\alpha, \beta)$.

$$\mathcal{P}_{\phi=\theta_a}(\theta_a, \theta_b) = \int d\alpha d\beta \, P_\phi(\alpha, \beta)\pi_{\theta_a}(\alpha)\pi_{\theta_b}(\beta)$$

$$= \frac{1}{16}\int d\alpha d\beta \, \delta(\alpha - \beta - \pi)\Big[\delta(\alpha - \phi - \frac{\pi}{4}) + \delta(\alpha - \phi + \frac{\pi}{4}) + \delta(\alpha - \phi - \frac{3\pi}{4})$$
$$+ \delta(\alpha - \phi + \frac{3\pi}{4})\Big] \times \Big[1 + m\cos(\alpha - \theta_a)\Big]\Big[1 + m\cos(\beta - \theta_b)\Big]\delta(m - \sqrt{2})$$

$$= \frac{1}{16}\int d\alpha \, \Big[\delta(\alpha - \theta_a - \frac{\pi}{4}) + \delta(\alpha - \theta_a + \frac{\pi}{4})\delta(\alpha - \theta_a - \frac{3\pi}{4}) + \delta(\alpha - \theta_a + \frac{3\pi}{4})\Big]$$
$$\times \Big[1 + m\cos(\alpha - \theta_a)\Big]\Big[1 - m\cos(\alpha - \theta_b)\Big]\delta(m - \sqrt{2})$$

$$= \frac{1}{16}\Big[\left(1 + m\cos\frac{\pi}{4}\right)\left(1 - m\cos(\theta_a - \theta_b + \frac{\pi}{4})\right)$$
$$+ \left(1 + m\cos\frac{\pi}{4}\right)\left(1 - m\cos(\theta_a - \theta_b - \frac{\pi}{4})\right)$$
$$+ \left(1 + m\cos\frac{3\pi}{4}\right)\left(1 - m\cos(\theta_a - \theta_b + \frac{3\pi}{4})\right)$$
$$+ \left(1 + m\cos\frac{3\pi}{4}\right)\left(1 - m\cos(\theta_a - \theta_b - \frac{3\pi}{4})\right)\Big]\delta(m - \sqrt{2})$$

$$= \frac{1}{8}\Big[\big\{1 - \cos(\theta_a - \theta_b) + \sin(\theta_a - \theta_b)\big\} + \big\{1 - \cos(\theta_a - \theta_b) - \sin(\theta_a - \theta_b)\big\}\Big]$$

$$= \frac{1}{4}\Big[1 - \cos(\theta_a - \theta_b)\Big].$$

16 Repeat Prob. 15 but use $P_\phi(\alpha, \beta)$ for an arbitrary ϕ.

$$\mathcal{P}_\phi(\theta_a, \theta_b) = \int d\alpha d\beta \, P_\phi(\alpha, \beta)\pi_{\theta_a}(\alpha)\pi_{\theta_b}(\beta)$$

$$= \frac{1}{16}\int d\alpha d\beta \, \delta(\alpha - \beta - \pi)\Big[\delta(\alpha - \phi - \frac{\pi}{4}) + \delta(\alpha - \phi + \frac{\pi}{4}) + \delta(\alpha - \phi - \frac{3\pi}{4})$$
$$+ \delta(\alpha - \phi + \frac{3\pi}{4})\Big] \times \Big[1 + \sqrt{2}\cos(\alpha - \theta_a)\Big]\Big[1 + \sqrt{2}\cos(\beta - \theta_b)\Big].$$

Since

$$1 + \sqrt{2}\cos\left(\phi \pm \frac{\pi}{4} - \theta_a\right) = 1 + \cos\phi\Big[\cos\theta_a \pm \sin\theta_a\Big] \mp \sin\phi\Big[\cos\theta_a \mp \sin\theta_a\Big],$$

$$1 - \sqrt{2}\cos\left(\phi \pm \frac{\pi}{4} - \theta_b\right) = 1 - \cos\phi\Big[\cos\theta_b \pm \sin\theta_b\Big] \pm \sin\phi\Big[\cos\theta_b \mp \sin\theta_b\Big],$$

$$A(\phi) \equiv \Big[1 + \sqrt{2}\cos\left(\phi + \frac{\pi}{4} - \theta_a\right)\Big]\Big[1 - \sqrt{2}\cos\left(\phi + \frac{\pi}{4} - \theta_b\right)\Big]$$
$$+ \Big[1 + \sqrt{2}\cos\left(\phi - \frac{\pi}{4} - \theta_a\right)\Big]\Big[1 - \sqrt{2}\cos\left(\phi - \frac{\pi}{4} - \theta_b\right)\Big]$$

$$= 2\Big[1 - (\cos\theta_b - \cos\theta_a)(\cos\phi - \sin\phi) - \cos(\theta_a - \theta_b)\Big],$$
$$A(\phi + \pi) = 2\Big[1 + (\cos\theta_b - \cos\theta_a)(\cos\phi - \sin\phi) - \cos(\theta_a - \theta_b)\Big].$$

Since
$$\mathcal{P}_\phi(\theta_a, \theta_b) = \frac{1}{16}\Big[A(\phi) + A(\phi + \pi)\Big],$$

we have
$$\mathcal{P}_\phi(\theta_a, \theta_b) = \frac{1}{4}\Big[1 - \cos(\theta_a - \theta_b)\Big].$$

17. Find the joint probability for HV$_3$, Derive $\tilde{\mathcal{P}}(\theta_a, \theta_b)$ for $\phi = \theta_a$ by using
$$\tilde{\mathcal{P}}_\phi(\alpha, \beta) = \delta(\alpha - \beta - \pi)\frac{1}{2}\Big[\delta(\alpha - \phi) + \delta(\alpha - \phi - \pi)\Big],$$
$$\tilde{\pi}_{\theta_a}(\alpha) = \frac{1}{2}\Big[1 + \cos(\alpha - \theta_a)\Big],$$
$$\tilde{\pi}_{\theta_b}(\beta) = \frac{1}{2}\Big[1 + \cos(\beta - \theta_b)\Big].$$

$$\tilde{\mathcal{P}}_{\phi=\theta_a}(\theta_a, \theta_b) = \int d\alpha d\beta \; \tilde{\mathcal{P}}_\phi(\alpha, \beta)\tilde{\pi}_{\theta_a}(\alpha)\tilde{\pi}_{\theta_b}(\beta)$$
$$= \int d\alpha d\beta \frac{1}{2}\delta(\alpha - \beta - \pi)\Big[\delta(\alpha - \theta_a) + \delta(\alpha - \theta_a - \pi)\Big]$$
$$\times \frac{1}{2}\Big[1 + \cos(\alpha - \theta_a)\Big]\frac{1}{2}\Big[1 + \cos(\beta - \theta_b)\Big]$$
$$= \frac{1}{8}\int d\alpha \Big[\delta(\alpha - \theta_a) + \delta(\alpha - \theta_a - \pi)\Big]\Big[1 + \cos(\alpha - \theta_a)\Big]\Big[1 - \cos(\alpha - \theta_b)\Big]$$
$$= \frac{1}{8}\Big[2\big\{1 - \cos(\theta_a - \theta_b)\big\} + 0\big\{1 + \cos(\theta_a - \theta_b)\big\}\Big]$$
$$= \frac{1}{4}\Big[1 - \cos(\theta_a - \theta_b)\Big].$$

18. Find the joint probability for HV$_3$, $\tilde{\mathcal{P}}(\theta_a, \theta_b)$ but use $\tilde{\mathcal{P}}_\uparrow(\alpha, \beta) = \tilde{\mathcal{P}}_{\phi=0}(\alpha, \beta)$ and show that
$$\tilde{\mathcal{P}}_{\phi=0}(\theta_a, \theta_b) = \frac{1}{4}\Big[1 - \frac{1}{2}\cos(\theta_a - \theta_b) - \frac{1}{2}\cos(\theta_a + \theta_b)\Big].$$

$$\tilde{\mathcal{P}}_{\phi=0}(\theta_a, \theta_b) = \int d\alpha d\beta \; \tilde{P}_\uparrow(\alpha, \beta) \; \tilde{\pi}_{\theta_a}(\alpha) \; \tilde{\pi}_{\theta_b}(\beta)$$

$$= \int d\alpha d\beta \frac{1}{2}\delta(\alpha - \beta - \pi)\Big[\delta(\alpha) + \delta(\alpha - \pi)\Big]$$

$$\times \frac{1}{2}\Big[1 + \cos(\alpha - \theta_a)\Big] \frac{1}{2}\Big[1 + \cos(\beta - \theta_b)\Big]$$

$$= \frac{1}{8} \int d\alpha \Big[\delta(\alpha) + \delta(\alpha - \pi)\Big]\Big[1 + \cos(\alpha - \theta_a)\Big]\Big[1 - \cos(\alpha - \theta_b)\Big]$$

$$= \frac{1}{8}\Big[(1 + \cos\theta_a)(1 - \cos\theta_b) + \{1 + \cos(\theta_a + \pi)\}\{1 - \cos(\theta_b + \pi)\}\Big]$$

$$= \frac{1}{4}\Big[1 - \frac{1}{2}\cos(\theta_a - \theta_b) - \frac{1}{2}\cos(\theta_a + \theta_b)\Big].$$

For $\theta_a = 0$ or $\theta_b = 0$ we have quantum result. But unlike quantum mechanics (which is OK for $P_\uparrow$ and P_ϕ, etc) we must choose $P_\sigma(\alpha, \beta)$ to "go with" $\tilde{\pi}_\phi(\alpha)$, i.e., must choose $\phi = \theta_{a,b}$ to get the correct answer.

For an arbitrary ϕ, we have

$$\tilde{\mathcal{P}}_\phi(\theta_a, \theta_b) = \int d\alpha d\beta \; \tilde{P}_\phi(\alpha, \beta)\tilde{\pi}_{\theta_a}(\alpha)\tilde{\pi}_{\theta_b}(\beta)$$

$$= \int d\alpha d\beta \; \frac{1}{2}\delta(\alpha - \beta - \pi)\Big[\delta(\alpha - \phi) + \delta(\alpha - \phi + \pi)\Big]$$

$$\times \frac{1}{2}\Big[1 + \cos(\alpha - \theta_a)\Big] \frac{1}{2}\Big[1 + \cos(\beta - \theta_b)\Big]$$

$$= \frac{1}{8}\int d\alpha \Big[\delta(\alpha - \phi) + \delta(\alpha - \phi - \pi)\Big]\Big[1 + \cos(\alpha - \theta_a)\Big]\Big[1 - \cos(\alpha - \theta_b)\Big]$$

$$= \frac{1}{8}\Big[\{1 + \cos(\phi - \theta_a)\}\{1 - \cos(\phi - \theta_b)\}$$

$$+ \{1 - \cos(\phi - \theta_a)\}\{1 + \cos(\phi - \theta_b)\}\Big]$$

$$= \frac{1}{4}\Big[1 - \frac{1}{2}\cos(\theta_a - \theta_b) - \frac{1}{2}\cos(2\phi - \theta_a - \theta_b)\Big].$$

If $\phi = \theta_a$ or $\phi = \theta_b$,

$$\tilde{\mathcal{P}}_\phi(\theta_a, \theta_b) = \frac{1}{4}\Big[1 - \cos(\theta_a - \theta_b)\Big].$$

19. For the mixed representation discussed in Sec. IV we here introduce an operator delta function

$$\delta(\hat{\pi}_a - x_a) = \int \frac{d\xi}{2\pi} e^{-i\xi(\hat{\pi}_a - x_a)}.$$

Show that for $|\Psi\rangle = |\uparrow\rangle$

$$P_\uparrow(x_a) = \text{Tr}\Big\{\delta(\hat{\pi}_a - x_a)\rho\Big\}$$
$$= \frac{1}{2}(1 + \cos\theta_a)\delta(x_a - 1) + \frac{1}{2}(1 - \cos\theta_a)\delta(x_a).$$

$$\text{Tr}\Big\{\delta(\hat{\pi}_a - x_a)\rho\Big\} = \int \frac{d\xi}{2\pi} e^{i\xi x_a} \text{Tr}\Big\{e^{-i\xi\hat{\pi}_a}\rho\Big\},$$

$$e^{-i\xi\hat{\pi}_a} = 1 - i\xi\hat{\pi}_a - \frac{\xi^2}{2}\hat{\pi}_a^2 + \cdots$$
$$= 1 + \left[-i\xi - \frac{\xi^2}{2}\cdots\right]\hat{\pi}_a$$
$$= 1 - \hat{\pi}_a + e^{-i\xi}\hat{\pi}_a,$$

and

$$\text{Tr}\Big\{\hat{\pi}_a\rho\Big\} = \langle\uparrow|\left(\cos\frac{\theta_a}{2}|\uparrow\rangle + \sin\frac{\theta_a}{2}|\downarrow\rangle\right)\left(\cos\frac{\theta_a}{2}\langle\uparrow| + \sin\frac{\theta_a}{2}\langle\downarrow|\right)|\uparrow\rangle$$
$$= \cos^2\frac{\theta_a}{2}.$$

Therefore,

$$P_\uparrow(x_a) = \text{Tr}\Big\{\delta(\hat{\pi}_a - x_a)\rho\Big\}$$
$$= \int \frac{d\xi}{2\pi} e^{i\xi x_a}\left\{\left(1 - \cos^2\frac{\theta_a}{2}\right) + \cos^2\frac{\theta_a}{2}e^{-i\xi}\right\}$$
$$= \sin^2\frac{\theta_a}{2}\delta(x_a) + \cos^2\frac{\theta_a}{2}\delta(x_a - 1)$$
$$= \frac{1}{2}(1 + \cos\theta_a)\delta(x_a - 1) + \frac{1}{2}(1 - \cos\theta_a)\delta(x_a).$$

20. Show that for spin-singlet system

$$P(x_a, x_b) = \frac{1}{4}\Big[\Big(1 - \cos(\theta_a - \theta_b)\Big)\Big\{\delta(x_a)\delta(x_b) + \delta(x_a - 1)\delta(x_b - 1)\Big\}$$
$$+ \Big(1 + \cos(\theta_a - \theta_b)\Big)\Big\{\delta(x_a)\delta(x_b - 1) + \delta(x_a - 1)\delta(x_b)\Big\}\Big].$$

Note that

$$\langle\uparrow|\hat{\pi}_a|\downarrow\rangle = \langle\downarrow|\hat{\pi}_a|\uparrow\rangle = \sin\frac{\theta_a}{2}\cos\frac{\theta_a}{2}, \qquad \langle\downarrow|\hat{\pi}_a|\downarrow\rangle = \sin^2\frac{\theta_a}{2}.$$

This, as in Prob. 19, leads to

$$\langle\uparrow|\delta(\hat{\pi}_a - x_a)|\downarrow\rangle = \langle\downarrow|\delta(\hat{\pi}_a - x_a)|\uparrow\rangle = -\frac{1}{2}\sin\theta_a\Big[\delta(x_a) - \delta(x_a - 1)\Big]$$

$$\langle\downarrow|\delta(\hat{\pi}_a - x_a)|\downarrow\rangle = \frac{1}{2}(1 + \cos\theta_a)\delta(x_a) + \frac{1}{2}(1 - \cos\theta_a)\delta(x_a - 1).$$

We then have

$$P(x_a, x_b) = \mathrm{Tr}\Big\{\delta(\hat{\pi}_a - x_a)\delta(\hat{\pi}_b - x_b)\rho\Big\}$$

$$= \frac{1}{2}\Big(\langle\uparrow,\downarrow| - \langle\downarrow,\uparrow|\Big)\delta(\hat{\pi}_a - x_a)\delta(\hat{\pi}_b - x_b)\Big(|\uparrow,\downarrow\rangle - |\downarrow,\uparrow\rangle\Big)$$

$$= \frac{1}{2}\Big[\langle\uparrow|\delta(\hat{\pi}_a - x_a)|\uparrow\rangle\langle\downarrow|\delta(\hat{\pi}_b - x_b)|\downarrow\rangle - \langle\uparrow|\delta(\hat{\pi}_a - x_a)|\downarrow\rangle\langle\downarrow|\delta(\hat{\pi}_b - x_b)|\uparrow\rangle$$

$$- \langle\downarrow|\delta(\hat{\pi}_a - x_a)|\uparrow\rangle\langle\uparrow|\delta(\hat{\pi}_b - x_b)|\downarrow\rangle + \langle\downarrow|\delta(\hat{\pi}_a - x_a)|\downarrow\rangle\langle\uparrow|\delta(\hat{\pi}_b - x_b)|\uparrow\rangle\Big].$$

With

$$A \equiv \langle\uparrow|\delta(\hat{\pi}_a - x_a)|\uparrow\rangle\langle\downarrow|\delta(\hat{\pi}_b - x_b)|\downarrow\rangle + \langle\downarrow|\delta(\hat{\pi}_a - x_a)|\downarrow\rangle\langle\uparrow|\delta(\hat{\pi}_b - x_b)|\uparrow\rangle$$

$$= \frac{1}{4}\Big[(1 - \cos\theta_a)\delta(x_a) + (1 + \cos\theta_a)\delta(x_a - 1)\Big]\Big[(1 + \cos\theta_b)\delta(x_b) + (1 - \cos\theta_b)\delta(x_b - 1)\Big]$$

$$+ \frac{1}{4}\Big[(1 + \cos\theta_a)\delta(x_a) + (1 - \cos\theta_a)\delta(x_a - 1)\Big]\Big[(1 - \cos\theta_b)\delta(x_b) + (1 + \cos\theta_b)\delta(x_b - 1)\Big]$$

$$= \frac{1}{2}\Big[(1 - \cos\theta_a\cos\theta_b)\Big\{\delta(x_a)\delta(x_b) + \delta(x_a - 1)\delta(x_b - 1)\Big\}$$

$$+ \frac{1}{2}\Big[(1 + \cos\theta_a\cos\theta_b)\Big\{\delta(x_a)\delta(x_b - 1) + \delta(x_a - 1)\delta(x_b)\Big\}\Big],$$

and

$$B \equiv \langle\uparrow|\delta(\hat{\pi}_a - x_a)|\downarrow\rangle\langle\downarrow|\delta(\hat{\pi}_b - x_b)|\uparrow\rangle + \langle\downarrow|\delta(\hat{\pi}_a - x_a)|\uparrow\rangle\langle\uparrow|\delta(\hat{\pi}_b - x_b)|\downarrow\rangle$$

$$= \frac{1}{2}\sin\theta_a\sin\theta_b\Big\{\delta(x_a) - \delta(x_a - 1)\Big\}\Big\{\delta(x_b) - \delta(x_b - 1)\Big\},$$

$$P(x_a, x_b) = \frac{1}{2}[A - B]$$

$$= \frac{1}{4}\Big[(1 - \cos\theta_a\cos\theta_b - \sin\theta_a\sin\theta_b)\delta(x_a)\delta(x_b)$$

$$+ (1 - \cos\theta_a\cos\theta_b - \sin\theta_a\sin\theta_b)\delta(x_a - 1)\delta(x_b - 1)$$

$$+ (1 + \cos\theta_a\cos\theta_b + \sin\theta_a\sin\theta_b)\delta(x_a)\delta(x_b - 1)$$

$$+ (1 + \cos\theta_a\cos\theta_b + \sin\theta_a\sin\theta_b)\delta(x_a - 1)\delta(x_b)\Big]$$

$$= \frac{1}{4}\Big[\Big(1 - \cos(\theta_a - \theta_b)\Big)\Big\{\delta(x_a)\delta(x_b) + \delta(x_a - 1)\delta(x_b - 1)\Big\}$$

$$+ \Big(1 + \cos(\theta_a - \theta_b)\Big)\Big\{\delta(x_a)\delta(x_b - 1) + \delta(x_a - 1)\delta(x_b)\Big\}\Big],$$

and the joint probability is simply found as

$$\mathcal{P}(\theta_a, \theta_b) = \int dx_a dx_b P(x_a, x_b) x_a x_b$$
$$= \frac{1}{4}\Big[1 - \cos(\theta_a - \theta_b)\Big].$$

ACKNOWLEDGMENTS

This work was supported by the US Office of Naval Research, the National Science Foundation, the Welch Foundation, and the US Air Force.

REFERENCES

1. M. O. Scully, Phys. Rev. D **28**, 2477 (1983).
2. F. Belinfante, *A Survey of Hidden-Variable Theories* (Pergamon, New York, 1973).
3. E. Wigner, Phys. Rev. **40**, 749 (1932)
4. M. Hillery, R. O'cornell, M. Scully, and E. Wigner, Phys. Rep. **106**, 121 (1984), and the references therein.
5. M. O. Scully and K. Wódkiewicz, *Coherence and Quantum Optics VI*, edited by J. H. Eberly *et al.*, p1047 (Plenum, New York 1990).

III. TOPICS IN QUANTUM MECHANICS AND QUANTUM CHAOS

Spectral Properties of Classically Chaotic Systems

F. Leyvraz

Centro de Ciencias Físicas, Universidad Nacional Autónoma de México
Cuernavaca, Morelos, México

Abstract. We describe the connection between quantum systems which have a chaotic classical counterpart and random matrix theory. As is well-known, it consists in the fact that the statistical properties of the spectra of such systems in the semiclassical limit are equivalent to those of random matrix theory. Here, we first briefly review some propertie of random matrices, and then proceed to justify the above-mentioned connection in two different ways: First, according to the classic work of Berry, we show how the result can be derived from periodic orbit theory, of which we give a rapid overview; second, we show how the same result can be obtained with greater generality but in a more speculative manner using the concept of structural invariance.

I INTRODUCTION

In the following, I will talk about some parts of what is commonly called "Quantum Chaos". As time is short and as my competence is limited, I will have to make a very personal and of necessity biassed choice of the subjects I shall treat. Here I want to give a rapid overview of what I shall talk about and mention in passing my main omissions.

To start, we must in some way define quantum chaos. As there exist a variety of different opinions on the subject, I shall try to steer away from controversy. Chaos, in the classical sense of the word, is linked to sensitive dependence on initial conditions, primarily in bound systems (for open systems, there exists also a notion of chaos, but sensitive dependence on initial conditions is not any more the only issue.) Such sensitive dependence on initial conditions means that any minor error in the initial position or velocity gets amplified at an initially exponential rate, until it becomes of order one. This is the usual definition of classical chaos. What is fairly uncontroversial is, that such a definition will not carry over literally to quantum mechanics. The reasons can be stated in many ways: The simplest is probably to note that, since a bound system always has a discrete eigenvalue spectrum, the behaviour of the expectation values of all observables is quasi-periodic in time, thus precluding sensitive dependence on initial conditions. On the other hand, it could be argued that the quasiperiods can be made arbitrarily long in some appropiate limit, and that the phenomenon should be discarded, just as the phenomenon of Poincaré times in traditional Statistical Mechanics. While I do not wish to enter into this dispute, it should certainly be the case that classical behaviour (and hence, true chaotic behaviour) is recovered in the semiclassical limit of quantum mechanics. The crux of the matter, in fact, appears to be that the limit is not approached *uniformly* in time. That is, classical dynamics is well reproduced by quantum mechanics up to a certain time but, beyond this time, specifically quantum effects dominate, which work to suppress chaos. This time, however, increases without limit as one goes further into the semiclassical domain.

In any case, however, it appears inescapable that classically chaotic systems should behave differently from non-chaotic ones, both in classical and in quantum mechanics. This is a consequence of the Correspondence Principle mentioned above: Since the quantum system must, in the semiclassical limit, approach a very different type of behaviour in the two cases, the quantum mechanical system must show qualitative differences as well. This will, in fact, be the definition we shall use for "Quantum Chaos": We shall take it to be the study of those characteristics of quantum systems which are peculiar to those systems which have a chaotic classical limit.

But what, specifically, do I mean by properties? Within this course, the answer will be: *Exclusively spectral properties*, that is properties of the eigenvalues of the Hamiltonian. The reason for this restriction can readily be understood: Spectra are invariant under change of basis, so we need not specify any particular basis in which to perform the study. Diffferent chaotic systems can therefore be compared unambiguously. This is a severe limitation from a practical point of view, however, and it is as well to realize it from the start. Indeed, most quantities of physical interest, such as transition amplitudes, matrix elements and the like, usually involve *both* the spectrum and the wave functions in some specified, problem-dependent basis. Nevertheless, for the reasons stated above, we shall not have much to say on this subject.

A few words about the semiclassical limit, which I have talked about so much in the above remarks. In the case of an ordinary Hamiltonian of the type

$$H = -\frac{\hbar^2}{2m}\Delta + V, \qquad (1)$$

the semiclassical limit is the limit of large energies, at least if the system remains bound for all energies. Equally well, it could be defined as the limit $\hbar \to 0$ at fixed energy (while at first sight somewhat unphysical, this has the great advantage of keeping the classical mechanics constant as the semiclassical regime is approached, something the infinite energy limit often fails to do.) In either case, the volume of the accessible phase space goes to infinity when measured in the appropiate quantum units, that is $(2\pi\hbar)^d$, where d is the dimension of the configuration space. Thus, in any sense of the word, the semiclassical limit is a limit of large excitation numbers and of high-dimensional matrices with large bases necessary for the calculation of the eigenvectors. This already shows one possible reason for the interest in quantum chaos: It always concerns the computationally most intractable part of a quantum system.

Now that we have defined the subject of our talk, let us give a quick overview of what we shall see:

—*Random Matrix Theory*: This theory was developed in the fifties an sixties [1,2] to cope with the very specific difficulties posed by the high-lying resonances of the nucleus, which are accessed by low-energy neutron scattering. In this particular situation, it appeared unreasonable to use any given Hamiltonian to describe the system. Wigner then proposed that one should use an ensemble of matrices, that is one should allow all Hermitian matrices to be possible Hamiltonians, giving each a specific probability. Since we are interested in the semiclassical limit, we may limit ourselves to large matrices. In this case, it can be shown that various spectral properties hold *with probability one*. That is, any matrix taken at random with the above probabilities, will certainly display these spectral properties. We shall discuss these in greater detail in the next chapter; for the moment it may be enough to say that the probability of having two neighbouring eigenvalues at a distance S is such a property. This means that if I randomly take a large matrix and make a histogram of its nearest-neighbour distances, this function will always be the same, independently of the matrix chosen.

Under these circumstances, one might indeed hope, that systems for which we have no reasonable way to devise a Hamiltonian behave in this way. And indeed, in the fifties and sixties, it was confirmed that neutron resonances in nuclei did satisfy these spectral properties (see e.g. [1] for a full discussion). The new development that arose and gave rise to the branch of "Quantum Chaos" was the growing recognition in the beginning of the eighties [3–5] that the same spectral properties also hold for quantum systems whose classical equivalent is chaotic. This was quite far from evident: Under these circumstances, we are not dealing with systems of arbitrary complexity. Indeed, these systems are often quite simple: The prototypes of chaotic Hamiltonians are various kinds of billiards, that is, free motion inside a cavity of a given (simple) shape with perfectly reflecting walls. In quantum mechanics, this simply amounts to looking at the spectrum of the Laplacian within a bounded domain with Dirichlet boundary conditions. There can therefore be no doubt as to the Hamiltonian we are dealing with. Nevertheless, in the semiclassical limit, the dynamics are sufficiently complex to create the same statistical properties as those of random matrices.

—*Periodic Orbit Theory*: This leads us to the second part of these lectures: In order to explain this phenomenon, as well as for many other reasons which I shall not go into, a considerable formalism has been developed, aiming at finding the properties of spectra from the underlying classical dynamics [6,7]. Since it turns out that classical periodic orbits play a major role in this theory, it has received the name of periodic orbit theory. In principle, the theory could come very close to being a rigorous connection between the semiclassical part of the spectrum and a sum involving the set of all periodic orbits of the system. In fact, there is a special case [8]—the trace formula of Selberg—where this connection is indeed exact and it is not necessary to talk about semiclassics. The problem, however, is that the sum over periodic orbits is an object which is *at least* as unknown as the spectrum itself. Nevertheless, it turns out to be possible to make certain statements about the long-range behaviour of the spectrum based entirely on this connection, together with a long series of largely uncontrolled approximations. In these lectures, we will show that some part of the statistical properties of random matrix theory (RMT) can indeed be obtained from this periodic orbit theory.

—*Structural Invariance*: This is an entirely different approach to the selfsame problem [9–11]: Instead of attacking it semiclassically using periodic orbits, we give plausible arguments as to why, roughly speaking, the same statistical ideas that justify using RMT when the Hamiltonian is unknown, can also be used to justify it when the Hamiltonian is chaotic.

To this I should add a few words as to why all this is so important. It is not merely an academic issue about the exact meaning and range of validity of the Correspondence Principle in quantum mechanics. Indeed, and this is a crucial point, there exists no limitation to the specific case of quantum mechanics and the limit of classical mechanics. Rather, we could just as well study electromagnetic waves in the limit of geometrical optics, or acoustic waves in the limit of ray acoustics. In fact, a large body of experiment has grown out of this research, and the various phenomena one observes and predicts, are also found in microwave cavities [12], quartz and alunminum crystals [13] and soap membranes [23]. The only common denominator these systems have, is that they all show some type of wave behaviour, and they are all very highly excited. In all these examples, one can define dynamics that play an equivalent role to classical mechanics, and when this dynamic becomes chaotic, the usual predictions of RMT and periodic orbit theory are in fact observed. So this shows another reason why quantum chaos has been of such interest: Far from being, as the name might suggest, a very special issue, it is in fact very generally applicable to a large variety of systems which until now were quite untractable both numerically and analytically. In a different way, and perhaps even more relevantly, it has shown common ways of thought for many problems that were previously viewed as being largely unrelated.

The following indicate the most glaring omissions in this program: First, no mention is made of *disordered systems*, which are perhaps the most important application. These are systems in which a random potential is introduced to model the existence of impurities in a sample. This case actually involves averaging over different Hamiltonians, in a way quite different from the one we shall be dealing with. Here many results are known. In particular, using a technique known as supersymmetry, Efetov [15] was able to show that, in a certain well-defined regime, such systems obey random-matrix statistics. It should be added that the whole supersymmetric approach is extremely powerful and has been applied to a large number of problems, either in disordered systems or simply in RMT, with a great deal of success [15,16]. I will not go into this, however, both because I do not have the expertise and because the formalism is truly formidable.

Second, I will limit myself to bound systems in finite geometries. I shall not treat chaotic scattering, though it is amenable to a very similar treatment as the bound case. I shall also exclude such systems as the Anderson model or the Harper model, which have infinite extension without being scattering systems. As we shall see, in the semiclassical limit, this excludes such phenomena as Anderson localization. More generally, we shall not treat transient phenomena, such as tunneling (which disappears in the semiclassical limit for chaotic systems) or localization in finite quasi-one dimensional systems.

Third, as mentioned above, I will not talk at all about eigenfunction properties. This is due to the fact that I do not want to talk about issues that depend on the existence of a basis that can legitimately be singled out among all bases. Nevertheless, there are some very important phenomena which relate to wave functions. Perhaps the most significant is *scarring* [17]: It has frequently been observed that the wave function in coordinate space, or some appropriately defined form of phase space wave function, is fairly strongly concentrated along short unstable periodic orbits, even in fully chaotic systems, where a uniform distribution in phase space was, in fact, expected from random matrix theory.

II RANDOM MATRIX THEORY

In the following, we shall consider the totality of all $N \times N$ Hermitian matrices and put a reasonable probability density on this set. To this end, we put two requirements on the measure:
—First, since no basis should be given any preference in the absence of any further information, the measure should be invariant under unitary transformations, that is, if $\mathcal{A}$ is a set of hermitian matrices and U is a fixed unitary matrix, then

$$\mu(U^{-1}\mathcal{A}U) = \mu(\mathcal{A}), \qquad (2)$$

where $\mu(\mathcal{A})$ denotes the measure. This requirement is absolutely basic and reflects the fact that we have no reason whatever to give any basis preference over another.
—Second, we shall assume, for convenience, that all the different matrix elements are independent random variables, only subject to the constraint of forming a Hermitian matrix.

From these two assumptions, it is a (non-trivial) task to show that there is (up to a trivial scale factor) only one possible measure, namely the following: All matrix elements are given the same Gaussian distribution, with given variance and mean value zero. It is easily checked that this measure does indeed satisfy our requirements. Because of its unitary invariance, this ensemble is called the Gaussian Unitary Ensemble or GUE. A similar ensemble exists for real symmetric matrices, which is invariant under orthogonal transformations of the basis, which is called the Gaussian Orthogonal Ensemble (GOE). It is specified by each element of the matrix being Gaussian distributed with a variance σ except for the diagonal terms, which are distributed with a variance 2σ, the mean being again equal to zero.

In this form, however, it is not easy to see what the spectral properties of a typical matrix might be. To obtain this, we proceed through an apparently somewhat indirect route indicated by Dyson [18]. Let A be an aritrary matrix taken from the GUE (say) and let B be another. Let us form the intermediate matrix

$$A(\epsilon) = \sqrt{1-\epsilon^2} A + \epsilon B, \tag{3}$$

and consider the matrix A in its eigenbasis. This, by definition, changes nothing to the statistical properties of the matrix B, since those are invariant under basis changes. Call E_i the eigenvalues of A and $E_i + \Delta E_i$ the eigenvalues of $A(\epsilon)$. One then has, using second order perturbation theory:

$$\Delta E_i = \epsilon b_{ii} - \epsilon^2 E_i/2 + \epsilon^2 \sum_{k \neq i} \frac{|b_{ik}|^2}{E_i - E_k}. \tag{4}$$

We can now diagonalize $A(\epsilon)$ and repeat the process with another B, taken equally at random from the ensemble. In this way, we go through the whole ensemble and the matrices $A(\epsilon)$ are always correctly GUE distributed. Here we are using the fact that the matrix elements are independently Gaussian distributed and that the Gaussian distribution is invariant under addition: That is, the distribution of the sum of two independent Gaussian distributed random variable is again Gaussian distributed, with a variance given by the sum of the variances. But the prefactors in Eq. (3) are so chosen that the variance of the matrix elements remain unchanged.

Now let us describe the resulting movement of the E_i as the B matrices are successively added to the original A matrix. Since the B's are independent, the first term in Eq. (4 is like a diffusion or noise term: The eigenvalues execute a random walk due to the independent increments caused by first order perturbation theory. However, the second order terms contain systematic parts, which then may be of the same order as the first term. We may rewrite this equation as follows

$$dE_i = \left(-\frac{E_i}{2} + 2 \frac{\partial}{\partial E_i} \sum_{1 \leq k < l \leq N} |b_{kl}|^2 \ln|E_k - E_l| \right) d\lambda + b_{ii} \sqrt{d\lambda}. \tag{5}$$

Here N denotes the size of the matrices and $d\lambda$ is ϵ^2. Now, in order to obtain consistent orders of magnitude, we choose the variance of the ensemble to be $\sqrt{N}$. This may appear unnatural, but it has the advantage of normalizing the spectrum so that the mean level spacing is of order one. Therefore the average value of $|b_{kl}|^2$ is N, and if we define $d\tau = N d\lambda$ we obtain

$$dE_i = \left(-\frac{E_i}{2N} + 2 \frac{\partial}{\partial E_i} \sum_{1 \leq k < l \leq N} \ln|E_k - E_l| \right) d\tau + d\xi_i. \tag{6}$$

Here $d\tau$ is equal to $Nd\lambda$ and $d\xi_i(\tau)$ is the diffusion term which arises from the random increments b_{ii} arising from first order perturbation theory. As these are uncorrelated, the $d\xi_i(\tau)$ have the characteristics of white noise:

$$\langle d\xi_i(0)d\xi_j(\tau)\rangle = \delta_{ij}\delta(\tau)d\tau. \tag{7}$$

The eigenvalues therefore move under the influence both of a noise term and of a force which comes from a repulsive logarithmic interaction potential between all eigenvalues. The eigenvalues are further confined by a harmonic potential which does not let the energies go to infinity. The presence of the logarithmic repulsion is the crucial fact, though. This *eigenvalue repulsion* is responsible for an extremely strong degree of correlation between the eigenvalues of a random matrix. This may appear paradoxical but, as we have just seen, the eigenvalues of a large random matrix behave similarly to a one-dimensional gas of particles confined by a harmonic potential and interacting through a very long-range potential, namely a logarithmic one. Note also that the only difference between the two ensembles GUE and GOE is only in the respective strength of the diagonal term, which is larger by a factor of two in the GOE case. This causes a relatively stronger diffusion and hence lessens the eigenvalue repulsion. We shall make this more quantitative later.

Due to this analogy with a system of equilibrium classical statistical mechanics (Eq. (6) is in fact simply a Langevin equation) it is easy to believe that such properties as the two-point correlation function or the nearest-neighbour spacing distribution should be independent of the particular matrix chosen. This can in fact be shown rigorously [19], though I shall not do it here.

It should also be noted that the Langevin equation Eq. (6) in fact gives more than merely the equilibrium behaviour: The dynamics of the eigenvalues has been used to model the dependence of eigenvalues of a chaotic system on an enternal parameter. Universal results concerning velocity-velocity correlations of eigenvalues as a function of an arbitrary parameter has been obtained by supersymmetry techniques [14].

Let us now draw a few elementary consequences from the above physical picture. Note first that the eigenvalue density will depend slowly on x. Equating the force acting on one eigenvalue due to the sum total of the other eigenvalues to that force which comes from the confining potential, one obtains for the mean eigenvalue density $\bar{\rho}(x)$:

$$\int_{-\infty}^{\infty} \frac{\bar{\rho}(y)}{x-y} = \frac{x}{4N}. \tag{8}$$

This can be seen to have the following solution

$$\bar{\rho}(x) = \frac{1}{16\pi N}\sqrt{1 - \left(\frac{y}{4N}\right)^2}. \tag{9}$$

From this, we see that in order to obtain meaningful results concerning the two-point correlation function and any other quantities, we must first eliminate the "secular" dependence given by Eq. (8) . This is the more so, as this density, the so-called "semi-circle" law, has no connection whatever with any realistic density of states whatever. This elimination is performed in the following manner: one writes the exact eigenvalue density in the following way:

$$\rho(x) = \overline{\rho}(x) + \rho_{fl}(x). \tag{10}$$

Here $\rho_{fl}(x)$ is the term which accounts for the deviations from the semicircle law which are inevitably present in any one realization of the ensemble. If we now transform the spectrum E_i according to the rule

$$x_i = \int_{-\infty}^{E_i} \overline{\rho}(E)\,dE, \tag{11}$$

one obtains essentially a sequence with average spacing equal to one. As it turns out, the statistical properties of this sequence are constant, that is, the two-point function and other correlation prioerties of the sequence are unaffected by the one-particle density. This is in strong contrast to ordinary statistical mechanics, but can be understood, once we realize that the logarithmic interaction potential does not, by its very nature, introduce a new length scale. The only length scale in the problem is the mean average spacing. It is therefore natural, that all properties should be constant once we have used the transformation in Eq. (11) to scale this length scale to one. Again, the same observations are constantly made in Hamiltonian systems: The separation indicated in Eq. (10) is also used. For these Hamiltonians, the average density is given in semiclassical terms either by the so-called Weyl formula, which I shall not discuss further, or else through fairly low-degree polynomial fits to the integrated density of states. Again, the independence of the statistics after this procedure (technically known as "unfolding the spectrum") is well established.

The properties I have been talking about can therefore now be described in clear terms: They are the two-point correlation function as well as all higher order correlation functions of the spectrum. Due to the analogy with a statistical mechanical system and due to the absence of additional length scales apart from the average level spacing, all these functions are expected to be the same for all matrices from the ensemble. Furthermore, they are also expected to be, after unfolding as described in Eq. (11), the same for all regions of the spectrum. It remains for us to see what these properties actually are, and how they can be computed. I shall not go into the more technical details, but will just draw an outline of the procedure.

First, one needs the joint probability distribution function (jpdf) of the eigenvalues, that is, the equivalent of the Gibbs ensemble. In fact, since Eq. (6) is just an ordinary Langevin equation, it can be checked that the stationary state is given by a Gibbs ensemble, namely:

$$P(E_1, \ldots, E_N) = \exp\left[-\beta\left(\frac{1}{4N}\sum_{k=1}^{N} E_k^2 + \sum_{1 \leq k < l \leq N} \ln|E_k - E_l|\right)\right], \tag{12}$$

where we have taken out a global factor of two, for the sake of aggreement with common usage. The inverse temperature β is determined by the intensity of the noise term. It can be checked that $\beta = 2$ for the GUE and $\beta = 1$ for the GOE. This leads to

$$P(E_1,\ldots,E_N) = \exp\left(-\frac{\beta}{4N}\sum_{k=1}^{N} E_k^2\right) \prod_{1\leq k<l\leq N} |E_k - E_l|^\beta. \tag{13}$$

The logarithmic interaction has therefore turned into a product structure which is reminiscent of Vandermonde determinants. And indeed, one of the ways of computing integrals over this probability measure uses precisely this analogy, together with the theory of Pfaffians, which are in essence the square root of the determinant of an antisymmetric even-dimensional matrix. Another approach is yet more abstract and involves integrating over both commuting and anticommuting variables. Because of an analogous techninque occasionally used in field theory, this approach is known as the supersymmetric approach. Neither can possibly fall within the scope of these lectures and the interested reader is referred to [2] and [15].

As a final remark, let me state without further proof, the results obtained by these various methods. The n-point correlation functions are defined by

$$\rho_n(E_1,\ldots,E_n) = \frac{N!}{(N-n)!}\int dE_{n+1}\cdots dE_N\, P(E_1,\ldots,E_N). \tag{14}$$

Here, in order to avoid difficulties related to unfolding, we may take the E_i for $1 \leq i \leq n$ all within the center of the spectrum. Further, it can be proved (and is intuitively obvious from the analogy with statistical mechanics), that this correlation function approach $\rho(E)^n$ as the eigenvalues separate to distances significantly larger than a few mean level spacings. Exact formulae can be given for all these functions, both for finite values of N and in the limit of $N \to \infty$, where these simplify appreciably.

Of all these correlation functions, the most important is the two-point correlation function. The exact formula in the $N \to \infty$ limit is given by

$$\rho_2(x) = 1 - \left(\frac{\sin \pi x}{\pi x}\right)^2 \tag{15}$$

for the GUE. Here x represents the distance $|E_1 - E_2|$, which is the only variable it depends on in this limit. Further, I have set the mean level spacing equal to one in this formula. For the GOE the corresponding formula is

$$\rho_2(x) = 1 - \frac{\sin^2 \pi x}{(\pi x)^2} + (\text{Si}(\pi x) - \pi/2)\left(\frac{\cos \pi x}{\pi x} - \frac{\sin \pi x}{(\pi x)^2}\right), \tag{16}$$

where

$$\text{Si}(x) = \int_0^x \frac{\sin x}{x}dx. \tag{17}$$

Note the linear repulsion at short distances for the GOE, which is replaced by a quadratic one for the GUE. This can be directly read off the general form of the jpdf as given in Eq. (13). It can also be obtained from much simpler and more general considerations: Take for example the two by two hermitian matrix

$$\begin{pmatrix} a & b \\ b^* & c \end{pmatrix}. \tag{18}$$

The eigenvalues $\lambda_{1,2}$ satisfy the equations

$$\lambda_1 + \lambda_2 = a + c \qquad \lambda_1 \lambda_2 = ac - |b|^2, \tag{19}$$

from which follows

$$(\lambda_1 - \lambda_2)^2 = (\lambda_1 + \lambda_2)^2 - 4\lambda_1\lambda_2 = (a-c)^2 + 4|b|^2. \tag{20}$$

It follows that, for $|\lambda_1 - \lambda_2|$ to be less than ϵ it is necessary that both $|a - c|$ and $|b|$ should be less than ϵ. In the Hermitian case, this involves *three* independent events (counting both the real and imaginary part of b), so that this probability is of order ϵ^3, whereas in the symmetric case it is of order ϵ^2. From this follows by differentiation that the probability of two eigenvalues being *at* a distance ϵ is as stated above.

Of equal, if not larger, importance is the behaviour of the two-point function at large distances. We see that the large x behaviour is always of the order x^{-2}, where it is necessary, in the case of the GUE, first to average over oscillations. The respective constants are $1/(2\pi^2)$ for the GUE and $1/\pi^2$ for the GOE. Note that the entire difference at large distances boils down to a factor of two, but that this is far from being the case generally. This will become relevant in the next chapter, where we shall be able to derive the factor of two from semiclassical considerations but, unfortunately, nothing more.

In order to visualize this large x behaviour, the following statistic is often used: Consider an interval of length L placed at random over the spectrum. On average, the number of eigenvalues it contains is L, since the mean level spacing is equal to one. We now define the *variance* of this number as $\Sigma_2(L)$. One can compute this from the two-point function in an elementary way: It is simply the average number of eigenvalue *pairs* in an interval of length L minus L^2, that is:

$$\Sigma_2(L) = L + \int_0^L dx \int_0^L dx' \, (\rho_2(x - x') - 1). \tag{21}$$

The various terms arise as follows: The first L comes from the average number of pairs of *identical* eigenvalues in the interval and the second counts all the other pairs. After a quick change of variables (to average and difference) one gets

$$\Sigma_2(L) = 2 \int_0^L (L - w)(\rho_2(w) - 1) dw + L. \tag{22}$$

It turns out, however, that

$$\int_0^\infty (1 - \rho_2(x))dx = \frac{1}{2}, \qquad (23)$$

so that Eq. (21) reduces to

$$\Sigma_2(L) = 2\int_0^L w(1 - \rho_2(w))dw, \qquad (24)$$

We note for future reference the large L behaviour of $\Sigma_2(L)$:

$$\Sigma_2(L) = C \ln L + O(1) \qquad (L \gg 1), \qquad (25)$$

where $C = 1/\pi^2$ for the GUE and $2/\pi^2$ for the GOE. This should be contrasted, say, with the behaviour of a random sequence: there the number of eigenvalues in an interval has normal fluctuations around L, so that $\Sigma_2(L)$ equals L always. There is, therefore, a very strong long-range siffness in the spectrum of a random matrix. It should be carefully noted that this is a far more specific phenomenon than level repulsion at short distances. The latter can be caused by virtually any mechanism coupling eigenvalues among each other, as appeared clearly in the two by two case. The former is therefore a much surer fingerprint of GOE or GUE behaviour.

A further statistic, which has enormous popularity, is the nearest neighbour spacing distribution. It does not depend merely on the two-point function. Rather, it is an infinite sum over all n point functions, which has no useful closed analytical form. However, the following approximations are amazingly good:

$$P(S) \approx \frac{\pi}{2} \exp(-\pi S^2/4) \qquad \text{(for GOE)},$$
$$P(S) \approx \frac{32}{\pi^2} \exp(-4S/\pi) \qquad \text{(for GUE)}, \qquad (26)$$

where $P(S)$ denotes the probability that two neighbouring eigenvalues are at a distance S. These formulae are called "the Wigner surmise" and are simply derived by assuming the appropiate level repulsion at the origin, assuming Gaussian decay at large S and using normalization and the average level spacing to fix the undetermined constants. The accuracy of this approximation is quite surprising. A straightforward computation shows that these formulae are in fact exact for two by two matrices (excercise: use the above formulae to work it out!)

In general, therefore, when one states that a given hamiltonian "has GOE statistics", one usually means that the two-point function or some equivalent measure ($\Sigma_2(L)$ or the Fourier spectrum, of which I have not spoken) have a form compatible with the above formulae for the GOE. Further, one usually checks for the nearest-neighbour distribution against the Wigner surmise of Eq. (26)

I should also point out, that under the hypothesis of basis independence and independence of the matrix elements made at the outset, there exist only three ensembles, originally characterized by Cartan [24], namely the GOE, GUE and Gaussian Symplectic Ensemble (GSE). The latter relates to spin systems and to a type of Hermitian quaternionic matrices. I shall not talk about these, as they play a fairly specialized role in the theory so far (time-reversal invariant systems of particles with half-integer spin and having significant spin-orbit coupling). However, it should be clear that the possibilities for constructing ensembles become much more numerous once the hypothesis of basis independence or that of the statistical independence of the matrix elements is given up. The following are a few examples:

—The Poisson Ensemble: One considers a completely fixed basis and takes all diagonal matrices, where the eigenvalues are independent identically distributed random variables. Here $\Sigma_2(L)$ is linear and there is no eigenvalue repulsion. $P(S)$ is equal to e^{-S}. An interesting variant consists in taking first a diagonal matrix and then subjecting this diagonal matrix to a random change of basis. The resulting matrices are then full, but their entries are by no means independent. In fact, the eigenvalue statistics is clearly unaffected by the random change of basis. One therefore has an ensemble of matrices which is basis-independent by construction, but which has spectral properties very different from those of the Gaussian ensembles.

—Block matrix ensembles: One considers a fixed block matrix structure and chooses any of the Gaussian ensembles to fill the blocks. For example, we might put two GOE's on the diagonal and zero in the off-diagonal blocks. This gives rise to a double GOE spectrum, which allows nearby eigenvalues, but still is fairly rigid at large distances, with $\Sigma_2(L)$ growing logarithmically for $L \gg 1$.

—Perturbed ensembles: One can also look at the sum of any of the above ensembles with a weak global GOE or GUE. This immediately creates eigenvalue repulsion, but has essentially no effect on the long-range rigidity of the spectrum.

—Ensemble of banded matrices: In these ensembles, the matrix elements are drawn independently from a distribution which depends on their opsition in the matrix. Typically, H_{ij} is drawn from a Gaussian ensemble with a given variance if $|i - j| < b$, where b is the band width and zero otherwise. In the limit $N \to \infty$, and b fixed, this clearly reduces to the Poisson ensemble described above. On the other hand, it is found that if $b \gg \sqrt{N}$, it reduces to a GOE or GUE as the case may be. These ensembles can therefore be used to describe transition from Poisson to GOE behaviour. Specifically, they are used a great deal to model systems showing localization. For a review of various such ensembles see [25].

Finally, for future applications, let me make an important remark: All the above ensembles deal with Hermitian or symmetric matrices, that is, they correspond to *Hamiltonians*. It is also possible to make similar ensembles of unitary matrices, namely the so-called Circular Unitary Ensemble (CUE) and the Circular Orthogonal Ensemble (COE). These consist of the set of all unitary matrices and the set of all symmetric unitary matrices respectively. These correspond to discrete steps of a time evolution in quantum mechanics. These correspond to discrete classical maps, and it turns out that the eigenphases of the quantization of a classical discrete map which generates a chaotic dynamics upon iteration, have statistical properties corresponding to those ensembles. Let us proceed to define them more precisely.

These ensembles are in fact characterized entirely by a requirement related to that of basis invariance. For the unitary ensemble it amounts to saying that the measure is invariant under multiplication to the left or to the right by any unitary matrix. For the COE, the requirement is, that the measure should be invariant under the tranformation

$$V \to U^t V U. \tag{27}$$

These transformations are the most general that maintain the symmetric character of the matrix V, just as multiplication to the left or right by an arbitrary unitary matrix is the most general transformation that keeps the unitary character of the matrix invariant. We shall exploit this idea at length in the fourth chapter. For the moment, we simply note that the eigenvalues of a unitary map lie on the unit circle. These eigenphases then have locally the same behaviour in the circular ensembles as in the Gaussian ones. This is just another way of stating that the eigenvalue repulsion is the same in both types of ensembles. The only thing that varies is the mode of confining the eigenvalues: In the Gaussian ensembles, it is done by means of a harmonic confining potential, whereas the geometry of the circle on which the eigenvalues of a unitary map must lie sees to confinement in a natural manner, that is quite free of arbitrariness. Obviously, there is no unfolding problem here either, so that overall these ensembles have a very "natural" appearance indeed.

III PERIODIC ORBIT THEORY: A SHORT INTRODUCTION

In the last lecture, we discussed at length the properties of random matrices. In this lecture, I wish to reproduce an argument substantially due to Berry [5] to derive the fact that $\Sigma_2(L)$ has the correct behaviour for a chaotic system. Before we do this, however, I wish to explain in detail what is the presumed connection between chaotic behaviour and RMT.

If a system is chaotic, has no discrete symmetries, and is not invariant under time reversal, then its short-range spectral fluctuations are described by those of a GUE. By this, I mean that the various statistical properties of the GUE hold with good accuracy. Specifically, I mean that not only the two-point function coincides at large distances with the GUE prediction, but also that it coincides well at short and intermediate distances. Further, the GUE prediction for the nearest-neighbour spacing distribution is satisfied and the higher order correlation functions agree with the theoretical predictions as well. All of this is borne out by numerical work [26].

If the system is time-reversal invariant (TRI), on the other hand, then the matrix is always symmetric in the position representation, so that one expects GOE statistics. Indeed, these have been extensively verified, as TRI systems are actually the most common in Nature as well as the easiest to simulate [4,27].

If the system has discrete symmetries, then it naturally falls into block matrix form, where the invariant subspaces of the symmetry give rise to this structure. In each block we therefore put a GOE or a GUE as the case may be. These are chosen to be inpendent, so that the block matrix models described in the previous lecture should be applied. Of course, a weakly broken discrete symmetry gives rise to a perturbed block matrix ensemble. All these predictions have been amply verified numerically.

There is one last fine point which we should mention: The rule in the case when there are both discrete symmetries and TRI is a bit subtle. The issue at hand is whether the invariant subspaces of the disrete symmetries are, or not, themselves invariant under the time-reversal operation. In the former case, there is no problem and we have a model of GOE blocks which describes the system. On the other hand, if they are not, then it is seen that the eigenvalues belonging to one invariant subspace are degenerate with those that belong to the corresponding time-reversed subspace. Under those circumstances, the non-TRI invariant subspaces have degenerate eigenvalues but GUE statistics. This might be viewed as a reflection of the non-TRI nature of the specific subspace to which the dynamics is limited [20,21].

Finally, though I shall insist much less on this point, integrable systems are usually associated with a Poisson spectrum, that is, a random set of eigenvalues. This rule has a fair number of exceptions, however, mainly of a number-theoretic nature, among which the harmonic oscillator is the most prominent. It is easily understood as a limiting case of the previous rule relating to invariant subspaces for discrete symmetries: In the case of continuous symmetries (which correspond to integrable systems), we might assign independent ensembles to each of an infinity of invariant subspaces, thereby generating a random spectrum. While this rule is more controversial, it is nevertheless the standard model used to analyze integrable or close to integrable systems.

To prove all this is far beyond the state of current knowledge. However, using the techniques of periodic orbit theory to be described below, we shall show that $\Sigma_2(L)$ for $L \gg 1$ behaves in the way expected from the above rules, except for the last point, which we shall not consider. It has, however, also been shown using this approach [21].

In order to simplify matters somewhat, we will only consider *discrete* dynamics. That is, we consider a fixed canonical map from a compact phase space onto itself. We limit ourselves to maps which are of the type of a time evolution, that is, bijective maps. The limitation to compact phase spaces may at first appear unnatural since, in ordinary phase space, momenta are usually defined as unbounded variables. However, for a bound system, both positions and momenta are limited to a bounded set of values for all times, so that phase space can always be chosen to be compact. It will in fact turn out that this restriction is vital: Without it, phenomena such as Anderson localization might occur and would altogether invalidate our conclusions.

To take a concrete example, let us consider the bounce map for a convex billiard. A billiard consists of a region in which a particle moves freely, with elastic reflection at the boundary. Each time the boundary is hit by the particle, we examine the position and velocity of the particle after the bounce. Since this is to all effects a Poincaré map, it is canonical in the appropiate coordinates. As can be shown, the position can be parametrized by arc length and the velocity by the cosine of the angle that the trajectory makes with the surface (or, in other words, with the tangential component of momentum). If we now define $l(s, s')$ as the length of the straight line connecting the points with coordinates s and s', a map is defined by the following equations:

$$p = -\frac{\partial l}{\partial s}, \qquad p' = \frac{\partial l}{\partial s'}. \tag{28}$$

Here the p variables correspond to the cosine of the angle of incidence, as explained above. It is easily checked that this map gives exactly the next bounce as a function of the first one. The dynamics of the billiard is thus entirely contained in this function. As to the geometry of the phase space, it is particularly simple: The arc length runs along a simple closed curve, so that this is equivalent to a circle. On the other hand, the cosine runs fron -1 to 1, with both ends representing physically distinct states. The phase space is therefore a cylinder with a boundary.

If we now wish to iterate the map described by Eq. (28), which is, by the way, bijective by construction, we proceed as follows: The relevant equations are the following

$$p = -\frac{\partial l}{\partial s}(s, s'), \qquad p' = \frac{\partial l}{\partial s'}(s, s'),$$
$$p' = -\frac{\partial l}{\partial s'}(s', s''), \qquad p'' = \frac{\partial l}{\partial s''}(s', s''). \tag{29}$$

From this follows that s' must be chosen (once s and s'' are given) in such a way that $l(s, s') + l(s', s'')$ is an extremum. Thus we define the generating function of the map iterated twice as

$$l_2(s, s'') = \underset{s'}{\text{ext}}\, (l(s, s') + l(s', s'')), \tag{30}$$

which is intuitively clear: A two-bounce orbit is obtained by taking the above extremum for the combined length. As this extremum can be multiple (and in general will be), we should introduce an index α to differentiate the various branches of $l_2(s, s'')$. However, this cannot really introduce any multivaluedness, since the original bounce map is bijective. Upon iteration, these branches proliferate, uncontrollably, as there is an ever increasing number of different ways of getting from s to s_N in N steps. This, together with the difficulty of making any general statements about the set of all orbits connecting two points, lies at the root of the trouble with the semiclassical trace formulae we are about to derive.

To go over to quantum mechanics, we first introduce a Hilbert space. Let us take as configuration space the circle. Our Hilbert space then consists of those functions on the circle with momenta bounded between -1 and 1. This means that the Fourier components are between $-1/\hbar$ and $1/\hbar$. Taking $\hbar$ to be $1/(2\pi N)$, we obtain the space of trigonometric polynomials of order N. This is in fact a general feature of compact phase spaces: The dimension of the corresponding Hilbert space is always finite for any value of $\hbar$ and corresponds to the number of phase space cell volumes $(2\pi\hbar)^d$ which "fit" into the total phase space volume. Here d represents the number of degrees of freedom or, in other words, the dimension of the configuration space. In this setting, we can now give a rigorous meaning to the "semiclassical limit", saying simply that it is the limit $N \to \infty$.

In order to quantize the map, we must find a unitary map that will approximately behave in a similar manner. In the following, we follow the ordinary convention of denoting position by q and the generating function by $S(q, q')$ We describe the unitary map by an integral kernel $K(q, q')$. While this may appear unnatural in view of the fact that the dimension of the Hilbert space is finite and one could in consequence do everything with sums, it turns out to be quite appropiate to stick to the integral notation when $N \gg 1$. To determine U_C, we solve for the following conditions

$$U_C q U_C^{-1} = q',$$
$$U_C p U_C^{-1} = p'. \tag{31}$$

From these equations one deduces

$$\langle \tilde{q}|q'U_C|q\rangle = \langle \tilde{q}|U_C q|q\rangle = q\langle \tilde{q}|U_C|q\rangle,$$
$$\langle \tilde{q}|p'U_C|q\rangle = \langle \tilde{q}|U_C p|q\rangle = \frac{\hbar}{i}\frac{\partial}{\partial q}\langle \tilde{q}|U_C|q\rangle. \tag{32}$$

Calling $K_C(\tilde{q}, q)$ the quantity $\langle \tilde{q}|U_C|q\rangle$ and evaluating the left-hand side of Eq. (32), one gets the following system of (generalized) differential equations

$$q'(\tilde{q}, -\frac{\hbar}{i}\frac{\partial}{\partial \tilde{q}})K_C(\tilde{q}, q) = qK_C(\tilde{q}, q),$$
$$p'(\tilde{q}, -\frac{\hbar}{i}\frac{\partial}{\partial \tilde{q}})K_C(\tilde{q}, q) = \frac{\hbar}{i}\frac{\partial}{\partial q}K_C(\tilde{q}, q). \tag{33}$$

These are the so-called MM equations [22] For the solution of these equation, we now put a WKB Ansatz of the form

$$K_C(\tilde{q}, q) = A(q, \tilde{q})e^{iS(q,\tilde{q})/\hbar}, \tag{34}$$

from which follow the Hamilton–Jacobi equations for $S(q, \tilde{q})$

$$q'(\tilde{q}, \frac{\partial S}{\partial \tilde{q}}) = q,$$
$$p'(\tilde{q}, \frac{\partial S}{\partial \tilde{q}}) = \frac{\partial S}{\partial q}. \tag{35}$$

From this finally follows that $S(q, \tilde{q})$ can be taken as the generating function of the original canonical transformation, so that we get

$$K_C(q, q') = \exp\left[\frac{iS(q', q)}{\hbar}\right] \left|\det \frac{\partial^2 S}{\partial q' \partial q}(q', q)\right|^{1/2}. \tag{36}$$

Note that within the approximation of stationary phase, this expression multiplied by an appropiate normalization constant which we shall always drop, gives a unitary operator, as indeed it should. Unitarity is not exact, however. Furthermore, within the same approximation, it is readily seen that the composition of two unitary maps is aproximately the unitary map corresponding to the composition of the two canonical transformations.

Now that we have a semiclassical formula in the simplest case, we must first generalize it to the case of many branches and then obtain a semiclassical formula for the trace of U_C as a function of the classical features of C.

Let us first generalize to the case of multiple $S_\alpha(q, q')$. One readily sees that the following Ansatz works, using similar techniques as above:

$$K_C(q, q') = \sum_\alpha e^{i\nu_\alpha \pi/4} e^{iS_\alpha(q', q)/\hbar} \left|\det \frac{\partial^2 S_\alpha}{\partial q' \partial q}(q', q)\right|^{1/2}. \tag{37}$$

Here ν_α is the difference between the number of positive and of negative eigenvalues of the matrix of second derivatives of $S(q', q)$. These arise from standard stationary phase approximation and are necessary for the unitarity of the operator to hold: The terms belonging to different indices in the product would otherwise not give correct results.

We are interested in the spectral properties of the eigenphases, that is, the properties of such quantities as

$$\Phi(\phi) = \sum_{j=1}^{N} \chi_L(2\pi(\phi_j - \phi)/N), \tag{38}$$

where $\chi_L(x)$ is equal to one for $0 \leq x \leq L$ and zero otherwise. This is simply the number of phases in an interval of length L, where the factor $2\pi/N$ serves to unfold the eigenphases, that is, to give them an average density of one. $\Sigma_2(L)$ is then the mean value of the square of $\Phi(\phi)$, once the mean value has been sutracted from $\Phi(\phi)$ that is, if

$$\Phi(\phi) = \sum_{m=-\infty}^{\infty} c_m e^{im\phi}, \tag{39}$$

then $\Sigma_2(L)$ is given by

$$\Sigma_2(L) = 2 \sum_{m=1}^{\infty} |c_m|^2. \tag{40}$$

It now remains to compute the c_m.

To this end, we note that

$$\text{Tr}\, U_C^k = \sum_{j=1}^N e^{ik\phi_j} \tag{41}$$

so that $\Phi(\phi)$ can be expressed as an infinite sum over various $\text{Tr}\, U_C^k$; to be specific, after developing $\chi_L(x)$ in a Fourier series, one obtains

$$\chi_L(x) = \sum_{m=-\infty}^{\infty} e^{\pi i mL/N} \frac{\sin \pi mL/N}{\pi m} e^{2\pi i mx/N}. \tag{42}$$

From this we find that

$$Phi(\phi) = \sum_{m=-\infty}^{\infty} e^{-im\phi} e^{\pi i mL/N} \frac{\sin \pi mL/N}{\pi m} \text{Tr}\, U_C^m. \tag{43}$$

From this follows the general formula

$$\Sigma_2(L) = 2\sum_{m=1}^{\infty} \frac{\sin^2 \pi mL/N}{(\pi m)^2} |\text{Tr}\, U_C^m|^2. \tag{44}$$

We are therefore in a position to evaluate the $\Sigma_2(L)$ of the quantization of an arbitrary map, if we are able to evaluate the traces of high powers of U_C. How high exactly? Eq. (44) indicates that we need at least to go up to m of order N/L. But we are typically interested in the behaviour of $\Sigma_2(L)$ for L of order one, which imposes the evaluation of very high order traces. As we shall see, if we were only interested in extremely large L, we could readily evaluate Eq. (44) without making any significant approximations. But unfortunately, this is not relevant, because the RMT regime we are interested in only concerns fairly small energy distances and hence fairly small values of L. For the larger values of L for which the evaluation of Eq. (44) is unproblematic, the behaviour also ceases to be universal and it has nothing to do any more with RMT.

Now we need to evaluate the trace. From what we have already done, this is fairly easy: Assume first that C is only given by one generating function S. One finds

$$\text{Tr}\, U_C = \int \langle q|U_C|q\rangle = \int e^{iS(q,q)/\hbar} \left|\det \frac{\partial^2 S}{\partial q \partial q'}(q,q)\right|^{1/2}. \tag{45}$$

If we now apply stationary phase to this integral, we see that the condition for stationarity is

$$\frac{\partial S}{\partial q}(q,q) + \frac{\partial S}{\partial q'}(q,q) = 0. \tag{46}$$

From this follows that the initial momentum to go from q to q is the same as the final momentum to go from q to q. This states that (q, p) is a *fixed point* of C. In the language of mechanics, if C represents the time evolution of a system, (q, p) is a *periodic orbit*. Thus we see that the traces we are interested in turn out to be sums over all periodic orbits. This is the main result of this section, and the various formulae we are about to write down are, in one form or another, the basis of everything that is called periodic orbit theory.

Continuing with the stationary phase approximation, we obtain from Eq. (46) by a straightforward generalization to the case of many branches

$$\text{Tr}\, U_C = \sum_{\alpha\gamma} e^{iS_{\alpha,\gamma}/\hbar} e^{i\nu_{\alpha,\gamma}\pi/4} |\det(1 - DC_\gamma)|^{-1/2}. \tag{47}$$

Here α denotes the branch and γ the periodic orbit. $S_{\alpha,\gamma}$ is value of the generating function to which the periodic orbit belongs at that particular point. As such it is simply the action of the periodic orbit and does not depend any more on the branch, but it is simply a characteristic of the fixed point γ. Similarly, the exponent $\nu_{\alpha,\gamma}$ are actually given by the difference between the number of positive and negative eigenvalues of S_α at the point γ. It is less trivial, but true, that these also only depend on the characteristics of the periodic orbit. They are known as the *Maslov indices* and reflect the nature of the extremum that the action has at the periodic orbit γ. Finally, DC_γ represents the Jacobian of the map C at the point γ.

Note carefully that we have made the assumption here that the fixed points of C were isolated. This assumption makes it impossible to treat completely integrable systems in the way just described. On the other hand, for such systems the WKB theory is so well developed that it can be said to give an essentially complete description of the quantization of integrable systems, though on quite different lines than the ones we are following. The above approach, on the other hand, is perfectly suited to fully chaotic systems, where the periodic orbits are isolated, unstable and more or less well separated.

If we now wish to go to higher m, there remain only minor adjustments: One does not call a fixed point of C^m a periodic orbit any more. Rather, this name is reserved for the *set* of points of the form $C^k x_\gamma$, where x_γ is a fixed point of C^m and $0 \leq k < m$. All the points of a given orbit are in every respect equivalent: They have identical Jacobian, Maslov indices and actions, so they can all be lumped together with a prefactor indicating the number of distinct points on the orbit. This is at most m, of course, but it may be less, since the orbit might be a *repeated* periodic orbit. We call this number m_γ. This then finally gives:

$$\text{Tr}\, U_C^m = \sum_\gamma m_\gamma |\det(1 - DC_\gamma)|^{-1/2}\, e^{i\nu_\gamma \pi/4} e^{iS_\gamma/\hbar}. \tag{48}$$

This is the so-called trace formula. As it stands, we should now pause to reflect on the approximations it contains: First, the quantization of C is only approximate, but it is correct in the semiclassical limit. Second, we have often used the stationary phase approximation, but it is also a well-controlled approximation, the error of which is well-known and goes to zero as $\hbar \to 0$. However, all this is only true *at fixed m*. If m is allowed to grow, then there is no reason to believe that Eq. (48) will hold. This is because the number of periodic orbits grows and these become ever more closely intertwined, so that the approximation of summing separately over each stationary point rapidly becomes untenable. Under no circumstances can we hope Eq. (48) to be even qualitatively correct beyond $m \approx N$, since then a typical periodic orbit has N points, meaning that it visits *each phase cell at least once*, which is ridiculous. In fact, there exist reasons to doubt the formula considerably earlier, but we shall be optimistic and use it up to this extreme limit.

We need the norm of the trace, for large m. Let us see if this brings any simplifications in the case of a fully chaotic system. For such a system, the number of isolated periodic orbits grows exponentially. It is easy to see that repeated orbits are exponentially rare under such circumstances, so we can set m_γ equal to m. Further, since m is large we have $S_\gamma \gg \hbar$. This implies that the phases $e^{iS_\gamma/\hbar}$ are nearly random. Recall that if $e^{i\alpha_j}$ are completely random phases, then

$$\left| \sum_j a_j e^{i\alpha_j} \right|^2 = \sum_j |a_j|^2. \tag{49}$$

This is due to the cancellation of the off-diagonal terms, which is complete if one averages over the phases. Here, we shall make the same assumption for the S_γ. This approximation, known as the *diagonal approximation*, gives

$$|\operatorname{Tr} U_C^m|^2 \approx \sum_\gamma m^2 \left| \det(1 - DC_\gamma) \right|^{-1}. \tag{50}$$

To understand the nature of the right-hand side of Eq. (50), let us look at the following expression: If x stands for the pair (p,q), what is $\delta(x - C^m x)$? Clearly, it is a measure with support in all fixed points of C^m. Evaluating it in terms of periodic orbits one obtains

$$\int dx\, \delta(x - C^m x) = \sum_\gamma m_\gamma \left| \det(1 - DC_\gamma) \right|^{-1}. \tag{51}$$

This coincides with the r.h.s. of Eq. (50) up to a factor of m. Now, it is natural to postulate that the periodic orbits in a chaotic system are in some sense uniformly distributed. The assumption is that a long periodic orbit cannot be distinguished from an arbitrary orbit, so that the invariant measure $\delta(x - C^m x)$ should not really be very different from Liouville measure (Ozorio de Almeida, Hannay) This is the so-called uniformity postulate. For this reason we finally state

$$|\operatorname{Tr} U_C^m|^2 \approx m\mathcal{V}, \tag{52}$$

where $\mathcal{V}$ is the phase space volume, which comes up in many other of the normalizations we dropped.

Putting Eq. (52) into the equation for $\Sigma_2(L)$ one obtains

$$\Sigma_2(L) = 2 \sum_{m=1}^{cN} \frac{\sin^2 \pi m L/N}{\pi^2 m}. \tag{53}$$

The sum has been truncated at some unspecified multiple of N, because the trace formula is hopelessly unreliable beyond it. On the other hand, if L is large enough, we can see that cN terms of the series give a good estimate of $\Sigma_2(L)$.

To check that the normalization is correct, we verify the $m = 0$ term in the equation for $\Phi(\phi)$. Since $\text{Tr}\, U_C^0$ is N, the term indeed gives L, so that the normalization is correct. $\Sigma_2(L)$ can now be estimated by the integral

$$2 \int_0^c \frac{\sin^2 \pi L x}{\pi^2 x} dx \approx \frac{1}{\pi^2} \ln L. \tag{54}$$

This was the announced result. In the assumption that the S_γ are fully random, we have implicitly stated that no two different orbits should, so to speak by construction, have exactly identical actions. But this is exactly what happens if either a discrete symmetry or time reversal invariance is present. This is the case because, for the vast majority of long periodic orbits, acting via time reversal or a discrete symmetry yields a different orbit from the original one. Thus, in the TRI case, periodic orbits come in time-reversed pairs, which have exactly identical actions. This is easily seen to lead to a factor of 2 in $|\text{Tr}\, U_C^m|^2$: Indeed, we must sum $4|\det(1 - DC_\gamma)|^{-1}$, but the sum itself extends only over one orbit from each pair, that is over half the orbits. The factor 4 comes from the fact that degenerate orbits miust be summed coherently, whereas the diagonal approximation consists in summing all orbits incoherently.

How should these results be viewed? Considering the number of approximations involved, it is certainly spectacular that it should work so well. On the other hand, the technical aspect may appear disturbing. Further, and this is most troubling of all, we have no way of getting systematically any further. This is due to the well-known fact, that the approximation in treating the diagonal approximation are fairly mild. That is, nothing short of going *beyond* the diagonal approximation will give us a clue, say, as to why GOE is not the same as twice GUE. In the diagonal approximation, this equality holds. Therefore in order to do better, we should take correlations between the actions of classical periodic orbits. Much work has been done so far, but the results appear disappointing compared to the amount of work they require. A final puzzling remark: We do, at the end, get the RMT result that we desire, at least in the large L limit, but where are the random matrices? It seems as though we did not really *understand* the reason for the coincidence between the final result using periodic orbits and the RMT result. It appears as (nearly) a miraculous coincidence. One might have liked it better, if some element of the theory could be identified as "the random matrix" which eventually leads to the RMT result, but this is simply not the case in the theory we discussed above.

On the other hand, it has the indisputable merit of identifying in a very down-to-earth manner the exact features responsible for logarithmic growth of $\Sigma_2(L)$. In the following lecture, we will try to present an approach with the opposite drawbacks.

IV A DIFFERENT APPROACH: STRUCTURAL INVARIANCE

In the following, the approach we will use will be the following: Starting from a given canonical map C, we first try to construct an ensemble of canonical maps, such that C will be a characteristic element of the ensemble. For technical reasons, it will turn out that this is not entirely possible. More accurately, to construct this ensemble would require a measure which we do not know to exist. However, we will indeed be able to construct an ensemble of unitary matrices of which the quantization U_C will be a typical element. Thus, at a probabilistic level, we shall in some sense prove the connection between dynamics and RMT. Obviously, this approach is not limited to large distances between energies, nor to two-point properties. On the other hand, the reasoning will be distinctly less concrete and will only apply in a statistical sense and with probability one. In a sense, one might therfore argue that it is neatly complementary to the periodic orbit approach presented above.

First, let me present a general method to derive an ensemble of which a given object is a characteristic element. Let us consider a set of objects with given properties and a specific object A. Furthermore, let the various objects be transformed among each other through the action of a group $\mathcal{G}$. The object A has a certain number of these properties. We define $\mathcal{H}$, *the structural invariance group of A*, in the following manner: Let $\mathcal{H}$ be that subgroup of $\mathcal{G}$ which leaves all properties of A invariant. If we now act on A with all the transformations of $\mathcal{H}$, we generate a collection Ω of objects, which has as additional structure that of a homogeneous space under the group $\mathcal{H}$. If $\mathcal{H}$ additionally has an invariant measure, then a measure on Ω can be generated in a natural way as follows: To each subset Σ of Ω I can associate the set of transformations $\mathcal{T}$ in $\mathcal{H}$ which map A in Σ. The Haar measure of $\mathcal{T}$ can then be taken as the measure of Σ. This is obviously independent of the choice of A, because of the invariance of the Haar measure. Furthermore, it is the only measure on Ω that is invariant under the group action and it is therefore obviously singled out.

Let us first consider a trivial example: Consider the set of all infinite binary sequences of ones and minus ones. As relevant properties, we take the value of all finite range correlation functions. By this I mean the following: Consider the sequence (s_i). The nearest-neighbour correlation is then given by

$$C_1 = \lim_{N \to \infty} \frac{1}{2N} \sum_{k=-N}^{N} s_k s_{k+1} \qquad (55)$$

and more general correlations defined through the obvious generalization. As transformations acting on binary sequences, we consider all possible transformations.

Thus, a transformation taken at random maps an arbitrary binary sequence on another equally arbitrary one. If the original sequence has no non-zero correlations, then, at least with probability one, this property is respected by all transformations and the structural invariance group $\mathcal{H}$ is the full group. At this stage, however, we see that the group is highly pathological and presumably has no Haar measure. To avoid this we restrict ourselves, for example, to the set of *finite* sequences. From this follows that we can consider the original sequence as a representative element of the full ensemble of all binary sequences, in which all sequences are taken to be equally probable. On the other hand, if there is a clear predominance of ones, say, but no further correlations, then we can take $\mathcal{H}$ to be the set of all maps which permute the elements of the original sequence. The set Ω is then the set of all sequences which arise from the original one through permutation, all sequences being equally probable. Again, it is reasonable to assume that the original sequence is indeed a representative member of the ensemble. In this sense, whatever property holds with probability one in the ensemble that has been constructed in this fashion will also presumably hold for the original element. Note that this approach does not always work: For example, if I find a non-zero two-point correlation, it is at least not evident how to define a group $\mathcal{H}$ that respects this property. A better approach might then be to define a *non-invariant* measure on $\mathcal{G}$ that takes this correlation into account and that respects it with probability one. However, it will turn out that we will be able to define a structural invariance group in the majority of the cases we shall be dealing with.

Let us now turn to the specific case of canonical transformations. As always, we shall limit ourselves to the case of bijective canonical transformations on a compact phase space. The restriction to bijective maps is a natural one, since we are considering stroboscopic maps of time dependent systems or Poincaré maps of bound systems. As pointed out in the beginning of the previous chapter, the second limitation is vital for the result to be valid at all. Otherwise the phenomenon of Anderson localization would generally invalidate our conclusions. As is readily seen, compacity holds in the usual cases for Poincaré maps of bound systems, but must be introduced in an often artificial manner for periodically driven systems, since the momenta are usually taken to vary over an unbounded range. However, they can be made periodic or bounded through some other device, in which case our remarks will hold.

In order to apply the approach described above, we must define the various ingredients involved in the construction. The set of all objects will be the set of all bijective canonical transformations on a compact phase space Γ. The specific object we start from will be a given such transformation C_0. Since we are interested in the short-range properties of the spectrum, and these in turn depend on the high iterates of the map involved, we shall consider only such properties as remain invariant under arbitrary iteration of the map. As for the group $\mathcal{G}$, we shall take it to be the direct product of the group of all bijective canonical canonical transformations with itself, with the following action on the set of all bijective canonical transformations:

$$p_{U,V} : C \to UCV. \tag{56}$$

To complete the construction, it would be necessary to have a Haar measure on the group $\mathcal{G}$. As far as we know, the existence of such a measure has neither been proved nor shown to be impossible. While we believe that such a measure does indeed exist in the case of compact phase spaces, we shall take another way: After all, what we are really interested in are not the properties of the canonical maps, but rather of the corresponding quantum unitary maps. After having constructed Ω in the set of all classical maps, we can proceed to define Ω_Q to consist of the set of all unitary maps corresponding to a given quantization of the maps in Ω. In a similar way, we can translate the group $\mathcal{G}$ in the quantum domain, where it becomes simply $U(N) \times U(N)$, where N is the dimension of the Hilbert space. The finiteness of N is a crucial point here, and this is precisely the point at which we use the fact that the phase space Γ is compact. From this follows that there exists a (unique) Haar measure on the quantized version of $\mathcal{G}$ and hence, for every closed subgroup $\mathcal{H}$ as well.

There is an apparent problem with the above program, however: The "translation" from classical to quantum language is by no means unique. One cannot, therefore, assign to each classical canonical map a well-defined unitary map. Nevertheless, since we are only interested in the semiclassical limit, this problem is not really severe, since all different possible choices will be very close to each other. One might worry, further, that an approximate translation of the group $\mathcal{H}$ from the classical to the quantum domain might lead one to lose the group property. But this is not the case, since the group $\mathcal{H}$ is defined through the invariance of the properties of C_0. This definition can itself be carried over to the quantum domain and the group property is then trivially maintained.

Finally, it remains to see what are the possible special properties of C_0. Since we agree that the only relevant properties are those that remain invariant under arbitrary iterations of the map, we are left with very few possibilities. Here are the three that I am aware of:

—Time-reversal Invariance: If there exists an anti-canonical map T such that $T^2 = 1$ and such that

$$TC_0TC_0 = 1, \tag{57}$$

then the map C_0 is said to be time-reversal invariant (TRI). In this case, we take $\mathcal{H}$ to consist of the pairs $(C, TC^{-1}T)$. These form a group, and it is straightforward to calculate that it leaves the TRI property invariant. As such, it is the structural invariance group of an arbitrary C_0 which has no specific property apart from TRI. From this follows that Ω is simply the set of all TRI bijective canonical maps, as was in fact to be expected. On the other hand, we are not (yet!) able to define a measure on Ω, since neither $\mathcal{G}$ nor $\mathcal{H}$ have well-defined Haar measures, as far as is known.

This can be readily translated into quantum-mechanical language if we note that in the absence of spin, T can be represented as complex conjugation in quantum mechanics, so that $\mathcal{H}$ is given by the subgroup consisting of the pairs of the type (U, U^t), where U^t is the transpose of U.

—Discrete Symmetries: Let C_0 commute with a group G of canonical transformations Q. This means that the evolution given by C_0 has symmetries, given by the maps Q. This property is indeed preserved for arbitrary iterations of C_0. The structural invariance group $\mathcal{H}$ is then given by the subgroup of all pairs (C, C') such that both C and C' commute with G. Under these circumstances, the action of $\mathcal{H}$ on C_0 respects the property of commuting with Q and generates as a set Ω the set of all bijective canonical maps that commute with Q.

Translating into quantum mechanics, this means that the quantum version of $\mathcal{H}$ consists of all pairs of unitary matrices that commute with the quantization U_Q of the maps Q. This means that the unitary matrices acting to the left and right of U_C can be put in block diagonal form in such a way that each block belongs to a given irreducible representation of G. Thus, the ensemble generated has a well defined block structure, in which all blocks statistically independent. On the other hand, if a given block transforms according to a d-dimensional representation of G, the eigenvalues of the corresponding block will be d-fold degenerate.

—Non-primitive Period: It is possible that C_0 does not correspond to a primitive period, that is, it might happen that C_0 can be written in the form:

$$C_0 = D^k \tag{58}$$

for some $k > 1$. In this case, it is not clear to me how to define a structural invariance group that leaves this property invariant, but the remedy is clear enough: It is simpler then to quantize D in the first place. The spectral properties of U_D^k can then be related to those of U_{C_0}.

We can now go back to the first two properties, and it is a straightforward excercise to show that a TRI system will have a COE, whereas a system with a discrete symmetry will have eigenphases that are divided into statistically independent blocks according to the eigenspaces of U_Q. Let us show this latter point in some detail following closely the reasoning presented in [20].

As we have shown above, in the case of discrete symmetry, the structural invariance group $\mathcal{H}$ is given by pairs of unitary matrices acting to the right and to the left of another matrix, which is also block diagonal in the same basis. The Haar measure of $\mathcal{H}$ is clearly the product measure of the Haar measures of the various unitary groups restriced to each block. From this follows that the ensemble is the product of the various CUE's involved, which is the rule we had stated in the beginning of the previous chapter.

Similarly, we had pointed out a difficulty in the case where the existence of a discrete symmetry was combined with TRI. This can now be treated in the above manner: We must determine the structural invariance group $\mathcal{H}$ that respects *simultaneously* TRI and the symmetry Q. In this case, if we take everything in an eigenbasis of U_Q, it is not clear any more that we can describe T through simple complex conjugation. Define U_T to be the *anti-unitary* representation of T in quantum mechanics. A moment's thought shows that two cases are possible: First, U_T leaves a given eigenspace of U_Q invariant. Then, in this eigenspace, U_T can be represented as complex conjugation and the ensemble restricted to this eigenspace is indeed the COE, using the same kind of arguments that lead to the COE when no symmetry is present. Second, U_T acts as the combination of complex conjugation and interchanging the given eigenspace with its time-recersed counterpart. This is clearly only possible if Q is not itself TRI, that is, if

$$TQTQ \neq 1. \tag{59}$$

The group $\mathcal{H}$ is then given by the pairs $(U, U_T U^{-1} U_T)$. In the block diagonal form, the matrix U has entries

$$\begin{pmatrix} U_1 & 0 \\ 0 & U_2 \end{pmatrix} \tag{60}$$

whereas the matrix $U_T U^{-1} U_T$ gives

$$\begin{pmatrix} U_2^t & 0 \\ 0 & U_1^t \end{pmatrix}. \tag{61}$$

From this follows readily that the eigenvalues in both eigenspaces are degenerate, but that the ensemble for each of them is actually the CUE.

As pointed out above, this was verified numerically. The specific example considered was a fully chaotic billiard with threefold symmetry but without any symmetry axis. The rotation by $2\pi/3$ is clearly not TRI and it divides the Hilbert space in three invariant subspaces, generated by $e^{im\phi}$ for m equal to one, zero or minus one modulo three respectively. Whereas the second is clearly invariant under TRI, the other two are interchanged among each other. For these, indeed, GUE statistics was clearly observed, whereas in the TRI subspace the usual GOE was observed.

Finally, it should be pointed out that it is possible to make the transition from eigenphases to eigenvalues for bounded time-independent systems. The key is to study the eigenphase distribution of the Poincaré map as a function of energy. It can then be shown, using results due to Bogomolny, that the statistics of eigenvalues is exactly the same as that for eigenphases, with the only minor difference that eigenvalues need to be unfolded because of the secular dependence of the density of states on energy.

V CONCLUSIONS

Summarizing, we have discussed the behaviour of classical chaotic systems in the semiclassical limit and we have argued that there exist certain properties which have a very remarkable *universality*. In particular, we find that the short-range spectral correlations of such systems are identical to those of certain ensembles of random matrices. Which ensemble to take is determined by very general properties of the system, such as TRI and the absence or presence of discrete symmetries. Thus, it can be argued that the spectral statistics of classically chaotic systems are fully determined by these, at least for short energy ranges. We have seen that this corresponds to the fact that the periodic orbits of a chaotic system are "universal" in the sense that they simply fill phase space uniformly. In this sense, all chaotic systems are equivalent for large times. On the other hand, as can be easily seen from the results derived in Section III, at large energy ranges, corresponding to short times, there appear features quite distinctive of each specific system.

To show this result, we have used two entirely different approaches: In the one, we have shown how periodic orbit theory together with some statistical assumptions concerning long periodic orbits can get one the large distance behaviour of the universal part. This approach, however, is limited to two-point properties and to the large distance regime. thus, it does not appear feasible to derive such things as the quadratic repulsion in the non-TRI case at small distances, as opposed to merely linear repulsion when TRI is present.

On the other hand, we have presented a somewhat more speculative approach, based on statistical concepts. There we attempt to construct an ensemble of dynamics, of which the given dynamic would be a typical representative. From such considerations would follow, that the connections stated above hold, at the very least, with probability one. The major technical difficulty in this case is the absence of a (known) Haar measure on the group of all bijective canonical transformations on a compact phase space.

ACKNOWLEDGMENTS

It is a pleasure to thank Dr. T.H. Seligman for valuable comments, as well as DGAPA project IN105595 for financial support.

REFERENCES

1. Brody , T. A., et al., *Rev. Mod. Phys.* **53**, 385 (1981).
2. Mehta, M. L., *Random Matrices and Statistical Theory of Energy Levels*, Academic Press, New-York, 1991.
3. Casati, G., Guarneri, I., and Valz-Gris, F., *Lett. Nuovo Cimento* **28**, 279 (1980).
4. Bohigas, O., Giannoni, M.-J., and Schmit, C., *Phys. Rev. Lett.* **52**, 1 (1984); *J. Phys. Lett.* **45**, L1015 (1984).
5. Berry, M. V., *Proc. Roy. Soc. London* **400**, 229 (1985).
6. Gutzwiller, M. C., *Chaos in Classical and Quantum Mechanics*, Springer Verlag, 1990.
7. Balian, R. B., and Bloch, C., *Ann. Phys.* (N.Y.) **60**, 401 (1970); *Ann. Phys.* (N.Y.) **63**, 592 (1971); **64**, 271 (1971); *Ann. Phys.* (N.Y.) **69**, 76 (1972); *Ann. Phys.* (N.Y.), **85**, 514 (1974).

8. See e.g. Balazs, N., and Voros, A., *Phys. Reports* **143**, 109 (1986), for a clear presentation.
9. Leyvraz, F., and Seligman, T. H., *Phys. Lett. A* **168**, 348 (1992).
10. Seligman, T. H., in *Quantum Chaos*, Casati, G., and Chirikov, B.V., eds., Cambridge University Press, 1995.
11. Leyvraz, F., and Seligman, T. H. , in *Proceedings of the IV Wigner Symposium, Guadalajara*, Atakishyiev, N.M., Seligman, T.H., and Wolf, K.B., eds., World Scientific, 1995.
12. Gräf . H. D. *et al.*, *Phys. Rev. Lett.* **69**, 1296 (1992); Alt, H. *et al.*, *Nucl. Phys. A* **560**, 293 (1993); Sridhar, S., and Heller, E. J., *Phys. Rev. A* **46**, 1728 (1992).
13. Ellegaard, C. *et al.*, *Phys. Rev. Lett.* **75**, 1546 (1995); in *Proceedings of the IV Wigner Symposium, Guadalajara*, (Atakishyiev, N. M., Seligman, T. H., and Wolf, K. B., eds.) World Scientific, 1995.
14. Simons, B. D., Lee, P. A., and Altshuler, B. L., *Phys. Rev. Lett.* **70**, 4122 (1993); *Nucl. Phys. B* **409** (FS), 487 (1993); Beenakker, C., *Phys. Rev. Lett.* **70**, 4126 (1993).
15. Efetov, K. B. *Supersymmetry in Disorder and Chaos*, Cambridge University Press, 1996.
16. Guhr, T., in *Proceedings of the IV Wigner Symposium, Guadalajara*, Atakishyiev, N. M., Seligman, T. H., and Wolf, K. B. eds., World Scientific, 1995; *J. Math. Phys.* **34**, 2523 (1993); J. Math. Phys. **34**, 2541 (1993); *Nucl. Phys. A* **560**, 223 (1993); Guhr, T. , and Weidenmüller, H. A., *Ann. Phys.* (N.Y.) **199**, 412 (1990).
17. Heller, E. J., in *Quantum Chaos and Statistical Nuclear Physics*, Seligman, T. H., and Nishioka, H., eds., *Lecture Notes in Physics* **263**, Springer Verlag, 1986.
18. Dyson, F. J., *J. Math. Phys.* **3**, 1191 (1962).
19. Pandey, A., *Ann. Phys.* (N.Y.), **119**, 170 (1979).
20. Leyvraz, F., Schmit, C., and Seligman, T. H., *J. Phys. A***29**, L575 (1996).
21. Keating, J. P., and Robbins, J. M., *J. Phys. A* **30**, L177 (1997).
22. Mello, P. A., and Moshinsky, M., *J. Math. Phys.* **16**, 2017 (1975).
23. Méndez, R. A. *et al.*, *Am. J. Phys.* (in press).
24. Cartan, E., *Abh. Math. Sem. Univ. Hamburg* **11**, 116 (1935).
25. Izrailev, F. M., *Phys. Reports* **196**, 299 (1990).
26. Seligman, T. H., and Verbaarschot, J. J. V., *Phys. Lett. A* **108**, 183 (1985).
27. Bohigas, O., *et al.*, *Phys. Reports* **223**, 43 (1993).

Interference Phenomena in Electronic Transport Through Chaotic Cavities: An Information-Theoretic Approach

Pier A. Mello[†] and Harold U. Baranger[††]

[†] *Instituto de Física, Universidad Nacional Autónoma de México, 01000 México D.F., México*
[††] *Bell Laboratories- Lucent Technologies, 700 Mountain Ave. 1D-230, Murray Hill NJ 07974*

I INTRODUCTION

Scattering of waves by complex systems has captured the interest of physicists for a long time. For instance, the problem of multiple scattering of waves has been of great importance in optics [1,2]. Interest in this problem has been revived recently, both for electromagnetic waves [3] and for electrons [4], in relation with the phenomenon of localization, which gives rise to a great many fascinating effects.

Nuclear physics, with typical dimensions of a few fm ($1fm = 10^{-15}$ m), offers excellent examples of quantum-mechanical scattering by "complex" many-body systems dating as far back as the 1930's when compound-nucleus resonances were discovered [5]. The treatment in these cases is often frankly statistical because the details of the many-body problem are intractably complicated.

In the last two decades, electron transport in disordered metals has been intensively investigated [4,6–10], as has transmission of electromagnetic waves through disordered media [3,4,7]. The typical size scale is $1\mu m$ for electronic systems and $1\mu m$ to $0.1m$ in the electromagnetic case. Because these are also examples of scattering in complex environments, where the character of the disorder is not exactly known, a statistical approach which treats an ensemble of disordered potentials is natural.

Amazingly, features similar to those of these complex nuclear and disordered problems have also been found in certain "simple" systems. While the geometry of these systems is apparently very simple– quantum-mechanical scattering of just one particle by three circular disks in a plane, for instance– the classical dynamics is fully chaotic. Such systems have been studied by the "quantum chaos" community, in which the main question is how the nature of the classical dynamics influences the quantum properties [11,12]. In contrast to the complex cases, these systems are amenable to exact (numerical) calculation; when the results are analyzed statistically, however, the results are closely related to those of the complex systems [11,12]. Experimentally, two types of simple scattering systems have been studied in particular: electron transport through microstructures called "ballistic quantum dots", whose dimensions are of the order of 1μm, and microwave scattering from metallic cavities, with typical dimensions of 0.1m.

The "universal" statistical properties of wave-interference phenomena observed in systems whose dimensions span about 14 orders of magnitude turn out to depend on very general physical principles and constitute the central topic of this article. Although our main interest throughout the paper is electronic transport through ballistic chaotic cavities, or ballistic quantum dots, described in Sec. I.B , we wish to emphasize the generality of the ideas involved, presenting first in Sec. I.A their application in the field of nuclear physics– where some of them were first introduced– with a brief reference to microwave cavities at the end of that subsection.

I.A The atomic nucleus and microwave cavities

One of the most successful models in nuclear physics, called the optical model of the nucleus, was invented by Feshbach, Porter and Weisskopf in the 1950's [13,14]. That model, which works very nicely over a wide range of energies, describes the scattering of a nucleon by an atomic nucleus– a complicated many-body problem– in terms of two distinct time scales:

1. A *prompt* response arising from *direct processes*, in which the incident nucleon feels a mean field produced by the other nucleons. This response is described mathematically in terms of the *average* of the actual scattering amplitudes over an energy interval: these averaged amplitudes, also known as *optical* amplitudes, show a much slower energy variation than the original ones.

2. A *delayed*, or *equilibrated*, response, corresponding to the formation and decay of the compound nucleus. It is described by the difference between the exact and the optical scattering amplitudes: it varies appreciably with energy and is studied with *statistical* concepts using techniques known as random matrix theory [15,16].

Just as in the field of statistical mechanics time averages are very difficult to construct and hence are replaced by *ensemble averages* using the notion of *ergodicity*, in the present context too one finds it advantageous to study energy averages in terms of ensemble averages through an ergodic property [17–20].

The optical model not only works well in nuclear physics, but has also been applied successfully in the description of a number of chemical reactions, thus bringing us from the nuclear to the molecular size scale [21].

The connection of the above problems with the theory of waveguides and cavities was proposed very clearly by Ericson and Mayer-Kuckuk more than thirty years ago [22]: "Nuclear-reaction theory is equivalent to the theory of waveguides... We will concentrate on processes in which the incident wave goes through a highly complicated motion in the nucleus... We will picture the nucleus as a closed cavity, with reflecting but highly irregular walls."

In fact, recent experiments with microwave cavities have shown features similar to those that had been observed in the nuclear case. Importantly, the "irregular walls" anticipated from the nuclear case are *not* necessary in order to see these features: the analogy between nuclear reaction theory and the theory of waveguides holds for simple smooth cavities as long as the corresponding classical dynamics is chaotic [23-25]. This has been the focus of several recent experiments involving microwave scattering from metallic cavities [26-29]. These studies in simple chaotic systems were predated by extensive studies on scattering of microwaves by a disordered dielectric medium; particularly important are the experiments of A. Genack and collaborators [30]. It is also interesting to note that the use of statistical concepts to analyze electromagnetic scattering by waveguides started quite independently of the connection to the complex scattering of nuclear physics, in the context of radio wave propagation [31].

I.B Ballistic Mesoscopic Cavities

The term *mesoscopic* system refers to microstructures in which the phase of the single-electron wave function– in an independent-electron approximation– remains coherent across the system of interest [4,6-10]: this means that the phase-coherence length l_ϕ associated with processes that can change the environment– the other electrons, or the phonon field– to an orthogonal state exceeds the system size [6]. This is realized in the laboratory with systems whose spatial dimensions are on the order of 1μm or less and at temperatures ≤ 1 K. In so-called *ballistic cavities*, or *quantum dots*, the electron motion is in addition practically ballistic, except for specular reflection from the walls: thus the elastic mean free path $l_{\rm el}$ also exceeds the system size. In the most favorable material system, GaAs heterostructures, this condition can also be realized for cavities of size $\leq 1\mu$m [6-9].

Experimentally, an electrical current is established through the leads that connect the cavity to the outside and the potential difference across the cavity is measured, from which the conductance G is extracted. In an independent-electron picture one thus aims at understanding the quantum-mechanical single-electron scattering by the cavity, while the leads play the role of waveguides. It is the multiple scattering of the waves reflected by the various portions of the cavity that gives rise to interference effects. Three important experimental probes of the interference effects are an external magnetic field B, the Fermi energy ϵ_F, and the shape of the cavity: when these are varied the relative phase of the various partial waves changes and so the interference pattern changes. The changing interference pattern in turn causes the conductance to change; this sensitivity of G to small changes in parameters through quantum interference is called *conductance fluctuations*.

The connection between scattering by simple chaotic cavities [23-25] and mesoscopic systems was first made theoretically [32]. Subsequently, cavities in the shape of a stadium, for which the single-electron classical dynamics would be chaotic, were first reported in Ref. [33]. More recently, several other types of structures have been investigated including experimental *ensembles* of shapes [34-44]. Averages of the conductance, its fluctuations, and its full distribution were obtained over such an ensemble.

It is the aim of the present paper to provide a theoretical framework to describe this physical situation. To this end we will set up a scheme similar to that explained in Sec. I.A in connection with the scattering problem in nuclear physics. A complementary treatment based on semiclassical ideas has also been developed but will not be covered here [32,45].

The paper is organized as follows. In Sec. II we present the general ideas of quantum-mechanical scattering by a cavity, introducing the basic object we shall work with– the scattering or S matrix of the problem– and its connection with the conductance that is measured experimentally.

Sec. III treats an ensemble of systems in terms of an ensemble of S matrices. The notion of invariant measure is introduced in Sec. III.A. In Sec. III.B we first show that in the case of 1-dimensional S matrices the analytic properties of S in the complex-energy plane and the property of ergodicity, plus the value of the optical S matrix $\langle S \rangle$ introduced in Sec. I.A above, determine the ensemble uniquely. In the multichannel case this is no longer the case. We then resort to an *information-theoretic* argument to select the ensemble.

Sec. IV presents specific analytical results in the absence of direct processes, first for the weak-localization correction and conductance fluctuations and then for the statistical distribution of the conductance.

In Sec. V we obtain explicitly the conductance distribution in the presence of direct processes in the case of a cavity connected to two leads that can support one channel each.

Sec. VI compares our theoretical predictions with the numerical solution of the Schrödinger equation for a number of structures, in the presence and in the absence of direct processes.

In Sec. VII we compare our theory with the experimental data that were already mentioned above and find a number of discrepancies. To reconcile theory and experiment, we realize the necessity of introducing the energy smearing caused by non-zero temperature as well as the effect of processes that destroy the coherence of the wave function in the sample.

Finally, the conclusions of our work are presented in Sec. VIII. Some of the main results in this paper have appeared in condensed form in our previous publications [46–48].

II THE SCATTERING PROBLEM

II.A Scattering waves: Definition of the S-matrix

Consider a system of noninteracting electrons. Since we shall be dealing with cases in which spin-orbit coupling is negligible, we disregard the spin degree of freedom and just consider "spinless electrons" in what follows. We are interested in studying the scattering of an electron at the Fermi energy $\epsilon_F = \hbar^2 k_F^2/2m$ by the 2D microstructure shown schematically in Fig. 1. The microstructure consists of a

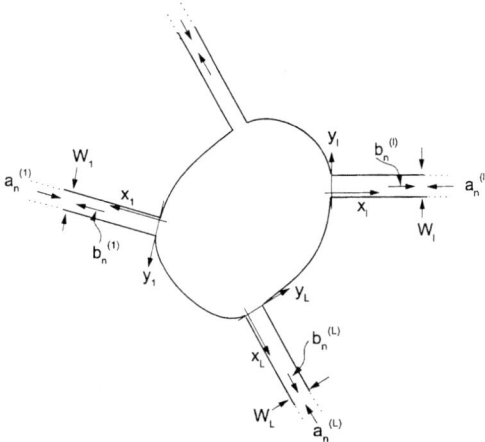

FIGURE 1. The 2D cavity studied in the text. The cavity is connected to the outside via L waveguides. The arrows inside the waveguides indicate incoming or outgoing waves as in Eq. (1). In waveguide l there can be N_l such incoming or outgoing waves: this is indicated in the figure by the amplitudes a_n^l, b_n^l, respectively, where $n=1,\cdots,N_l$.

cavity, connected to the outside by L leads, ideally of infinite length, that play the role of waveguides. The l-th lead ($l=1,\cdots,L$) has width W_l. We are interested in the (scattering) solutions of the Schrödinger equation *inside* such a structure, with the ideal boundary condition that the walls of the cavity and leads are completely impenetrable: hence the wave function must vanish there.

In each lead l we introduce a system of coordinates x_l, y_l, as indicated in Fig. 1. The x_l axis runs along the lead and points outwards from the cavity. The y_l axis runs in the transverse direction and is tangential to the cavity wall and its continuation across the lead; y_l takes on the values 0 and W_l on the two walls of the lead. In lead l and for $x_l > 0$ we have the elementary solutions to the Schrödinger equation

$$e^{\pm i k_n^{(l)} x_l} \chi_n(y_l) \tag{1}$$

where the positive (negative) sign is for outgoing (incoming) waves. Here the functions $\chi_n(y_l)$ are the solution of the transverse Hamiltonian which in the presence of a magnetic field may depend on $k_n^{(l)}$ [49]. The solution of the scattering problem consists in expressing the amplitude of the outgoing waves in terms of the incoming ones.

In the absence of a magnetic field, the explicit problem can be simply stated; we now present this case in detail, noting that it can be generalized to the $B \neq 0$ case [49]. For $B=0$, the functions $\chi_n(y_l)$ are

$$\chi_n(y_l) = \sqrt{\frac{2}{W_l}} \sin K_n^{(l)} y_l, \quad K_n^{(l)} = \frac{n\pi}{W_l}, \quad n = 1, 2, \ldots \qquad (2)$$

where $K_n^{(l)}$ is the "transverse wave number". The functions $\chi_n(y_l)$ vanish on the two walls of the lead and form a complete orthonormal set of functions for the variable y_l; i.e.

$$\langle \chi_n | \chi_n \rangle = \delta_{nm}. \qquad (3)$$

This "transverse quantization" is a consequence of the boundary condition on the walls of the leads. Each possibility defined by the integer n is named a *mode*, or *channel*. The "longitudinal" wave number $k_n^{(l)}$ satisfies the relation

$$[k_n^{(l)}]^2 + [K_n^{(l)}]^2 = k_F^2. \qquad (4)$$

If $K_n^{(l)} < k_F$, then $[k_n^{(l)}]^2 > 0$, $k_n^{(l)}$ is real, and the $e^{\pm i k_n^{(l)} x_l}$ occurring in Eq. (1) represent running waves along the leads: we thus have *running modes* or *open channels*. On the other hand, when $K_n^{(l)} > k_F$, then $[k_n^{(l)}]^2 < 0$ and $k_n^{(l)}$ is pure imaginary, thus giving rise to exponentially decaying waves along the leads: these modes are called *evanescent modes* or *closed channels*. If

$$N_l < k_F W_l / \pi < N_l + 1, \qquad (5)$$

there are N_l open channels in lead l. Very far away along the leads, i.e. as $x_l \to \infty$, only the contribution of the open channels contributes to the wave function. The most general form of the asymptotic wave function in lead l is thus the linear combination

$$\sum_{n=1}^{N_l} \left[a_n^{(l)} \frac{e^{-i k_n^{(l)} x_l}}{(\hbar k_n^{(l)} / m)^{1/2}} + b_n^{(l)} \frac{e^{i k_n^{(l)} x_l}}{(\hbar k_n^{(l)} / m)^{1/2}} \right] \chi_n(y_l). \qquad (6)$$

Note that the normalization of the plane waves is such that they give rise to *unit flux*.

We define the N_l-dimensional vector

$$\mathbf{a}^{(l)} = (a_1^{(l)}, \ldots, a_{N_l}^{(l)})^T \qquad (7)$$

that contains the N_l incoming amplitudes in lead l ($l = 1, \ldots, N_l$). Putting all the $\mathbf{a}^{(l)}$ ($l = 1, \ldots, L$) together, we form the vector

$$\mathbf{a} = (\mathbf{a}^{(1)}, \ldots, \mathbf{a}^{(L)})^T \qquad (8)$$

associated with the incoming waves in all the channels for all the leads. We can make similar definitions for the outgoing-wave amplitudes. The *scattering matrix*, or S *matrix*, is then defined by the relation

$$\mathbf{b} = S\mathbf{a}, \qquad (9)$$

connecting the incoming to the outgoing amplitudes. In terms of individual leads we can write

$$S = \begin{bmatrix} r_{11} & t_{12} & \cdots & t_{1L} \\ t_{21} & r_{22} & \cdots & t_{2L} \\ \vdots & \vdots & \ddots & \vdots \\ t_{L1} & t_{L2} & \cdots & r_{LL} \end{bmatrix}. \qquad (10)$$

Here, r_{ll} is an $N_l \times N_l$ matrix, containing the reflection amplitudes from the N_l channels of lead l back to the same lead; t_{lm} is an $N_l \times N_m$ matrix, containing the transmission amplitudes from the N_m channels of lead m to the N_l channels of lead l. The S matrix is thus a square matrix with dimensionality n given by

$$n = \sum_{l=1}^{L} N_l. \qquad (11)$$

While we have carried out the analysis explicitly for $B = 0$, the main results, Eqs. (1), (6), (9), and (10), also hold in the presence of a magnetic field.

Having chosen the unit-flux normalization for the plane waves in Eq. (6), *flux conservation* implies *unitarity* of the S matrix [1,50–52]; i.e.

$$SS^\dagger = I. \qquad (12)$$

In the absence of other symmetries, we have this requirement only. This is the *unitary* case, also designated in the literature as $\beta = 2$. In the presence of *time-reversal invariance* (TRI) (as in the absence of a magnetic field) and no spin, the S matrix, besides being unitary, is *symmetric* [1,50–53]:

$$S = S^T \qquad (13)$$

This is the *orthogonal* case, also designated as $\beta = 1$. The *symplectic* case ($\beta = 4$), arising in the presence of spin and time-reversal invariance, will not be touched upon in this presentation.

II.B The conductance

For a two-lead problem, which will occur most frequently in our analysis, the S matrix has the structure

$$S = \begin{bmatrix} r_{11} & t_{12} \\ t_{21} & r_{22} \end{bmatrix} \equiv \begin{bmatrix} r & t' \\ t & r' \end{bmatrix}. \qquad (14)$$

In the particular case $N_1 = N_2 = N$, i.e. when the two leads have the same number of channels N, the four blocks r, t, r', t' are $N \times N$ and the S matrix is $2N \times 2N$.

As we mentioned in the Introduction, we are interested in the electronic *conductance* of the microstructure. If we assume that the latter is placed between two reservoirs (at different chemical potentials) shaped as expanding horns with negligible reflection back to the microstructure, then the Landauer-Büttiker formula [57–59] expresses the conductance G in terms of the *scattering properties of the microstructure itself* as

$$G = \frac{e^2}{h} g, \quad g = 2T, \tag{15}$$

where the factor 2 arises from the two spin degrees of freedom and the "spinless conductance" T is the transmission coefficient $|t_{ab}|^2$ summed over initial and final channels:

$$T = \mathrm{Tr}(tt^\dagger). \tag{16}$$

II.C Polar representation of the S matrix

In the two-lead case with $N_1 = N_2 = N$, one can parametrize the S matrix in the so called "polar representation" as [54–56,46]

$$S = \begin{bmatrix} v_1 & 0 \\ 0 & v_2 \end{bmatrix} \begin{bmatrix} -\sqrt{1-\tau} & \sqrt{\tau} \\ \sqrt{\tau} & \sqrt{1-\tau} \end{bmatrix} \begin{bmatrix} v_3 & 0 \\ 0 & v_4 \end{bmatrix} = VRW. \tag{17}$$

Here, τ stands for the N-dimensional diagonal matrix of eigenvalues τ_a ($a=1,\cdots,N$) of the Hermitian matrix $tt^\dagger$. The v_i ($i=1,\cdots,4$) are arbitrary $N \times N$ unitary matrices for $\beta=2$, with the restrictions

$$v_3 = v_1^T, \quad v_4 = v_2^T \tag{18}$$

for $\beta = 1$. It is readily verified that any matrix of the form (17) satisfies the appropriate requirements of an S matrix for $\beta = 1, 2$. The converse statement, as well as the uniqueness of the polar representation, can also be proved using an argument similar to that in Ref. [55].

In the polar representation, we can write the total transmission T of Eq. (16) as

$$T = \sum_a \tau_a. \tag{19}$$

Thus the polar representation is natural for the study of conductance since it separates the magnitude of the transmission– the transmission eigenvalues $\{\tau_a\}$– from the irrelevant phase effects– the unitary matrices $\{v_i\}$.

III ENSEMBLES OF S MATRICES: AN INFORMATION-THEORETIC APPROACH

As we mentioned in the Introduction, we are interested in ensembles of systems, which will be represented as *ensembles of S matrices* endowed with a *probability measure*. For that purpose it is important to introduce first the notion of *invariant measure* for our S matrices: this we develop in the first subsection. The second subsection is devoted to the derivation of the probability measure for our ensemble of S matrices, using an information-theoretic point of view.

III.A The invariant measure

The invariant measure is the measure which equally weights all matrices which satisfy the unitarity and symmetry constraints. Intuitively, it corresponds to the most random distribution consistent with the constraints, or in other words the one with the least information. Mathematically, such a measure is defined by requiring that it remain invariant under an automorphism of a given symmetry class of matrices into itself [53,54]:

$$d\mu^{(\beta)}(S) = d\mu^{(\beta)}(S'). \tag{20}$$

For $\beta=1$, the transformed matrix S' is related to S by

$$S' = U_0 S U_0^T, \tag{21}$$

U_0 being an arbitrary, but fixed, unitary matrix. Clearly, Eq. (21) is an automorphism of the set of unitary symmetric matrices into itself. For $\beta=2$,

$$S' = U_0 S V_0 \tag{22}$$

where U_0 and V_0 are arbitrary fixed unitary matrices. For $\beta = 2$, the resulting measure is the well known Haar's measure of the unitary group and its uniqueness is well known [60,61]. Uniqueness for $\beta = 1$ was shown in Ref. [53]. Using the invariant measures, Eqs. (20) - (22), as the probability measures for ensembles of S-matrices defines the *Circular Orthogonal* and *Unitary Ensembles (COE, CUE)* for $\beta=1,2$, respectively.

Several explicit representations of the invariant measure are known, the classic one being in terms of the eigenphases and eigenvectors of the S-matrix. For our purposes, the polar representation of Eq. (17) is of particular interest because of its connection to the conductance properties of the cavity. We thus consider the two-equal-lead case, $N_1 = N_2 = N$, and express the invariant measure explicitly in this parametrization.

We first recall a well known result from differential geometry. Consider the expression for the arc element

$$ds^2 = \sum g_{\mu\nu}(x)\delta x_\mu \delta x_\nu, \qquad (23)$$

written in terms of *independent variables* and the metric tensor $g_{\mu\nu}(x)$. Assuming that ds^2 remains invariant under the transformation $x_\mu = x_\mu(x'_1, x'_2, \cdots)$, one can prove that the volume element

$$dV = |\det g(x)|^{1/2} \prod_\mu dx_\mu \qquad (24)$$

remains invariant under the same transformation [62].

We now go back to our random-S-matrix problem. We define the differential arc element as

$$ds^2 = \text{Tr}[dS^\dagger dS]. \qquad (25)$$

This expression remains invariant under the transformations (21) and (22). Substituting for S the form (17), one can extract the metric tensor; applying Eq. (24), one then finds the invariant measure. This is done in VIII; the result is [46,63]

$$d\mu^{(\beta)}(S) = P^{(\beta)}(\{\tau\}) \prod_a d\tau_a \prod_i d\mu(v^{(i)}), \qquad (26)$$

where $P^{(\beta)}(\{\tau\})$ denotes the joint probability density of the $\{\tau\}$, $d\mu(v^{(i)})$ is the invariant or Haar's measure on the unitary group $U(N)$ [61], and C_β is a normalization constant. From VIII we have

$$P^{(1)}(\{\tau\}) = C_1 \prod_{a<b} |\tau_a - \tau_b| \prod_c \frac{1}{\sqrt{\tau_c}}, \qquad P^{(2)}(\{\tau\}) = C_2 \prod_{a<b} |\tau_a - \tau_b|^2. \qquad (27)$$

Eqs. (26) and (27) explicitly specify the invariant measure. The factor involving the product over pairs gives the repulsion of the eigenvalues; notice that the repulsion is linear for the orthogonal case while quadratic in the unitary case, as typically occurs.

III.B The information-theoretic model

III.B.1 The one-channel case

To begin our discussion, consider first a physical problem that can be described by a 1×1 S matrix. This is the case, for instance, for a particle scattered by a 1D potential that is nonzero in the region $-a \leq x \leq 0$, to which an impenetrable wall is added at $x=-a$: the particle then lives in the semiinfinite domain $-a \leq x \leq \infty$. Another example is that of a 2D cavity, connected to the outside by only one lead that in turn supports only one open channel. From unitarity, S must be a complex number of unit modulus at every energy; i.e. $S(E) = e^{i\theta(E)}$. In the Argand diagram

Re(S)-Im(S), $S(E)$ is represented, for a given energy, by a point on the unitarity circle: that point is defined by the angle $\theta(E)$. As the energy changes, so does the representative point: this is what we may call, pictorially, the "motion" of $S(E)$ as a function of energy; it resembles the motion in phase space of the representative point of a classical system as a function of time.

We ask the following question: as we move along the energy axes, *what fraction of the time do we find θ lying inside a given interval $d\theta$?* Let us call $dP(\theta) = p(\theta)d\theta$ that fraction.

To answer this question, we first analyze how to construct energy averages. By this we always mean a *local* energy average, performed inside an interval I that contains many resonances and yet is small compared to an energy interval over which "gross structure" quantities showing a secular variation, like the average spacing Δ of resonances, vary appreciably. We then expect the dependence of the average in question on the center of the interval E_0, its width I, as well as the particular weighting function used to define the average, to be *weak*.

We devise an *idealized* situation: the argument E in $S(E)$ is extended all the way from $-\infty$ to $+\infty$, in such a way that *local averages are everywhere the same as inside I*: this idealization, which we call *stationarity*, will only represent well what goes on *locally* inside I in the actual system. We indicate an energy average of a quantity by placing a bar over it while an ensemble average is denoted by angular brackets.

Ref. [19] chooses, as the weighting function, a Lorentzian. Using the fact that $S(E)$ is *analytic* in the upper half of the complex energy plane (*causality*), it shows that

$$\overline{S^k} = \left[\overline{S}\right]^k, \tag{28}$$

i.e. *the average of the k-th power of S coincides with the k-th power of the average of S*. Thus, the quantity $\overline{S}$, referred to as the optical S matrix in Sec. I.A, plays a special role, in that the average of any power of S can be expressed in terms of it. Ref. [19] then finds the answer to the above-posed question: the fraction of time $p(\theta)$ spent by θ in a unit interval around θ in its journey along the energy axis is *uniquely* given by the expression

$$p(\theta) = \frac{1}{2\pi} \frac{1 - |\overline{S}|^2}{|S - \overline{S}|^2}, \quad S = e^{i\theta}, \tag{29}$$

and *depends only upon the average, or optical, S matrix $\overline{S}$*. This expression is also known as *Poisson's kernel* [54]. The conclusion is remarkable: it tells us that the *system specific details are irrelevant*, except for the optical S matrix.

As an example, consider a 1D δ-potential centered at $x = 0$ and a perfectly reflecting wall at $x = -a$. An energy stretch containing 100 resonances starting from $ka = 10,000$ (so that secular variations of gross-structure quantities can be neglected) was sampled to find the fraction of time that θ falls in a certain small

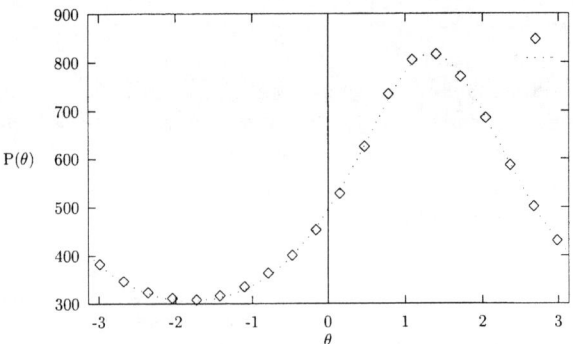

FIGURE 2. In the δ-potential model a stretch of energy containing 100 resonances, starting from $ka = 10,000$, was sampled to find the fraction of time that θ falls in a unit interval around θ: the result is indicated by diamonds. The curve is a plot of Poisson's kernel (29), with the value of $\overline{S}$ extracted from the numerical data, in order to have a parameter-free fit. The agreement is excellent. (From Ref. [65].)

interval around θ [65]. The result is compared in Fig. 2 with Poisson's kernel (29), where the value of the optical $\overline{S}$ was extracted from the numerical data themselves, so as to have a parameter-free fit. We observe that the agreement is excellent.

Consider now a collection of systems, described by an ensemble of S matrices endowed with a probability measure. As an idealization, suppose we further construct $S(E)$ as a *stationary random function of energy*, for $-\infty < E < \infty$. Then we know the conditions under which *ergodicity*, understood as equality of energy and ensemble averages (except for a set of zero measure), holds [64]. Let us then assume that our ensemble is ergodic.

The condition (28) arising from analyticity, together with ergodicity, implies the relation

$$\left\langle S^k \right\rangle = \langle S \rangle^k \tag{30}$$

between ensemble averages, often called the *analyticity-ergodicity (AE)* requirement. The ensemble measure is thus *uniquely* given by

$$dP_{\langle S \rangle}(S) = p_{\langle S \rangle}(\theta) d\theta, \qquad p_{\langle S \rangle}(\theta) = \frac{1}{2\pi} \frac{1 - |\langle S \rangle|^2}{|S - \langle S \rangle|^2}, \tag{31}$$

once $\langle S \rangle$ is specified. The ensemble depends parametrically upon the single complex number $\langle S \rangle$, any other information being irrelevant.

We note in passing that Eq. (30) implies that a function $f(S)$ that is analytic in its argument, and can thus be expanded in a power series in S, must fulfill the *reproducing property* [54,66]

$$f(\langle S \rangle) = \int f(S) dP_{\langle S \rangle}(S). \tag{32}$$

It is because the probability measure appears as the kernel of this integral equation that it is called Poisson's kernel.

III.B.2 *The multi-channel case*

We now consider S matrices of dimensionality n, that can describe, in general, a multi-lead problem with n channels altogether, as explained in Sec. II. The Argand diagram discussed above for $n=1$ has to be generalized to include the axes $\text{Re} S_{11}$, $\text{Im} S_{11}$, $\text{Re} S_{12}$, $\text{Im} S_{12}$, ...,$\text{Re} S_{nn}$, $\text{Im} S_{nn}$; S is restricted to move on the surface determined by unitarity ($SS^\dagger = I$) and, for $\beta = 1$, symmetry ($S = S^T$).

We assume E is far from thresholds and recall that again $S(E)$ is *analytic* in the upper half of the complex-energy plane. The study of the statistical properties of S is again simplified by idealizing $S(E)$, for real E, as a *stationary random-matrix function* satisfying the condition of *ergodicity*. The same argument as in the 1D case above shows that the AE requirement (30) is generalized to

$$\langle (S_{a_1 b_1})^{n_1} \cdots (S_{a_k b_k})^{n_k} \rangle = \langle S_{a_1 b_1} \rangle^{n_1} \cdots \langle S_{a_k b_k} \rangle^{n_k}. \tag{33}$$

Notice that this expression involves only S, and not S^* matrix elements. Similarly, if $f(S)$ is a function that can be expanded as a series of nonnegative powers of $S_{11}, \cdots, S_{nn}$ (analytic in S), we must have the reproducing property (32).

Our starting point is the invariant measure $d\mu_\beta(S)$ that was introduced in the last subsection. The average of S evaluated with that measure vanishes (shown explicitly in Sec. IV.A), so that the prompt, or direct, components described in the Introduction vanish. It is easy to check that the AE requirements (33) or, equivalently, the reproducing property (32), is satisfied exactly for the invariant measure. Ensembles that contain more information than the invariant one are constructed by multiplying the latter by appropriate functions of S. We relate the probability density $p^{(\beta)}_{\langle S \rangle}(S)$ to the differential probability through

$$dP^{(\beta)}_{\langle S \rangle}(S) = p^{(\beta)}_{\langle S \rangle}(S) d\mu_\beta(S) \tag{34}$$

and require the fulfillment of the AE conditions. It was shown in Ref. [66] that, for $n > 1$, the AE conditions and reality of the answer are not enough to determine the probability distribution uniquely. However, it was shown that the probability density (V_β is a normalization factor)

$$p_{\langle S \rangle}(S) = V_\beta^{-1} \frac{[\det(I - \langle S \rangle \langle S \rangle^\dagger)]^{(\beta n + 2 - \beta)/2}}{|\det(I - S \langle S \rangle^\dagger)|^{\beta n + 2 - \beta}}, \tag{35}$$

known again as Poisson's kernel, not only satisfies the AE requirements (33) [54], but the *information* $\mathcal{I}$ associated with it

293

$$\mathcal{I}[p] \equiv \int p_{\langle S \rangle}(S) \ln p_{\langle S \rangle}(S) d\mu(S) \tag{36}$$

is *less than or equal to that of any other probability density satisfying the AE requirements for the same* $\langle S \rangle$ [66]. Notice that, for $n=1$, Eq. (35) reduces to (31). Thus the information entering Poisson's kernel specifies : **1)** General properties: *i) flux conservation* (giving rise to unitarity of the S matrix), *ii) causality* and the related analytical properties of $S(E)$ in the complex-energy plane, and *iii)* the presence or absence of time-reversal (and spin-rotation symmetry when spin is taken into account), that determines the *universality class*: orthogonal, unitary (or symplectic) **2)** A specific property: the ensemble average $\langle S \rangle$ ($= \overline{S}$ under ergodicity), which controls the presence of *prompt*, or *direct processes* in the scattering problem. System-specific *details other than the optical S are assumed to be irrelevant.*

The fact that for $n > 1$ AE and reality do not fix the ensemble uniquely is not surprising. In general (the 1×1 case being exceptional) we expect, physically, the matrix $\langle S \rangle$ to be insufficient to characterize the full distribution when, in addition to the prompt and equilibrated components, there are other contributions associated with different time scales [18]. Out of all possibilities, though, the information-theoretic argument selects the one where the prompt and equilibrated components and the associated optical S are the only physically relevant quantities.

In addition to the completely general derivation of Poisson's kernel above, we present a concrete construction of this distribution following Ref. [67,68]. For the equilibrated part of the response, suppose there is an S-matrix S_0 which is distributed according to the circular ensembles. For the prompt response, imagine a scattering process S_1 occuring prior to the response S_0. The total scattering is the composition of these two parts. Specifically, imagine bunching the L leads of the cavity into a "superlead" containing n incoming and n outgoing waves. Along the superlead, between the cavity and infinity, we connect a scatterer of the appropriate symmetry class described by S_1. Since there are n incoming and n outgoing waves on either side of the scatterer, S_1 is $2n$-dimensional and can be written

$$S_1 = \begin{bmatrix} r_1 & t_1' \\ t_1 & r_1' \end{bmatrix}. \tag{37}$$

The composition of the two scattering processes yields a total S

$$S = r_1 + t_1'(1 - S_0 r_1')^{-1} S_0 t_1. \tag{38}$$

One can prove [54,68,67,69] the following statement: the distribution of S is Poisson's measure (35) with $\langle S \rangle = r_1$ if and only if the distribution of S_0 is the invariant measure. That is, Eq. (38) transforms between the problem with direct processes and the one without (for the one-energy distribution considered here). Also, one can show [54,69] that the distribution is independent of the choice of t_1 and t_1', as long as they belong to a unitary matrix S_1.

Note that throughout this work we use arguments which refer only to physical information expressible entirely in terms of the S-matrix. An alternate point of view is to express everything in terms of an underlying Hamiltonian which one then analyzes using statistical or information-theoretic assumptions. These two points of view give, in fact, the same results: one can prove [70,67,68] that, for $\langle S \rangle = 0$, a Gaussian Ensemble for the underlying Hamiltonian gives a Circular Ensemble for the resulting S. The argument was extended to $\langle S \rangle \neq 0$ in Refs. [67,68] using the transformation (38) above.

IV ABSENCE OF DIRECT PROCESSES

In this case the optical matrix $\langle S \rangle$ vanishes and Poisson's kernel (35) reduces to the invariant measure:

$$dP^{(\beta)}_{\langle S \rangle = 0} = d\mu^{(\beta)}(S). \tag{39}$$

We now derive the implications of this distribution for T, starting with the average and variance of T and then turning to its distribution.

IV.A Averages of products of S: Weak-localization and conductance fluctuations

Averages over the invariant measure of products of S matrix elements (invariant integration) can be evaluated using solely the properties of the measure, without performing any integration explicitly [20,71]. We first discuss the unitary case, because it is simpler, and then the orthogonal one.

IV.A.1 The case $\beta = 2$

To illustrate the procedure, consider first the average over the invariant measure of a single S matrix element, to be denoted as

$$\langle S_{a\alpha} \rangle^{(2)}_0 = \int S_{a\alpha} d\mu^{(2)}(S). \tag{40}$$

Even though it is trivial to recognize that this average vanishes— since the invariant measure gives the same weight to $S_{a\alpha}$ and to its negative— we present a more formal argument, which will be generalized later to more complicated averages. If U^0 is an *arbitrary* but *fixed* unitary matrix, we define the transformed $\tilde{S}$ as

$$S = U^0 \tilde{S}. \tag{41}$$

Introducing (41) in (40) we have

$$\langle S_{a\alpha} \rangle^{(2)}_0 = \sum_{a'} U^0_{aa'} \int \tilde{S}_{a'\alpha} d\mu^{(2)}(\tilde{S}) = \sum_{a'} U^0_{aa'} \langle S_{a'\alpha} \rangle^{(2)}_0 \tag{42}$$

where we have used the defining property (20) of the invariant measure and the definition of $\langle S_{a'a}\rangle_0$. In particular, if we take, as the arbitrary fixed matrix U^0

$$U^0_{aa'} = e^{i\theta_a}\delta_{aa'}, \tag{43}$$

we find

$$\langle S_{a\alpha}\rangle_0 = e^{i\theta_a}\langle S_{a\alpha}\rangle_0. \tag{44}$$

Since this expression should hold for arbitrary θ_a, we conclude that

$$\langle S_{a\alpha}\rangle_0 = 0. \tag{45}$$

The above argument can be generalized to prove that

$$\left\langle \left[S_{b_1\beta_1}\cdots S_{b_p\beta_p}\right]\left[S_{a_1\alpha_1}\cdots S_{a_q\alpha_q}\right]^*\right\rangle_0^{(2)} = 0, \tag{46}$$

unless $p=q$ and unless $\{a_1,\ldots,a_p\}$ and $\{b_1,\ldots,b_p\}$ constitute the same set of indices except for order, with the same condition for the sets $\{\alpha_1,\ldots,\alpha_p\}$, $\{\beta_1,\ldots,\beta_p\}$. In particular, consider $p=q=1$. Using the same argument as above, we find

$$\langle S_{b\beta}S^*_{a\alpha}\rangle_0^{(2)} = \sum_{a'b'} U^0_{bb'}\left(U^0_{aa'}\right)^* \langle S_{b'\beta}S^*_{a'\alpha}\rangle_0^{(2)}, \quad \forall U^0. \tag{47}$$

For U^0 given as in Eq. (43), we have

$$\langle S_{b\beta}S^*_{a\alpha}\rangle_0 = e^{i(\theta_b-\theta_a)}\langle S_{b\beta}S^*_{a\alpha}\rangle_0 \tag{48}$$

which vanishes unless $b=a$. Had we defined $\tilde{S}$ through right multiplication instead of left multiplication in (41), we would have concluded $\beta=\alpha$. The only nonvanishing possibility in Eq. (47) is thus

$$\langle |S_{a\alpha}|^2\rangle_0 = \sum_{a'} |U^0_{aa'}|^2 \langle |S_{a'\alpha}|^2\rangle_0, \quad \forall U^0. \tag{49}$$

For instance, the choice of the matrix that produces a permutation of the indices 1 and 2 as U^0 yields:

$$U^0 = \begin{bmatrix} 0 & 1 & 0 & \cdots & 0 \\ 1 & 0 & 0 & \cdots & 0 \\ 0 & 0 & 1 & \cdots & 0 \\ \vdots & \vdots & \vdots & \ddots & \vdots \\ 0 & 0 & 0 & \cdots & 1 \end{bmatrix}, \quad \langle |S_{1\alpha}|^2\rangle = \langle |S_{2\alpha}|^2\rangle. \tag{50}$$

Similarly

$$\left\langle |S_{1\alpha}|^2 \right\rangle_0 = \cdots = \left\langle |S_{n\alpha}|^2 \right\rangle_0. \tag{51}$$

The final result for the average intensity follows from unitarity

$$\sum_{a=1}^{n} \left\langle |S_{a\alpha}|^2 \right\rangle_0 = 1 \implies \left\langle |S_{a\alpha}|^2 \right\rangle_0^{(2)} = \frac{1}{n}. \tag{52}$$

As an application, we calculate the *average conductance* when we have a cavity connected to the outside by means of two leads supporting N_1 and N_2 open channels (so that $n = N_1 + N_2$):

$$\langle T \rangle_0^{(2)} = \sum_{a=1}^{N_1} \sum_{b=1}^{N_2} \left\langle |t_{ab}|^2 \right\rangle_0^{(2)} = \frac{N_1 N_2}{N_1 + N_2} = \left[\frac{1}{N_1} + \frac{1}{N_2} \right]^{-1}, \tag{53}$$

which is the series addition of the two conductances N_1 and N_2. (The superscript on the angular brackets indicates $\beta = 2$.)

With similar arguments one finds [71]:

$$\left\langle |S_{12}|^2 |S_{34}|^2 \right\rangle_0^{(2)} = \frac{1}{n^2 - 1}, \tag{54}$$

$$\left\langle |S_{12}|^2 |S_{13}|^2 \right\rangle_0^{(2)} = \frac{1}{n(n+1)}, \tag{55}$$

$$\left\langle |S_{12}|^4 \right\rangle_0^{(2)} = \frac{2}{n(n+1)}. \tag{56}$$

Here, 1,2,3,4 stand for any quartet of different indices, so that n in each case must be large enough to accomadate as many indices as necessary.

As an application, we calculate the second moment of the conductance as

$$\langle T^2 \rangle_0^{(2)} = \sum_{a,c=1}^{N_1} \sum_{b,d=1}^{N_2} \left\langle |t_{ab}|^2 |t_{cd}|^2 \right\rangle_0^{(2)} = \frac{N_1^2 N_2^2}{(N_1 + N_2)^2 - 1}. \tag{57}$$

The variance of the conductance is then given by

$$[\text{var}(T)]_0^{(2)} = \frac{N_1^2 N_2^2}{(N_1 + N_2)^2 \left[(N_1 + N_2)^2 - 1 \right]}. \tag{58}$$

Notice that in the limit $N_1, N_2 \to \infty$ with $N_1/N_2 = K$ fixed

$$[\text{var}(T)]_0^{(2)} \to \frac{K^2}{(K+1)^4}, \tag{59}$$

a constant which depends only on the ratio N_1/N_2 and on no other details of the cavity. Since this limit corresponds to increasing the width of the waveguides and so of the full system, the fact that the result is constant is the analog of the so-called *universal conductance fluctuations (UCF)* well-known for quasi-1D disordered systems [4,6]. In particular, for $K = 1$, $\text{var}(T) \to 1/16$, slightly less than the quasi-1D value of $1/15$.

IV.A.2 The case $\beta = 1$

Just as above, we first illustrate the procedure through the average over the invariant measure of a single S matrix element [Eq. (40) written for $\beta = 1$] which vanishes trivially. We introduce the transformed $\widetilde{S}$, again a unitary symmetric matrix, through

$$S = U^0 \widetilde{S} (U^0)^T. \tag{60}$$

Substituting (60) in the integral definition of $\langle S_{ab} \rangle_0^{(1)}$ and using the defining property of the invariant measure (20) we have

$$\langle S_{ab} \rangle_0^{(1)} = \sum_{a'} U^0_{aa'} U^0_{bb'} \int \widetilde{S}_{a'b'} d\mu^{(1)}(\widetilde{S}) = \sum_{a'} U^0_{aa'} U^0_{bb'} \langle S_{a'b'} \rangle_0^{(1)}. \tag{61}$$

In particular, if we take as the arbitrary fixed matrix U^0 the one given by Eq. (43), we find

$$\langle S_{ab} \rangle_0 = e^{i(\theta_a + \theta_b)} \langle S_{ab} \rangle_0 \implies \langle S_{ab} \rangle_0 = 0 \tag{62}$$

since θ_a, θ_b are arbitrary.

Just as for $\beta = 2$, the above argument can be generalized to prove that

$$\left\langle \left[S_{a_1 b_1} \cdots S_{a_p b_p} \right] \left[S_{\alpha_1 \beta_1} \cdots S_{\alpha_q \beta_q} \right]^* \right\rangle_0^{(1)} = 0, \tag{63}$$

unless $p = q$ and unless $\{a_1, b_1, \ldots, a_p, b_p\}$ and $\{\alpha_1, \beta_1, \ldots, \alpha_p, \beta_p\}$ constitute the same set of indices except for order. In particular, for $p = q = 1$ we find

$$\left\langle S_{ab} S^*_{\alpha\beta} \right\rangle_0^{(1)} = \sum_{a'b'\alpha'\beta'} U^0_{aa'} U^0_{bb'} \left(U^0_{\alpha\alpha'} U^0_{\beta\beta'} \right)^* \left\langle S_{a'b'} S^*_{\alpha'\beta'} \right\rangle_0^{(1)}, \forall U^0. \tag{64}$$

For instance, for U^0 given as in Eq. (43), we have

$$\left\langle S_{ab} S^*_{\alpha\beta} \right\rangle_0^{(1)} = e^{i(\theta_a + \theta_b - \theta_\alpha - \theta_\beta)} \left\langle S_{ab} S^*_{\alpha\beta} \right\rangle_0^{(1)}, \tag{65}$$

which vanishes unless $\{a, b\} = \{\alpha, \beta\}$ or $\{\beta, \alpha\}$. As a particular case, take $a = b = \alpha = \beta = 1$ and, as U^0, a matrix with the structure

$$U^0 = \begin{bmatrix} U^0_{11} & U^0_{12} & 0 & \cdots & 0 \\ U^0_{21} & U^0_{22} & 0 & \cdots & 0 \\ 0 & 0 & 1 & \cdots & 0 \\ \vdots & \vdots & \vdots & \ddots & \vdots \\ 0 & 0 & 0 & \cdots & 1 \end{bmatrix}. \tag{66}$$

We find

$$\left\langle |S_{11}|^2 \right\rangle_0 = \sum_{a'b'} U^0_{1a'} U^0_{1b'} \left(U^0_{1a'} U^0_{1b'} \right)^* \left\langle S_{a'b'} S^*_{a'b'} \right\rangle_0$$

$$+ \sum_{a' \neq b'} U^0_{1a'} U^0_{1b'} \left(U^0_{1b'} U^0_{1a'} \right)^* \left\langle S_{a'b'} S^*_{b'a'} \right\rangle_0$$

$$= \left[\left|U^0_{11}\right|^4 + \left|U^0_{12}\right|^4 \right] \left\langle |S_{11}|^2 \right\rangle_0 + 4 \left|U^0_{11}\right|^2 \left|U^0_{12}\right|^2 \left\langle |S_{12}|^2 \right\rangle_0. \tag{67}$$

Squaring the unitarity relation $|U^0_{11}|^2 + |U^0_{12}|^2 = 1$ and substituting in Eq. (67) we finally obtain

$$\left\langle |S_{11}|^2 \right\rangle^{(1)}_0 = 2 \left\langle |S_{12}|^2 \right\rangle^{(1)}_0. \tag{68}$$

This result is very important. It states that time-reversal invariance (TRI) has the consequence that *the average of the absolute value squared of a diagonal S-matrix element is twice as large as that of an off-diagonal one*, under the invariant measure. By unitarity, the specific value of each one of these averages is given by

$$\left\langle |S_{aa}|^2 \right\rangle^{(1)}_0 = \frac{2}{n+1}, \quad \left\langle |S_{a \neq b}|^2 \right\rangle^{(1)}_0 = \frac{1}{n+1}. \tag{69}$$

Just as in the above case $\beta = 2$, we calculate as an application the *average conductance* when our cavity is connected to the outside by two leads with N_1 and N_2 open channels ($n = N_1 + N_2$):

$$\langle T \rangle^{(1)}_0 = \sum_{a=1}^{N_1} \sum_{b=1}^{N_2} \left\langle |t_{ab}|^2 \right\rangle^{(1)}_0 = \frac{N_1 N_2}{N_1 + N_2 + 1}. \tag{70}$$

Here, the extra 1 in the denominator as compared with Eq. (53) is the *weak-localization correction* (WLC), a symmetry effect resulting from TRI. We can rewrite Eq. (70) separating out the WLC term as

$$\langle T \rangle^{(1)}_0 = \frac{N_1 N_2}{N_1 + N_2} - \frac{N_1 N_2}{(N_1 + N_2)(N_1 + N_2 + 1)}. \tag{71}$$

In particular, for $N_1 = N_2 = N$ and for $N \to \infty$, corresponding to a large system, the WLC term tends to the universal number $-1/4$.

Using a similar procedure, one finds the results (n being again the dimensionality of the S matrix) [20]

$$\left\langle |S_{12}|^2 |S_{34}|^2 \right\rangle^{(1)}_0 = \frac{n+2}{n(n+1)(n+3)}, \tag{72}$$

$$\left\langle |S_{12}|^2 |S_{13}|^2 \right\rangle^{(1)}_0 = \frac{1}{n(n+3)}, \tag{73}$$

$$\left\langle |S_{12}|^4 \right\rangle_0^{(1)} = \frac{2}{n(n+3)}. \tag{74}$$

A comment like that made right after Eq. (56) applies here as well. Just as in Eq. (57), we now find for the second moment of the conductance

$$\left\langle T^2 \right\rangle_0^{(1)} = \frac{N_1 N_2 \left[N_1 N_2 \left(N_1 + N_2 + 2 \right) + 2 \right]}{(N_1 + N_2)(N_1 + N_2 + 1)(N_1 + N_2 + 3)} \tag{75}$$

and for its variance

$$[\mathrm{var}(T)]_0^{(1)} = \frac{2 N_1 N_2 \left(N_1 + 1 \right) \left(N_2 + 1 \right)}{(N_1 + N_2)(N_1 + N_2 + 1)^2 (N_1 + N_2 + 3)} \to 2 \frac{K^2}{(K+1)^4} \tag{76}$$

where in the limit $N_1, N_2 \to \infty$ with $N_1/N_2 = K$, a fixed number. Note that in this universal limit, the variance here is exactly twice as large as for $\beta = 2$, Eq. (59), another result of time-reversal invariance. For the particular case $K=1$, the limiting value of the variance is $1/8$.

IV.B The distribution of the conductance in the two-equal-lead case

In the last section we characterized the conductance of a cavity through the first two moments of its transmission. If the distribution of the conductance is Gaussian, this is sufficient to characterize the full distribution. In fact, it can be shown that in the large-size universal limit, $N \to \infty$, the distribution of the conductance is indeed Gaussian [72].

In general, however, the probability density of T will not be Gaussian, and it is of interest, then, to derive results for this density. For this purpose, the polar representation of Sec. II.C is particularly useful since the conductance is directly related to the $\{\tau\}$ whose joint probability distribution we know. Specifically, the distribution of the transmission T of Eq. (19) can be obtained by direct integration of the $P^{(\beta)}(\{\tau\})$ of Eq. (27):

$$w^{(\beta)}(T) = \int \delta(T - \sum_a \tau_a) P^{(\beta)}(\{\tau_a\}) \prod_a d\tau_a. \tag{77}$$

We consider a few examples below.

IV.B.1 The case $N=1$

In this case we have only one τ_a, that we may call τ, so that $T=\tau$, and $0 \leq T \leq 1$. Eq. (27) then gives

$$w^{(1)}(T) = \frac{1}{2\sqrt{T}}, \qquad w^{(2)}(T) = 1. \tag{78}$$

For $\beta = 1$, we thus have a higher probability to find small T's than $T \sim 1$: this is clearly a symmetry effect, a result of TRI that favors backscattering and hence low conductances.

IV.B.2 The case $N = 2$

Now $T = \tau_1 + \tau_2$, and $0 \leq T \leq 2$. In VIII we show that

$$w^{(1)}(T) = \begin{cases} \frac{3}{2}T, & 0 < T < 1 \\ \frac{3}{2}\left(T - 2\sqrt{T-1}\right), & 1 < T < 2 \end{cases} \tag{79}$$

and

$$w^{(2)}(T) = 2\left[1 - |1 - T|\right]^3 \tag{80}$$

For $\beta = 1$, notice the square-root cusp at $T = 1$. We find, once again, a higher probability for $T < 1$ than for $T > 1$ to occur. On the other hand, for $\beta = 2$, $w(T)$ is again symmetric around $T = 1$.

IV.B.3 The case $N = 3$

Now $T = \tau_1 + \tau_2 + \tau_3$, and $0 \leq T \leq 3$. For $\beta = 2$ one finds

$$w^{(2)}(T) = \begin{cases} \frac{9}{42}T^8, & 0 < T < 1 \\ -\frac{2781}{14} + \frac{6588}{7}T - 1818T^2 + 1836T^3 \\ \quad - 1035T^4 + 324T^5 - 54T^6 + \frac{36}{7}T^7 - \frac{3}{7}T^8, & 1 < T < \frac{3}{2} \end{cases} \tag{81}$$

and the distribution is symmetric about $T = 1.5$. As mentioned above, $w^{(\beta)}(T)$ gradually approaches a Gaussian distribution.

IV.B.4 Arbitrary N

In this case it is straightforward to obtain the dependence of the tail of the distribution in the region $0 < T < 1$. In this region the constraint that $\tau < 1$ does not enter; a calculation given in VIII shows that

$$w_N^{(\beta)}(T) \propto T^{\beta N^2/2 - 1}. \tag{82}$$

V PRESENCE OF DIRECT PROCESSES

In order to treat cases involving direct processes, as in billiards in which short paths produce a prompt response, we now need Poisson's kernel (35) in its full generality. We discuss below some analytical results for the distribution of the spinless conductance T in the case of a cavity connected to the outside by means of two leads supporting one open channel each ($N_1 = N_2 = 1$), giving rise to a 2-dimensional S matrix. There is only one τ in this case and it is its distribution that we seek, since $T = \tau$. While the expressions that we derive are somewhat cumbersome, they are used for comparing to numerical results in Sec V where plots of several examples are displayed.

We write the optical S-matrix $\overline{S}$, a subunitary matrix, as

$$\overline{S} = \begin{bmatrix} x & w \\ z & y \end{bmatrix}, \tag{83}$$

where the entries are, in general, complex numbers.

V.A The case $\beta=2$

From Eq. (35) we write the differential probability for the S matrix as

$$dP_{\overline{S}}^{(2)}(S) = \frac{[\det(I - \overline{S}\,\overline{S}^\dagger)]^n}{|\det(I - S\,\overline{S}^\dagger)|^{2n}} d\mu_0^{(2)}(S), \tag{84}$$

where

$$d\mu_0^{(2)}(S) = \frac{d\mu^{(2)}(S)}{V}, \qquad \int d\mu_0^{(2)}(S) = 1. \tag{85}$$

We are interested here in the case $n=2$.

In the polar representation (17) with $n=2$, the S matrix has the form

$$S = \begin{bmatrix} e^{i\alpha} & 0 \\ 0 & e^{i\beta} \end{bmatrix} \begin{bmatrix} -\sqrt{1-\tau} & \sqrt{\tau} \\ \sqrt{\tau} & \sqrt{1-\tau} \end{bmatrix} \begin{bmatrix} e^{i\gamma} & 0 \\ 0 & e^{i\delta} \end{bmatrix}, \tag{86}$$

the invariant measure (85) being [see Eqs. (26), (27)]

$$d\mu_0(S) = d\tau \frac{d\alpha\, d\beta\, d\gamma\, d\delta}{(2\pi)^4}. \tag{87}$$

1) As a particular case, suppose the optical S matrix is diagonal, so that there is *only direct reflection* and no direct transmission: in Eq. (83) we choose $w=z=0$. Substituting Eqs. (83), (86), (87) in (84) we find

$$dP_{X,Y}^{(2)}(\tau,\varphi,\psi) = \frac{(1-X^2)^2(1-Y^2)^2}{\left|\left(e^{-i\varphi}+X\sqrt{1-\tau}\right)\left(e^{-i\psi}-Y\sqrt{1-\tau}\right)-XY\tau\right|^4}\frac{d\tau d\varphi d\psi}{(2\pi)^2} \quad (88)$$

where $\varphi = \alpha + \gamma$, $\psi = \beta + \delta$, $X = |x|$, $Y = |y|$. The distribution of the conductance T is thus

$$w_{X,Y}^{(2)}(T) = (1-X^2)^2(1-Y^2)^2 \qquad (89)$$

$$\times \left\langle \frac{1}{\left|\left(e^{-i\varphi}+X\sqrt{1-T}\right)\left(e^{-i\psi}-Y\sqrt{1-T}\right)-XYT\right|^4} \right\rangle_{\varphi,\psi}$$

where $\langle \cdots \rangle_{\varphi,\psi}$ denotes an average over the variables φ, ψ. The result is $(0 < T < 1)$

$$w_{X,Y}^{(2)}(T) = K\frac{A - B(1-T) + C(1-T)^2 + D(1-T)^3}{[E - 2F(1-T) + G(1-T)^2]^{5/2}}, \qquad (90)$$

where

$$K = (1-X^2)^2(1-Y^2)^2, \qquad A = (1-X^4Y^4)(1-X^2Y^2)$$
$$B = (X^2+Y^2)(1-6X^2Y^2+X^4Y^4) + 4X^2Y^2(1+X^2Y^2)$$
$$C = (1+X^2Y^2)(6X^2Y^2-X^4-Y^4) - 4X^2Y^2(X^2+Y^2)$$
$$D = (X^2+Y^2)(X^2-Y^2)^2, \qquad E = (1-X^2Y^2)^2$$
$$F = (1+X^2Y^2)(X^2+Y^2) - 4X^2Y^2, \qquad G = (X^2-Y^2)^2.$$

This result reduces to 1 when $X = Y = 0$, as it should. A particularly interesting case is that of "equivalent channels", i.e. $X = Y$, in which the above expression reduces to

$$w_{X,X}^{(2)}(T) = (1-X^2)\frac{(1-X^4)^2 + 2X^2(1+X^4)T + 4X^4T^2}{[(1-X^2)^2 + 4X^2T]^{5/2}}. \qquad (91)$$

The structure of this result is clear if we notice that

$$w_{X,X}^{(2)}(0) = \left[\frac{1+X^2}{1-X^2}\right]^2 > 1, \qquad w_{X,X}^{(2)}(1) = \frac{(1-X^2)(1+X^4)}{(1+X^2)^3} < 1,$$

and hence $w_{X,X}^{(2)}(0) > w_{X,X}^{(2)}(1)$, so that small conductances are emphasized, as expected, because of the presence of direct reflection and no direct transmission.

2) The case of *only direct transmission* and no direct reflection is obtained by setting $x = y = 0$ in Eq. (83). The conductance distribution is obtained from Eq. (90) with the replacement $X \to W = |w|$, $Y \to Z = |z|$, $T \to 1 - T$. In the equivalent-channel case we now obtain a conductance distribution that emphasizes large conductances.

3) The case of a general optical S matrix, Eq. (83), has also been worked out and the result expressed in terms of quadratures: because of its complexity, it will not be quoted here.

V.B The case $\beta=1$

This case is more complicated than that for $\beta=2$ and we have only succeeded in treating analytically some particular cases. Take, for instance, a diagonal $\overline{S}$, i.e. $w=z=0$ in Eq. (83). With the same notation as above, we find

$$w^{(1)}_{X,Y}(T) = \left|\left(1-X^2\right)\left(1-Y^2\right)\right|^{3/2} \frac{1}{2\sqrt{T}} \tag{92}$$

$$\times \left\langle \frac{1}{\left|\left(e^{-i\varphi}+X\sqrt{1-T}\right)\left(e^{-i\psi}-Y\sqrt{1-T}\right)-XYT\right|^3} \right\rangle_{\varphi,\psi},$$

a result that has to be integrated numerically. When $X=Y=0$, the distribution (92) reduces to $1/2\sqrt{T}$, as it should. It is interesting to notice that for $Y=0$ the above result can be integrated analytically, to give

$$w^{(1)}_{X,0}(T) = \frac{(1-X^2)^{3/2}}{2\sqrt{T}} \,_2F_1\left(3/2, 3/2; 1; X^2(1-T)\right), \tag{93}$$

$_2F_1$ being a hypergeometric function [73].

VI COMPARISON WITH NUMERICAL CALCULATIONS

The information-theoretic approach that we have been discussing is expected to be valid for cavities in which the classical dynamics is completely chaotic, a property that refers to the *long time* behavior of the system. It is in such structures that the long time response is ergodic and equilibrated, and so one can expect that maximum entropy considerations will play a role. In this section we examine particular cavities numerically in order to determine to what extent the information-theoretic approach really holds. The structures that we consider are all "billiards"– they consist of hard walls surrounding a cavity with constant potential– with two leads. We start by considering particularly simple structures, then treat structures in which the absence of direct processes is assured, then move on to structures having particularly obvious direct processes.

VI.A Simple structures

In the "quantum chaos" literature– the study of how quantum properties depend on the nature of the classical dynamics in a system– several billiards are used as standard examples of closed chaotic systems. The two most studied are the Sinai billiard– the region enclosed between a square and a circle centered in the square– and the stadium billiard– two half-circles joined by straight edges. The classical dynamics in these two billiards is known to be completely chaotic. For a test case open system, then, it is natural to take one of these billiards and attach two leads. The open stadium billiard was studied previously for this reason [26,32,75–77]. Here we directly compare results for this system to the predictions of the information-theoretic approach.

The numerical methods used in these calculations are covered in detail in Ref. [74]. Briefly, the procedure consists of the following three steps. First, discretize the Hamiltonian onto a square mesh using the simplest finite-difference scheme. Solving the Schrödinger equation is then reduced to solving a set of linear equations. Second, find the Green function at the desired energy from one lead to the other using a recursive procedure and outgoing-wave boundary conditions. This procedure essentially uses the sparseness of the finite-difference matrix to efficiently solve the linear equations. Third, note that the transmission amplitude can be obtained from this Green function by simply projecting onto the transverse wave-functions in the lead. The main parameter in these calculations, ka, is the size of the mesh compared to the wavelength. In the results shown here, ka is always less than 0.8 and $ka < 0.5$ in most cases. For these values, the anisotropy of the Fermi surface is small; the non-parabolicity of the dispersion is larger, but does not concern us here since we treat transport at a fixed energy.

Two simple quantities to calculate both numerically and theoretically are the average and variance of the conductance. The analytic results for the invariant measure are in Sec. IV.A. Fig. 3 shows the numerical results for the two asymmetric open stadium shown on the side– asymmetric half-stadiums are used in order to avoid the complications of reflection symmetry. The top panel shows the deviation of the average transmission from the classical value of the transmission. This classical value was obtained numerically by tracing trajectories through the cavity. For fully equilibrated scattering, the classical value is $N/2$ where N is the number of channels in each lead, but in Fig. 3 $T_{\text{classical}}/N$ is 0.60 (0.58) for the upper (lower) cavity. The bottom panel shows the variance of T. While the numerical results are similar in magnitude to the predictions, the agreement is clearly not very good.

Before proceeding with our discussion of the conductance in these cavities, we step back to perform the most common test of random matrix theory. In the context of closed systems, it is natural to consider the statistics of the energy levels. The degree to which these statistics agree with the Wigner-Dyson statistics derived from random matrix theory is often used as the prime indication of the validity of the theory for a given system. In the context of the S-matrix, the analog is to look at the statistics of the eigenphases: in the large N limit their statistics is also Wigner-Dyson [15,16]. In fact, studies of eigenphases of chaotic scattering systems were carried out prior to any interest in the conductance [23,24]. The statistics of the eigenphases can be characterized by three representative quantities [15,16]. First, the mean density of eigenphases measures the uniformity of the system; for the circular ensembles (CE) it is constant. Second, the nearest neighbor distribution highlights the repulsion at short scales; it is approximately the Wigner surmise

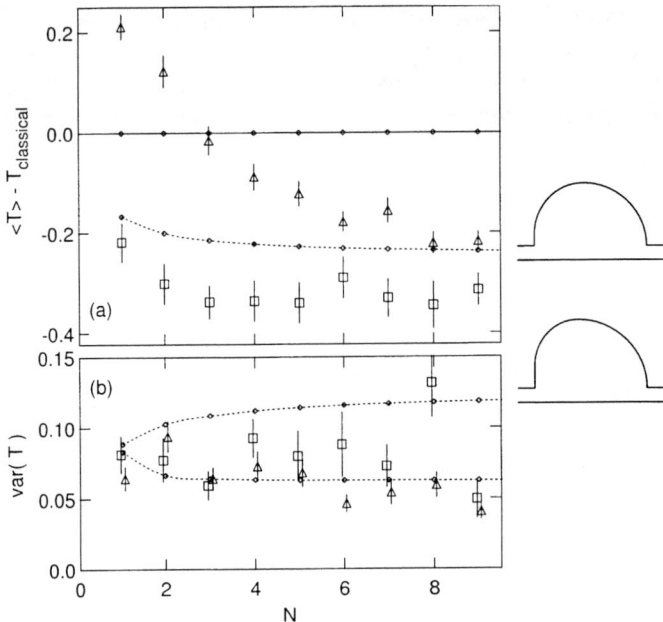

FIGURE 3. The magnitude of (a) the quantum correction to the classical conductance and (b) the conductance fluctuations as a function of the number of modes in the lead N. The two asymmetric stadium-like billiards shown at the right were used; the average of the results for the two cavities is shown. The numerical results for $B=0$ (squares with statistical error bars) and $B \neq 0$ (triangles) are compared to the predictions of the COE (points, dotted line) and CUE (points, dashed line). The agreement is poor. For both billiards, 25 energies were sampled for each value of N, and for non-zero field $BA/\phi_0 = 2, 4$ where A is the area of the cavity.

for the CE. Third, the variance of the number of phases within a certain range L, denoted $\Sigma^2(L)$, indicates the rigidity of the spectrum at large scales; it grows logarithmically with L for the CE.

These three quantities are shown for the simple open stadium in Fig. 4. The agreement with the predictions of the CE is good for all three, in contrast to the results for the conductance above, despite N not being very large. Since the conductance involves the transmission coefficient which is a property of the wavefunctions, this indicates that the distribution of the eigenvectors is more sensitive to deviations from the CE than the distribution of eigenphases. Thus the eigenphase statistics cannot be taken as a definitive indication of the validity of the CE: it is perfectly possible to have excellent eigenvalue statistics while having poor eigenvector statistics [78].

Returning to the properties of the conductance, we believe that the deviations from the CE seen in the numerics (Fig. 3) are caused by the presence of short paths

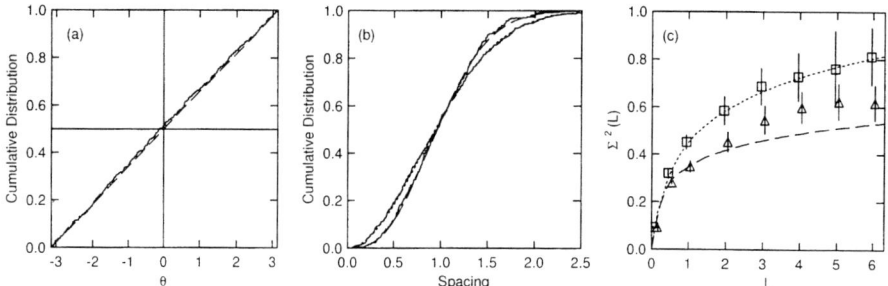

FIGURE 4. Statistics of the eigenphases of the S-matrix for the second stadium shown in Fig. 3 with $N=9$. (a) Cumulative distribution of the eigenphase density (solid line) compared to the CE (dashed). (b) Cumulative distribution of the difference between nearest-neighbor phases for both $B=0$ and $BA/\phi_0 = 4$ compared to the COE (dotted) and CUE (dashed). The spacing is normalized to the mean separation. (c) Variance of number of phases in an interval L for both $B=0$ (squares) and $BA/\phi_0 = 4$ (triangles) compared to the COE (dotted) and CUE (dashed). All three statistics agree with the prediction of the circular ensembles– constant density, a spacing distribution given by the Wigner surmise, and a logarithmically increasing variance– despite the poor agreement for the transmission in Fig. 3.

in these simple structures. This means that the response is not fully equilibrated. The two most obvious types of short paths in these structures are the direct paths between the leads and the whispering gallery paths– those that hit only the half-circle. Short paths will be included in the analysis in Sec. VI.B below. One way to minimize the effect of these short paths is to make the openings to the leads very small so that the probability of being trapped for a long time increases. Presumably the CE will apply to any completely chaotic billiard in the limit that the openings are very small (the number of modes in each lead should remain constant). However, such a structure is difficult to treat numerically, except in the very small N limit, which, in fact, we discuss in Sec. VI.B below.

VI.B Absence of direct processes

In order to compare with the predictions of the circular ensembles, we wish, therefore, to study structures in which the most obvious direct paths are absent. To this end, we have introduced "stoppers" into the stadium to block both the direct and the whispering gallery trajectories: examples are shown in Fig. 5. In order to study the statistical properties, the conductance as a function of energy is calculated. Because the energy variation is on the scale of $\hbar\gamma_{\rm esc}$ (the escape rate from the cavity) [24,32], it is much more rapid than the spacing between the modes in the leads ($\hbar v_F/W$). Thus many independent samplings of the conductance at a fixed number of modes may be obtained. In addition, we vary slightly the position

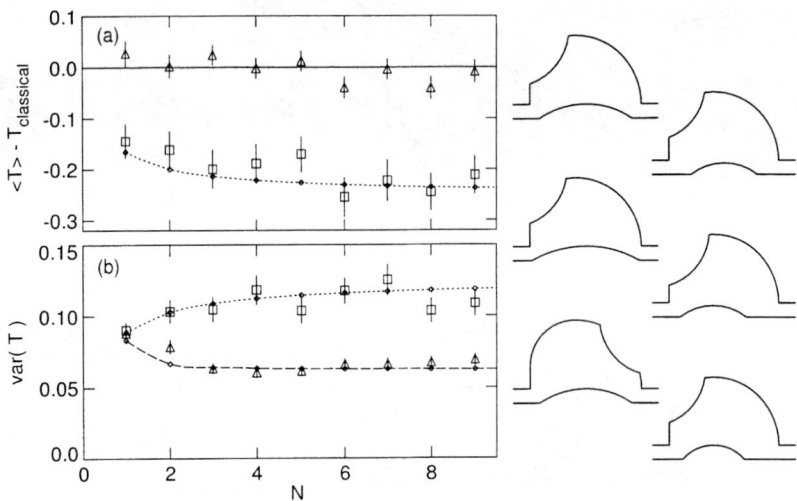

FIGURE 5. The magnitude of (a) the quantum correction to the classical conductance and (b) the conductance fluctuations as a function of the number of modes in the lead N. The numerical results for $B=0$ (squares with statistical error bars) agree with the prediction of the COE (points, dotted line), while those for $B \neq 0$ (triangles) agree with the CUE (points, dashed line). The six cavities shown on the right were used; the average of the numerical results is plotted. Note that each cavity has stoppers to block both the direct and whispering gallery trajectories. For non-zero field, $BA/\phi_0 = 2, 4$ where A is the area of the cavity. (After Ref. [46].)

of the stoppers so as to change the interference effects and collect better statistics. The numerical results in Fig. 5 used 50 energies for each N (all chosen away from the threshold for the modes) and the 6 different stopper configurations shown; the classical transmission probabilities for these cavities ranged from 0.46 to 0.51 with a mean of 0.49. In addition, for non-zero magnetic field two values were used. We see that the agreement with the CE is now very good for both the mean and the variance, both for $B = 0$ and for non-zero B. This supports our view that the deviations in the simple structure Fig. 3 are caused by short paths.

While the agreement of the mean and variance of the conductance with the CE gives a strong indication of the validity of the information-theoretic model for real cavities, a much more dramatic prediction of the model is the strongly non-Gaussian distribution of T for a small number of modes. The analytic results for the CE were given in section IV. These are compared to the numerical results for $N = 1, 2$ in Fig. 6, using the same data as in Fig. 5. Note that the data is consistent with a square-root singularity in the case $N = 1$ $B = 0$ and with cusps in the two $N = 2$ cases. Thus we see that even for this much more stringent test, the agreement between the behavior of real cavities and the CE is excellent.

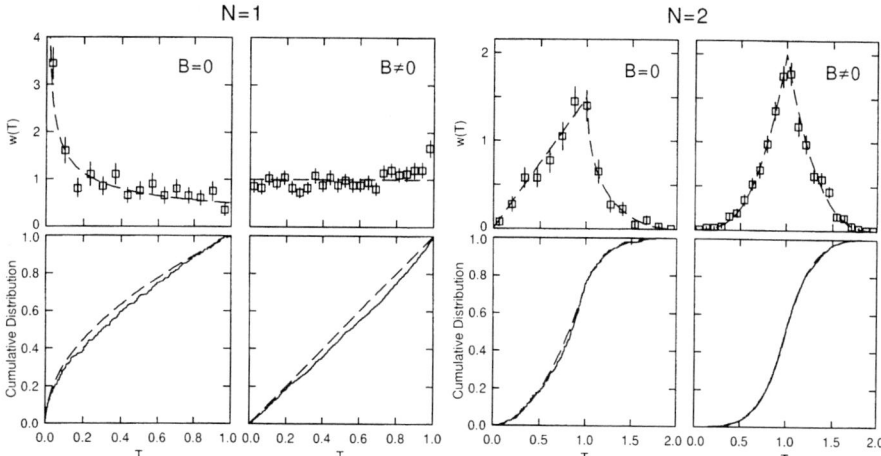

FIGURE 6. The distribution of the transmission intensity at fixed $N = 1$ or 2 in both the absence and presence of a magnetic field, compared to the analytic COE and CUE results. The panels in the first row are histograms; those in the second row are cumulative distributions. Note both the strikingly non-Gaussian distributions and the good agreement between the numerical results and the CE in all cases. The cavities and energy sampling points used are the same as those in Fig. 5; for $B \neq 0$, $BA/\phi_0 = 2, 3, 4,$ and 5 were used. (After Ref. [46].)

VI.C Presence of direct processes

We now want to look at more general structures than those used in the last section for comparing to the circular ensemble results. In particular, we will remove the stoppers that blocked short paths, and compare the numerical results with the predictions of Poisson's kernel, following Ref. [48]. We will not, however, consider the most general structure: Poisson's kernel is expected to hold in situations where there are *two* widely separated time scales, a prompt response and an equilibrated response. Thus we will study structures where we expect this to be true. Since we have only obtained explicit results in the case $N = 1$ (see Sec. V), we will study this case numerically as well.

We have computed the conductance for several billiards shown in Fig. 7. Statistics were collected by (1) sampling in an energy window larger than the energy correlation length but smaller than the interval over which the prompt response changes, and (2) using several slightly different structures. Typically we used 200 energies in $kW/\pi \in [1.6, 1.8]$ (where W is the width of the lead) and 10 structures found by changing the height or angle of the convex "stopper". Note that the stopper here is used to increase statistics, not block short paths. As in the absence of direct processes, since we are mostly averaging over energy, we rely on ergodicity to compare the numerical distributions to the ensemble averages of the

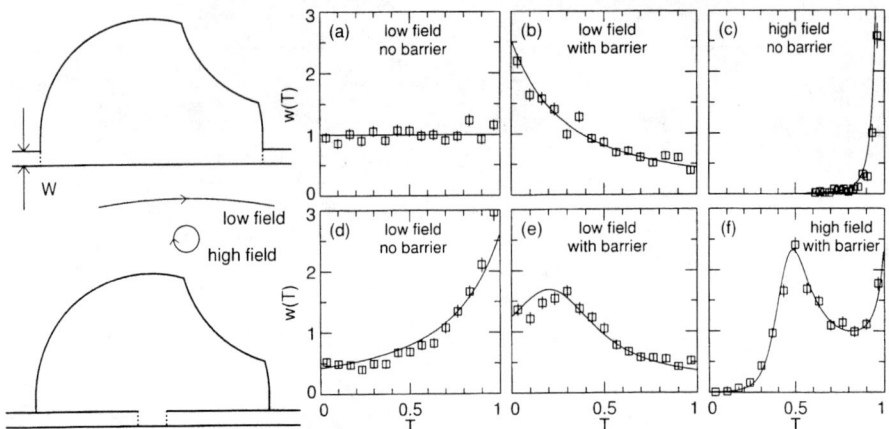

FIGURE 7. The distribution of the transmission coefficient for $N = 1$ in a simple billiard (top row) and a billiard with leads extended into the cavity (bottom row). The magnitude of the magnetic field and the presence or absence of a potential barrier at the entrance to the leads (marked by dotted lines in the sketches of the structures) are noted in each panel. Cyclotron orbits for both fields, drawn to scale, are shown on the left. The squares with statistical error bars are the numerical results; the lines are the predictions of the information-theoretic model, parametrized by an optical S-matrix extracted from the numerical data. The agreement is good in all cases. (From Ref. [48].)

information-theoretic model. The optical S-matrix was extracted directly from the numerical data and used as $\langle S \rangle$ in Poisson's kernel; in this sense the theoretical curves shown below are *parameter free*.

We first consider the billiard shown in Fig. 7 at low magnetic field ($BA/\phi_0 = 2$ where A is the area of the cavity; by low magnetic field we mean that the cyclotron radius r_c is much larger than the size of the cavity ($r_c = 55\,W$ shown to scale). In this case $w(T)$ is nearly uniform [Fig. 7(a)], and $\langle S \rangle$ is small because direct trajectories are negligible in this large structure. We thus obtain good agreement with the invariant-measure prediction of a constant distribution (78).

In order to increase $\langle S \rangle$ we make one of three changes: (1) introduce potential barriers at the openings of the leads into the cavity (dashed lines in structures of Fig. 7), (2) increase the magnetic field, or (3) extend the leads into the cavity. In the first case, the barriers are chosen so that the bare transmission of each barrier is 1/2 (determined by calculation in an infinite lead). They cause direct reflection and skew the distribution towards small T [Fig. 7(b)]. Since the reflection from the barrier is immediate while the transmitted particles are trapped for a long time, one has two very different response times. Second, the large magnetic field ($BA/\phi_0 = 80$) corresponds to r_c just larger than the width of the lead ($r_c = 1.4\,W$). The field increases one component of the direct transmission– the one corresponding to skipping orbits along the lower edge- and skews the distribution towards large T [Fig. 7(c)]. Third, extending the leads into the cavity increases the direct transmission in both directions and also skews the distribution towards large T [Fig. 7(d)]. We have done this rather than consider a smaller cavity since in our case the equilibrated component is trapped for a long time, yielding a clear separation of scales.

In each of these cases, the numerical histogram is compared to the information-theoretic model (solid lines) in which the numerically obtained $\overline{S}$ is inserted. In panels (b)-(d) the curve plotted is the analytic expression of Eq. (90) and the corresponding one for direct transmission. Note the excellent agreement with the information-theoretic model.

Since the long-time classical dynamics in each of the three structures (a), (b) and (d) is chaotic, these results show that a wide variety of behavior is possible for chaotic scattering, the invariant-measure description applying only when there is a single characteristic time scale.

In case (c), the dynamics is not completely chaotic because of the small cyclotron radius, and so one would not expect the circular ensemble to apply. In Ref. [46] we found that increasing the magnetic flux through the structure beyond a few flux quanta spoiled the agreement with the circular ensemble; we now know that a nonzero $\langle S \rangle$ is generated and that the present model describes the data very well. The excellent agreement found here with a flux as high as 80 suggests extending the analysis to the quantum Hall regime.

By combining several of the modifications used above, different $\langle S \rangle$ and so different distributions can be produced. First, by using extended leads with barriers at their ends, one can cause both prompt transmission and reflection; the result is shown in Fig. 7(e). Finally, increasing the magnetic field in this structure produces a large average transmission and a large average reflection. The resulting $w(T)$, Fig. 7(f), has a surprising two-peak structure: one peak near $T=1$ caused by the large direct transmission and another near $T=1/2$. For cases (e) and (f), four intervals of 50 energies each were treated independently (since the four intervals show slightly different $\langle S \rangle$'s) and the four sets of data were then superimposed [79]. Even in these two unusual cases, the prediction of the information-theoretic model is in excellent agreement with the numerical results.

VII COMPARISON WITH EXPERIMENTAL DATA: DEPHASING EFFECTS

The random S-matrix theory of quantum transport through ballistic chaotic cavities was seen in the last section to be in good agreement with numerical simulations for structures in which the assumptions of the theory are expected to hold: that agreement includes the average conductance, its variance and probability density.

The ultimate test of the theory should, however, be comparisons with experiment. A naïve comparison indicates poor agreement. For instance, for $N_1 = N_2 = N = 2$, both the theoretical weak-localization correction (WLC) and variance are larger than the experimental results [38]. The theoretical WLC (including spin factor) is $-2/5 = -0.40$ from Eq. (71), while the experimental one is ~ -0.31. The theoretical var(g) is $4/15 \approx 0.27$ for $\beta = 2$ [Eq. (58)] and $72/175 \approx 0.41$ for $\beta = 1$ [Eq. (76)], giving the ratio $[\text{var}(g)]_0^{(1)} / [\text{var}(g)]_0^{(2)} = 54/35 \approx 1.54$. The experimental results for the variance, on the other hand, are ~ 0.015 for $B \neq 0$ and ~ 0.034 for $B = 0$, giving the ratio 2.27. In addition, the measured probability density [38] is close to a Gaussian distribution, which differs from the prediction of Eqs. (79), (80). More recent experimental data for $N = 1$ [44] show a distribution which is clearly asymmetric for $B = 0$, favoring low conductances as required by weak localization, but the asymmetry is by no means as pronounced as that of Eq. (78) which has a square-root singularity at the origin.

To reconcile these discrepancies, we must realize that inherent in the discussion of the previous sections is the assumption that one can neglect processes which destroy the coherence of the wave function inside the sample and neglect energy smearing caused by non-zero temperature. Even if the phase-breaking length l_ϕ is larger than the geometrical size of the cavity, for sufficiently narrow leads the electron may spend enough time inside the sample to feel the effect of phase-breaking mechanisms. The discussion of these effects and their relevance for the description of the experimental data is the subject of the present section.

We simulate the presence of phase-breaking events through a model invented by M. Büttiker [80], where, in addition to the physical leads 1,2 attached to reservoirs at chemical potentials μ_1, μ_2, a "fake lead" 3 connects the cavity to a phase-randomizing reservoir at chemical potential μ_3. The idea is that any particle which exits the cavity through lead 3 is replaced by a particle from the reservoir; since the replacement particle comes from a reservoir, it is incoherent with respect to the exiting particle, hence the phase-breaking. Requiring the current in lead 3 to vanish determines μ_3; the two-terminal dimensionless conductance is then found to be

$$g \equiv G/(e^2/h) = 2 \left[T_{21} + \frac{T_{23}T_{31}}{T_{32} + T_{31}} \right] . \tag{94}$$

In this equation, the factor of 2 accounts for spin explicitly and T_{ij} is the transmission coefficient for "spinless electrons" from lead j to lead i. This expression for g has a natural interpretation: the first term is the coherent transmission and the second term is the sequential transmission from 1 to 2 via 3. In terms of the S matrix, T_{ij} is

$$T_{ij} = \sum_{a_i=1}^{N_i} \sum_{b_j=1}^{N_j} \left| S_{ij}^{a_i b_j} \right|^2, \tag{95}$$

where N_i is the number of channels in lead i; $S_{ij}^{a_i b_j}$ is the matrix element in the position a_i, b_j of the ij block of the S matrix, rows and columns in this block being labeled from 1 to N_i and from 1 to N_j, respectively. The total number of channels will be designated by

$$N_T = \sum_i N_i. \tag{96}$$

Occasionally, we shall use the notation N_ϕ to designate the number of *phase-breaking channels* N_3 in the fake lead 3. These N_ϕ channels are physically related to the phase-breaking scattering rate γ_ϕ via the relation $N_\phi/(N_1 + N_2) \approx \gamma_\phi/\gamma_{\rm esc}$, where $\gamma_{\rm esc}$ is the escape rate from the cavity. This "fake lead" model has been used for various studies of the effect of phase-breaking in mesoscopic systems [81].

We now make the assumption that the *total* $N_T \times N_T$ scattering matrix S obeys the distribution (39) given by the invariant measure [47,82]. Through this assumption, the effect of the third lead is felt somehow uniformly in space rather than at any given point. This even-handed statistical treatment makes the "fake lead" approach more physically reasonable. A further generalization to the case of non-ideal coupling of the fake lead to the cavity is studied in Ref. ([83]) but will not be presented here. In the following, we confine the analytical discussion to the large N_ϕ limit, then present the results of numerical random-matrix theory calculations, and finally compare again to experiment.

VII.A Large N_ϕ

We find below the average and variance of the conductance when $N_\phi \gg 1$ [47]. As we shall see, we need, for that purpose, the average and the covariance of the transmission coefficients T_{ij} introduced above.

From Eqs. (52) and (69), the average of T_{ij} ($i \neq j$) is given by (as usual, $\beta = 1, 2$)

$$\langle T_{ij} \rangle_0^{(\beta)} = \frac{N_i N_j}{N_T + \delta_{\beta 1}}, \tag{97}$$

so that

$$\langle T_{21} \rangle_0^{(\beta)} = \frac{N_2 N_1}{N_T + \delta_{\beta 1}} \approx \frac{N_2 N_1}{N_3} + O\left(\frac{1}{N_3^2}\right) \tag{98}$$

$$\langle T_{31} \rangle_0^{(\beta)} = \frac{N_3 N_1}{N_T + \delta_{\beta 1}} \approx N_1 \left[1 - \frac{N_1 + N_2 + \delta_{\beta 1}}{N_3} + O\left(\frac{1}{N_3^2}\right)\right] \tag{99}$$

$$\langle T_{32} \rangle_0^{(\beta)} = \frac{N_3 N_2}{N_T + \delta_{\beta 1}} \approx N_2 \left[1 - \frac{N_1 + N_2 + \delta_{\beta 1}}{N_3} + O\left(\frac{1}{N_3^2}\right)\right], \tag{100}$$

where the $\approx$ sign refers to the situation $N_3 \gg 1$ while $N_1, N_2 = O(1)$. Turning to the covariance, using Eqs. (54)-(56) and (97), one finds for $\beta = 2$

$$\langle \delta T_{ij} \delta T_{kl} \rangle_0^{(2)} = \frac{N_i N_j}{N_T^2 (N_T^2 - 1)} \left[N_T^2 \delta_{ik} \delta_{jl} - N_k N_T \delta_{jl} - N_l N_T \delta_{ik} + N_k N_l \right]. \quad (101)$$

Likewise, using Eqs. (72)-(74) for $\beta = 1$ yields

$$\begin{aligned}
\langle \delta T_{ij} \delta T_{kl} \rangle_0^{(1)} &= \frac{1}{N_T (N_T + 1)^2 (N_T + 3)} \\
&\times \{ N_i N_j (N_T + 1)(N_T + 2)(\delta_{ik}\delta_{jl} + \delta_{il}\delta_{jk}) + 2 N_i N_k \delta_{ij}\delta_{kl} \\
&+ 2 N_i N_k N_l \delta_{ij} + 2 N_i N_j N_k \delta_{kl} + 2 N_i N_T (N_T + 1)\delta_{ijkl} + 2 N_i N_j N_k N_l \\
&- (N_T + 1)[2 N_i N_l \delta_{ijk} + 2 N_i N_k \delta_{ijl} + 2 N_i N_j (\delta_{ikl} + \delta_{jkl}) \\
&+ N_i N_j N_l (\delta_{ik} + \delta_{jk}) + N_i N_j N_k (\delta_{il} + \delta_{jl})]\}.
\end{aligned} \quad (102)$$

Here, a δ with two or more indices vanishes unless all its indices coincide, in which case its value is 1. These expressions are valid for arbitrary N_i, N_j, and also for $i = j$ if T_{ii} is interpreted as the reflection coefficient R_{ii} from lead i back to itself.

The conductance of Eq. (94) can be expressed as

$$\begin{aligned}
g &= 2 \left[\overline{T}_{21} + \delta T_{21} + \frac{(\overline{T}_{23} + \delta T_{23})(\overline{T}_{31} + \delta T_{31})}{(\overline{T}_{32} + \delta T_{32} + \overline{T}_{31} + \delta T_{31})} \right] \\
&= 2 \left[\overline{T}_{21} + \delta T_{21} + \frac{\overline{T}_{23} \overline{T}_{31}}{\overline{T}_{32} + \overline{T}_{31}} \frac{(1 + \delta T_{23}/\overline{T}_{23})(1 + \delta T_{31}/\overline{T}_{31})}{1 + (\delta T_{32} + \delta T_{31})/(\overline{T}_{32} + \overline{T}_{31})} \right],
\end{aligned} \quad (103)$$

where we have written

$$T_{ij} = \overline{T}_{ij} + \delta T_{ij}. \quad (104)$$

For convenience, we use interchangeably a bar or the bracket $\langle \cdots \rangle_0^{(\beta)}$ to indicate an average over the invariant measure.

From Eqs. (101) and (102) we find, for the variance of T_{ij} ($i \neq j$)

$$\langle (\delta T_{ij})^2 \rangle_0^{(2)} = \frac{N_i N_j (N_T - N_i)(N_T - N_j)}{N_T^2 (N_T^2 - 1)} \quad (105)$$

and

$$\langle (\delta T_{ij})^2 \rangle_0^{(1)} = \frac{N_i N_j [(N_T + 1 - N_i)(N_T + 1 - N_j) - (N_T + 1) + N_i N_j]}{N_T (N_T + 1)^2 (N_T + 3)}. \quad (106)$$

For two leads only, i.e. for $N_\phi = 0$, these last two expressions reduce to Eqs. (58) and (76), respectively. On the other hand, for $N_\phi \gg 1$ and $i = 1, 2$

$$\frac{\left[\langle (\delta T_{i3})^2 \rangle_0^{(\beta)} \right]^{1/2}}{\overline{T}_{i3}} \sim O\left(\frac{1}{N_\phi} \right). \quad (107)$$

Thus, $\delta T_{23}/\overline{T}_{23}$ and $\delta T_{13}/\overline{T}_{13}$ are small quantities in Eq. (103), and we can make the expansion

$$g = 2\left(\overline{T}_{21} + \delta T_{21}\right)$$
$$+ \frac{\overline{T}_{23}\overline{T}_{31}}{\overline{T}_{32} + \overline{T}_{31}}[1 + \frac{\delta T_{23}}{\overline{T}_{23}} + \frac{\delta T_{31}}{\overline{T}_{31}} - \frac{\delta T_{32} + \delta T_{31}}{\overline{T}_{32} + \overline{T}_{31}} + \frac{\delta T_{23}\delta T_{31}}{\overline{T}_{23}\overline{T}_{31}}$$
$$+ \frac{(\delta T_{23} + \delta T_{31})^2}{\left(\overline{T}_{32} + \overline{T}_{31}\right)^2} - \frac{\delta T_{23}(\delta T_{32} + \delta T_{31})}{\overline{T}_{23}\left(\overline{T}_{32} + \overline{T}_{31}\right)} - \frac{\delta T_{31}(\delta T_{32} + \delta T_{31})}{\overline{T}_{31}\left(\overline{T}_{32} + \overline{T}_{31}\right)} + \cdots] \qquad (108)$$

The average of this conductance for $N_\phi \gg 1$ is obtained by using Eq. (107):

$$\langle g \rangle_0^{(\beta)} = 2\left\{\overline{T}_{21} + \frac{\overline{T}_{23}\overline{T}_{31}}{\overline{T}_{32} + \overline{T}_{31}}\left[1 + O\left(\frac{1}{N_\phi^2}\right)\right]\right\}. \qquad (109)$$

Substituting Eqs. (98)-(100), we have

$$\langle g \rangle_0^{(\beta)} = 2\frac{N_2 N_1}{N_2 + N_1}\left[1 - \frac{\delta_{\beta 1}}{N_\phi}\right] + O\left(\frac{1}{N_\phi^2}\right) \qquad (110)$$

which gives the weak-localization correction

$$\delta \langle g \rangle \equiv \langle g \rangle_0^{(1)} - \langle g \rangle_0^{(2)} = -2\frac{N_2 N_1}{N_2 + N_1}\frac{1}{N_\phi} + O\left(\frac{1}{N_\phi^2}\right). \qquad (111)$$

Turning to the variance, we subtract the average conductance from Eq. (108) and obtain to lowest order in $1/N_\phi$

$$\frac{1}{2}\delta g \equiv \frac{1}{2}\left[g - \langle g \rangle_0^{(\beta)}\right] \approx \delta T_{21} + \frac{\overline{T}_{23}\overline{T}_{31}}{\overline{T}_{32} + \overline{T}_{31}}\left[\frac{\delta T_{23}}{\overline{T}_{23}} + \frac{\delta T_{31}}{\overline{T}_{31}} - \frac{\delta T_{32} + \delta T_{31}}{\overline{T}_{32} + \overline{T}_{31}}\right]. \qquad (112)$$

and obtain, for $\beta = 1, 2$, to lowest order in $1/N_\phi$

$$\frac{1}{2}\delta g \approx \delta T_{21} + \frac{N_1}{N_2 + N_1}\delta T_{23} + \left(\frac{N_2}{N_2 + N_1}\right)^2 \delta T_{31} - \frac{N_2 N_1}{(N_2 + N_1)^2}\delta T_{32}. \qquad (113)$$

Squaring this last expression and averaging, we obtain the variance of the conductance in terms of the $\langle \delta T_{ij} \delta T_{kl} \rangle_0$. We must now substitute the variances and covariances given in Eqs. (101) and (102) to obtain our final result:

$$[\text{var}(g)]_0^{(\beta)} = \left[\frac{2 N_2 N_1}{(N_2 + N_1) N_\phi}\right]^2 \frac{2}{\beta}\left[1 + (2 - \beta)\frac{N_2^3 + N_1^3}{N_2 N_1 (N_2 + N_1)^2}\right] + \cdots. \qquad (114)$$

The ratio of variances for $\beta = 1$ and $\beta = 2$ is

$$\frac{[\text{var}(g)]_0^{(1)}}{[\text{var}(g)]_0^{(2)}} = 2\left[1 + \frac{N_2^3 + N_1^3}{N_2 N_1 (N_2 + N_1)^2}\right] + \cdots. \qquad (115)$$

We observe from this last equation, first, that this ratio is independent of N_ϕ to leading order, and, second, that the ratio of variances is larger than 2, and as high as 3 for $N_2 = N_1 = N = 1$. For comparison, in the absence of phase-breaking and for $N_2 = N_1$, that ratio lies between 1 and 2.

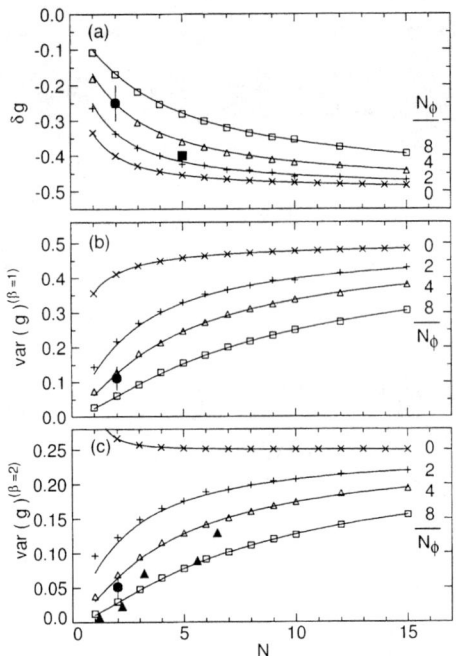

FIGURE 8. Magnitude of quantum transport effects as a function of the number of channels in the leads, $N_1 = N_2 = N$, for $N_\phi = 0$, 2, 4, and 8. (a) The weak-localization correction. (b) The variance for the orthogonal case ($B = 0$). (c) The variance for the unitary case (nonzero B). Open symbols are numerical results (20,000 matrices used, statistical error is the symbol size). Solid lines are interpolation formulae. Solid symbols are experimental results of Refs. [34] (squares), [35] (triangles), and [38] (circles) corrected for thermal averaging. The introduction of phase-breaking decreases the "universality" of the results but leads to good agreement with experiment. (From Ref. [47].)

VII.B Arbitrary N_ϕ

In order to evaluate effects of phase-breaking when N_ϕ is not large, we numerically evaluate the random-matrix theory, generating random $N_T \times N_T$ unitary or orthogonal matrices and computing g from Eq. (94). Fig. 8 shows the weak-localization correction and the variance of the conductance as the number of modes in the leads is varied for several fixed N_ϕ. This result is relevant to experiments at fixed temperature in which the size of the opening to the cavity is varied. Though δg and var$g$ are nearly independent of N in the perfectly coherent limit– the "universality" discussed above– phase-breaking channels cause variation. Thus the universality can be seen only if $N_\phi \ll N$; otherwise, the behavior is approximately linear.

The results of Sec. IV, which correspond to $N_\phi = 0$, and those of the last section,

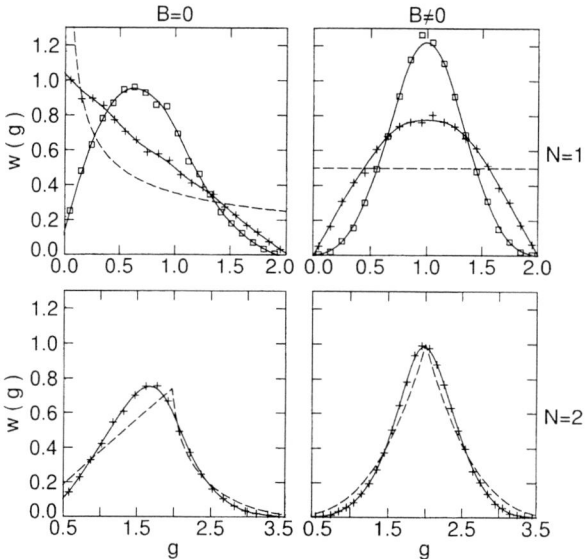

FIGURE 9. Probability density of the conductance in the orthogonal (first column) and unitary (second column) cases for $N=1$ (first row) and $N=2$ (second row). Increasing the phase-breaking from zero (dashed lines, analytic) to $N_\phi = 1$ (plus symbols, numerical) to $N_\phi = 2$ (squares, numerical) smooths the distribution. (From Ref. [47].)

obtained for $N_\phi \gg 1$, suggest an approximate interpolation formula for arbitrary values of N_ϕ. For the case $N_2 = N_1 = N$, Ref. [47] proposes

$$\delta \langle g \rangle \approx -\frac{N}{2N + N_\phi} \qquad (116)$$

for the WLC and

$$[\mathrm{var}(g)]_0^{(\beta)} \approx \left\{ \left[[\mathrm{var}(g)]_{0,N_\phi=0}^{(\beta)}\right]^{-1/2} + \left[[\mathrm{var}(g)]_{0,N_\phi \gg 1}^{(\beta)}\right]^{-1/2} \right\}^{-2} \qquad (117)$$

for the variance. These interpolation formulae are compared with the numerical simulations in Fig. 8: the agreement is good, the only significant deviation being for $N=1$ and small N_ϕ.

In addition to the mean and variance, the probability density of the conductance, $w(g)$, can be evaluated. Fig. 9 shows the distribution in the weak phase-breaking regime with a small number of modes in the lead. From the variance and mean, we know that the phase-breaking will narrow the distribution and make it more symmetric because the weak-localization correction goes to zero. Not surprisingly, the phase-breaking in addition smooths the distribution, and the extreme non-Gaussian structure in $w(g)$ is washed out.

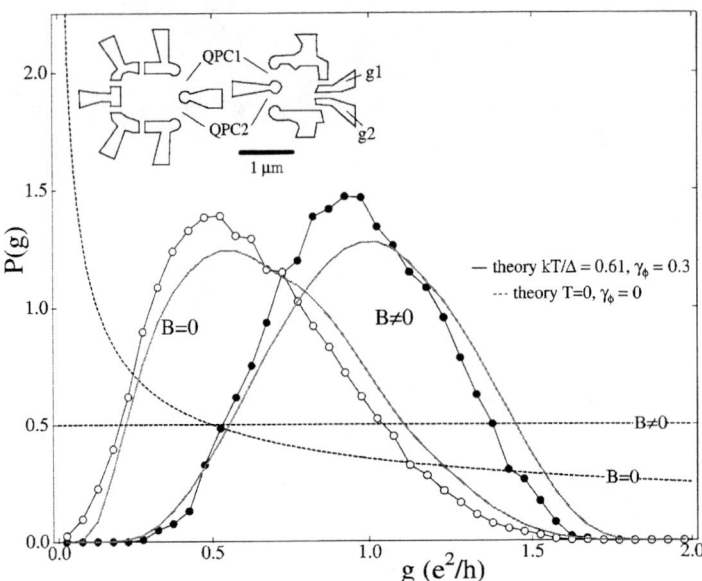

FIGURE 10. Experimental conductance distributions for both $B=0$ (open circles) and 40 mT (filled circles) for a 0.5 μm^2 device at 100 mK with $N=1$. These are compared to theoretical curves for both zero temperature (dashed lines) and non-zero temperature (solid lines). Though the effect of non-zero temperature is substantial, the measured distribution at $B=0$ is clearly not Gaussian. Inset: pattern of gates defining the dot. (After Ref. [44].)

VII.C Experiment

Now that we have a tool for evaluating the effects of phase-breaking, we return to a comparison of the information-theoretic model with experiments. First, for the mean and variance of the conductance, in Fig. 8 are also shown as black dots the results of three experiments from Refs. [34,35,38]. Since the theoretical results take into account through N_ϕ the effect of finite temperature insofar as dephasing is concerned, without including thermal smearing, the experimental variance has been corrected to compensate for this latter effect [84]. In the case of Ref. [38] measurements of all three quantum transport quantities were made, and so this data can be used to test the consistency of the theory. Notice that a value of $N_\phi = 4$-8 allows the simultaneous fit of the WLC and the variance for both $\beta = 1$ and 2. This indicates that the theory and experiment are indeed in good agreement.

Second, the full distribution of the conductance has been measured in the $N=1$ case and compared to the random-matrix theory including phase-breaking [44]. The results are shown in Fig. 10. In the $B=0$ case, the experimental distribution is clearly not Gaussian and is skewed towards small values of g, as expected.

On the other hand, the distribution is not nearly as dramatic as the completely coherent theory suggests. An important recent development in relation with the fictitious lead model described above has been carried out in Ref. [83], with the motivation of describing a spatially more uniform distribution of phase-breaking events: the fictitious lead is considered to support an infinite number of modes, each with vanishing transmission, allowing a continuous value for the dephasing rate. The experimental results are analyzed in Ref. [44] with this improved model, and the result is also shown in Fig. 10. It is seen that the shape of the conductance distribution is reproduced well with a dephasing rate obtained independently from the WLC. This comparison is strong support for the validity of the theory. The numerical value of the variance at various temperatures shows a discrepancy, though; in particular, the observed ratio of variances for $\beta = 1$ and $\beta = 2$ is considerably larger than that given by the model: this is as yet unexplained.

VIII CONCLUSION

In this paper we developed a statistical theory aimed at the description of the quantum-mechanical scattering of a particle by a cavity, whose geometry is such that the classical dynamics of the system is chaotic. We studied, as our main application, the electronic transport through ballistic microstructures, in an independent-particle approximation.

The theory, which was developed in the past within nuclear physics to describe the scattering of a nucleon by a nucleus, describes the regime in which there are two distinct time scales, associated with a prompt and an equilibrated response. The prompt response is described in terms of the energy average $\overline{S}$– also known as the optical S matrix– of the actual S matrix. Through the notion of ergodicity, $\overline{S}$ is calculated as the average $\langle S \rangle$ over an ensemble of similar systems, represented mathematically by an ensemble of S matrices belonging to the universality class in question: orthogonal, unitary or symplectic. In addition, the ensemble satisfies the analyticity (causality)-ergodicity requirements and the optical $\langle S \rangle$ evaluated over it has a specified (matrix) value. The ensemble discussed in the text is the one that carries minimum information, or maximum entropy, satisfying these conditions. It is thus meant to describe those situations in which any other additional information is irrelevant. In this procedure one constructs the one-energy statistical distribution of S using only the above physical information– expressible in terms of S itself– without ever invoking any statistical assumption for the underlying Hamiltonian which never enters the analysis.

From the resulting S-matrix distribution, known as Poisson's kernel, properties of the quantum conductance have been derived: its average, its fluctuations, and its full distribution in certain cases, both in the absence and presence of direct processes. We obtain good agreement with the results of the numerical solution of the Schrödinger equation for cavities in which the assumptions of the theory hold: either ones in which prompt response is absent (Sec. IV), or ones in which there are two widely separated time scales (Sec. V). As for the comparison with experimental data, agreement was found once temperature smearing and dephasing effects were taken into account, at least within a phenomenological model.

The effect of time-reversal symmetry– determining the universality class β– has been seen to be of fundamental importance. One other symmetry was not touched upon in this article, because of lack of space: *spatial reflection symmetries*. While such symmetries are not relevant for disordered systems– the traditional subject of mesoscopic physics– they are possible in ballistic systems in which one can control the scattering geometry. How such symmetries affect the interference contribution to transport was studied in Refs. [85,86]. For reflection symmetries, S is block diagonal in a basis of definite parity with respect to that reflection, with a circular ensemble in each block (if $\langle S \rangle = 0$). The key point is that the conductance may couple the different parity-diagonal blocks of S, and thus the resulting quantum transport properties are a nontrivial generalization of the circular ensemble results. These effects are discussed in Ref. [85,86] for structures presenting a "left-right", "up-down", "four-fold", and "inversion" spatial symmetry and compared with the results of numerical solutions of the Schrödinger equation. From an experimental point of view, it would be nice to have a confirmation of our results, both in microstructures as well as in microwave cavities, the latter ones being perhaps simpler to handle.

We should point out that the theory presented here is not applicable when there are other relevant time scales in the problem. One example is the case of a disordered quasi-1D system, where the diffusion time across the system is also an important time scale. In the past, this extra piece of information was taken into account explicitly by expressing the transfer matrix for the full system as the product of the transfer matrices for the individual slices that compose the system [87–89,4,7,10]. An example of an intermediate situation is that of a finite number of cavities connected in series. Whether more than two time scales are physically relevant for *single cavities* as well is not clear at present. It seems likely that the deviations of the numerical results from the theory for the simple structures in Sec. VI.A are for this reason [90]. Also, it is conceivable that additional time scales are at the root of the discrepancies discussed in Ref. [85] in relation with the WLC. An extension of the maximum-entropy theory presented above to include other time scales is not known at present.

We also stress that the theory presented here is applicable to the statistics of functions of the S matrix at a single value of the energy. The joint statistical distribution of the S matrix at two or more energies has escaped, so far, an analysis within the philosophy described above (some aspects of the two-point problem have been studied assuming an underlying Hamiltonian described by a Gaussian ensemble as in Refs. [91,92]). An approach coming close to that philosophy was initiated in Ref. [93] for the simplest quantity of a two-point character: the statistical distribution of the time delay arising in the scattering process– a quantity involving the energy derivative of S– motivated by the study of the electrochemical capacitance of a mesoscopic system. It was found that another piece of information is needed for the description: the statistical distribution of the K-matrix of resonance widths. Ref. [93] deals with 1×1 S-matrices; subsequent work derived the distribution of time delays for N channels [94]. In a further development, Ref. [95] finds a transformation that relates the k-point distribution of the n-dimensional S-matrix for the case $\langle S \rangle \neq 0$ to that for $\langle S \rangle = 0$, thus relating the problems in the presence and in the absence of direct processes.

Finally, we remark that the information-theory described in the text makes use of the standard Boltzmann-Shannon definition of information and entropy. Other definitions have been presented in the literature: see, for instance, Ref. [96], and the references contained therein. What would be the physics behind the use of other definitions of entropy and how our results would be modified is not known at present.

ACKNOWLEDGMENTS

The authors are grateful for helpful conversations with P. W. Brouwer, R. A. Jalabert, V. A. Gopar, C. M. Marcus, M. Martínez, and T. H. Seligman. PAM acknowledges partial support of CONACyT, under contract No. 2645P-E. Both authors wishes to thank the hospitality of The Aspen Center for Physics, where part of this paper was developed and discussed.

EVALUATION OF THE INVARIANT MEASURE

We evaluate dS in the polar representation (17) and then the arc element ds^2 of Eq. (25) keeping V and W independent, as is the case for $\beta = 2$. We use this same algebraic development and set $W = V^T$ in the proper place to analyze the $\beta = 1$ case. Differentiating S of Eq. (17) we obtain

$$dS = (dV)RW + V(dR)W + VR(dW) = V\left[(\delta V)R + dR + R(\delta W)\right] \qquad (118)$$

where we have defined the matrices

$$\delta V = V^\dagger dV, \qquad \delta W = (dW)W^\dagger, \qquad (119)$$

which are *antihermitian*, as can be seen by differentiating the unitarity relations $V^\dagger V = I$, $WW^\dagger = I$. The arc element of Eq. (25) is thus

$$\begin{aligned}
ds^2 &= \mathrm{Tr}\{\left[R^T(\delta V)^\dagger + dR^T + (\delta W)^\dagger R^T\right] \cdot \left[(\delta V)R + dR + R\delta W\right]\} \\
&= \mathrm{Tr}[R^T(\delta V)^\dagger(\delta V)R + R^T(\delta V)^\dagger dR + R^T(\delta V)^\dagger R(\delta W) \\
&\quad + (dR^T)(\delta V)R + (dR^T)dR + (dR^T)R(\delta W) \\
&\quad + (\delta W)^\dagger R^T(\delta V)R + (\delta W)^\dagger R^T dR + (\delta W)^\dagger R^T R(\delta W)].
\end{aligned} \qquad (120)$$

The various differentials occurring in the previous equation can be expressed as

$$dR = \frac{1}{2} \begin{bmatrix} d\tau/\sqrt{\rho} & d\tau/\sqrt{\tau} \\ d\tau/\sqrt{\tau} & -d\tau/\sqrt{\rho} \end{bmatrix}, \tag{121}$$

$$\delta V = \begin{bmatrix} v_1^\dagger dv_1 & 0 \\ 0 & v_2^\dagger dv_2 \end{bmatrix} = \begin{bmatrix} \delta v_1 & 0 \\ 0 & \delta v_2 \end{bmatrix}, \tag{122}$$

$$\delta W = \begin{bmatrix} (dv_3)v_3^\dagger & 0 \\ 0 & (dv_4)v_4^\dagger \end{bmatrix} = \begin{bmatrix} \delta v_3 & 0 \\ 0 & \delta v_4 \end{bmatrix}, \tag{123}$$

where we have used the abbreviation

$$\rho = 1 - \tau. \tag{124}$$

We now calculate the various terms in (120). The first plus ninth terms give

$$\text{Tr}\left[(\delta V)^\dagger(\delta V) + (\delta W)^\dagger(\delta W)\right] = \text{Tr}\sum_{i=1}^{4}(\delta v_i)^\dagger \delta v_i. \tag{125}$$

The second plus fourth terms give

$$\text{Tr}\left[R^T(\delta V)^\dagger dR + (dR^T)(\delta V)R\right] \tag{126}$$
$$= \text{Tr}[\sqrt{\rho}(\delta v_1)^\dagger d\sqrt{\rho} + \sqrt{\tau}(\delta v_2)^\dagger d\sqrt{\tau}$$
$$+\sqrt{\tau}(\delta v_1)^\dagger d\sqrt{\tau} + \sqrt{\rho}(\delta v_2)^\dagger d\sqrt{\rho} + h.c.] = 0,$$

where h.c. stands for Hermitian conjugate. We have used the identity

$$\text{Tr}\left[D(\delta v)^\dagger D' + D'(\delta v)D\right] = 0, \tag{127}$$

where D and D' denote any two real diagonal matrices and δv an antihermitian one. Using the identity

$$\text{Tr}\left[(\delta v_i)^\dagger D(\delta v_j) D\right] = \text{Tr}\left[D(\delta v_j)^\dagger D(\delta v_i)\right], \tag{128}$$

we find for the third plus seventh terms

$$\text{Tr}\left[R^T(\delta V)^\dagger R(\delta W) + (\delta W)^\dagger R^T(\delta V)R\right]$$
$$= 2\text{Tr}[\sqrt{\rho}(\delta v_1)^\dagger \sqrt{\rho}(\delta v_3) + \sqrt{\tau}(\delta v_2)^\dagger \sqrt{\tau}(\delta v_3)$$
$$+\sqrt{\tau}(\delta v_1)^\dagger \sqrt{\tau}(\delta v_4) + \sqrt{\rho}(\delta v_2)^\dagger \sqrt{\rho}(\delta v_4)]. \tag{129}$$

The fifth term gives

$$\text{Tr}(dR^T)(dR) = \frac{1}{2}\sum_a \frac{(d\tau_a)^2}{\tau_a \rho_a}. \tag{130}$$

Finally, the sixth plus eighth terms give

$$\text{Tr}\left[(dR^T)R(\delta W) + h.c.\right]$$
$$= \frac{1}{2}\text{Tr}\left[-(d\tau)\delta v_3 + (d\tau)\delta v_3 + (d\tau)\delta v_4 - (d\tau)\delta v_4 + h.c.\right] = 0. \tag{131}$$

Substituting these expressions in (120), we find

$$ds^2 = 2\text{Tr}\left\{\frac{1}{2}\sum_{i=1}^{4}(\delta v_i)^\dagger (\delta v_i) + \frac{(d\tau)(d\tau)}{4\tau\rho} + \sqrt{\rho}(\delta v_1)^\dagger \sqrt{\rho}(\delta v_3)\right.$$
$$\left. + \sqrt{\tau}(\delta v_2)^\dagger \sqrt{\tau}(\delta v_3) + \sqrt{\tau}(\delta v_1)^\dagger \sqrt{\tau}(\delta v_4) + \sqrt{\rho}(\delta v_2)^\dagger \sqrt{\rho}(\delta v_4)\right\}. \tag{132}$$

The antihermitian matrix δv_i can be expressed as

$$\delta v_i = \delta a_i + i\delta s_i, \tag{133}$$

where δa_i is real antisymmetric and δs_i is real symmetric. Substituting in the expression for ds^2 and rearranging terms, we find

$$ds^2 = \sum_a \left\{ \sum_{i=1}^{4}((\delta s_i)_{aa})^2 + 2\rho_a[(\delta s_1)_{aa}(\delta s_3)_{aa} + (\delta s_2)_{aa}(\delta s_4)_{aa}] \right.$$
$$\left. + 2\tau_a[(\delta s_2)_{aa}(\delta s_3)_{aa} + (\delta s_1)_{aa}(\delta s_4)_{aa}] + \frac{(d\tau_a)^2}{2\tau_a\rho_a} \right\}$$
$$+ 2\sum_{a<b}\left\{\sum_{i=1}^{4}((\delta a_i)_{ab})^2 + \sum_{i=1}^{4}((\delta s_i)_{ab})^2 \right. \tag{134}$$
$$+ 2\sqrt{\rho_a\rho_b}[(\delta a_1)_{ab}(\delta a_3)_{ab} + (\delta a_2)_{ab}(\delta a_4)_{ab} + (\delta s_1)_{ab}(\delta s_3)_{ab} + (\delta s_2)_{ab}(\delta s_4)_{ab}]$$
$$\left. + 2\sqrt{\tau_a\tau_b}[(\delta a_2)_{ab}(\delta a_3)_{ab} + (\delta a_1)_{ab}(\delta a_4)_{ab} + (\delta s_2)_{ab}(\delta s_3)_{ab} + (\delta s_1)_{ab}(\delta s_4)_{ab}]\right\}.$$

The case $\beta=1$

In this case, $v_3 = v_1^T$, $v_4 = v_2^T$ so $\delta v_3 = \text{delta}v_1)^T$, $\delta v_4 = \text{delta}v_2)^T$, and hence

$$\delta a_3 = -\delta a_1, \qquad \delta s_3 = \delta s_1, \qquad \delta a_4 = -\delta a_2, \qquad \delta s_4 = \delta s_2. \tag{135}$$

Substituting in (134), we have

$$ds^2 = 2\sum_a \left\{((\delta s_1)_{aa})^2 + ((\delta s_2)_{aa})^2 + 2\tau_a (\delta s_1)_{aa} (\delta s_2)_{aa}\right.$$
$$\left. + \rho_a \left[((\delta s_1)_{aa})^2 + ((\delta s_2)_{aa})^2\right] + \frac{(d\tau_a)^2}{4\tau_a(1-\tau_a)}\right\}$$
$$+ 4 \sum_{a<b} \left\{((\delta a_1)_{ab})^2 + ((\delta a_2)_{ab})^2 + ((\delta s_1)_{ab})^2 + ((\delta s_2)_{ab})^2\right.$$
$$+ 2\sqrt{\tau_a \tau_b}[(\delta s_1)_{ab} (\delta s_2)_{ab} - (\delta a_1)_{ab} (\delta a_2)_{ab}]$$
$$\left. + \sqrt{\rho_a \rho_b} \left[((\delta s_1)_{ab})^2 + ((\delta s_2)_{ab})^2 - ((\delta a_1)_{ab})^2 - ((\delta a_2)_{ab})^2\right]\right\}, \quad (136)$$

or

$$\frac{1}{2}ds^2 = \sum_a \left\{(1+\rho_a)\left[((\delta s_1)_{aa})^2 + ((\delta s_2)_{aa})^2\right] + 2\tau_a (\delta s_1)_{aa} (\delta s_2)_{aa} + \frac{(d\tau_a)^2}{4\tau_a \rho_a}\right\}$$
$$+ 2\sum_{a<b} \left\{(1+\sqrt{\rho_a \rho_b})\left[((\delta s_1)_{ab})^2 + ((\delta s_2)_{ab})^2\right]\right. \quad (137)$$
$$+ (1-\sqrt{\rho_a \rho_b})\left[((\delta a_1)_{ab})^2 + ((\delta a_2)_{ab})^2\right]$$
$$\left. + 2\sqrt{\tau_a \tau_b}[(\delta s_1)_{ab} (\delta s_2)_{ab} - (\delta a_1)_{ab} (\delta a_2)_{ab}]\right\}.$$

In (137), $(\delta s_1)_{aa}$, $(\delta s_2)_{aa}$ and τ_a ($a=1, \cdots, N$) contribute N independent variations each, while $(\delta s_1)_{ab}$, $(\delta s_2)_{ab}$, $(\delta a_1)_{ab}$, $(\delta a_2)_{ab}$ ($a<b$) contribute $N(N-1)/2$ each, giving a total of

$$\nu = 3N + 4\frac{N(N-1)}{2} = 2N^2 + N, \quad (138)$$

which is the correct number of independent parameters for a $2N$-dimensional unitary symmetric matrix.

The metric tensor appearing in Eq. (23) has a simple block structure, consisting of 1×1 and 2×2 blocks along the diagonal, as follows. There are N 2-dimensional blocks with rows and columns labeled $(\delta s_1)_{aa}$, $(\delta s_2)_{aa}$:

$$\begin{array}{c} \quad\quad (\delta s_1)_{aa} \quad (\delta s_2)_{aa} \\ \begin{array}{c}(\delta s_1)_{aa} \\ (\delta s_2)_{aa}\end{array} \left[\begin{array}{cc} 1+\rho_a & \tau_a \\ \tau_a & 1+\rho_a \end{array}\right] \end{array} \quad (139)$$

and N 1-dimensional blocks $1/4\rho_a\tau_a$ labeled by $d\tau_a$, giving, altogether, the contribution

$$\prod_{a=1}^N \frac{(1+\rho_a)^2 - \tau_a^2}{4\rho_a \tau_a} = \prod_{a=1}^N \frac{1}{\tau_a} \quad (140)$$

to det g. There are $N(N-1)/2$ 2-dimensional blocks with rows and columns labeled $(\delta s_1)_{ab}$, $(\delta s_2)_{ab}$,

$$\begin{array}{c} (\delta s_1)_{ab} \quad (\delta s_2)_{ab} \\ \begin{array}{c} (\delta s_1)_{ab} \\ (\delta s_2)_{ab} \end{array} \left[\begin{array}{cc} 1 + \sqrt{\rho_a \rho_b} & \sqrt{\tau_a \tau_b} \\ \sqrt{\tau_a \tau_b} & 1 + \sqrt{\rho_a \rho_b} \end{array} \right] \cdot 2, \end{array} \quad (141)$$

and also $N(N-1)/2$ 2-dimensional blocks with rows and columns labeled $(\delta a_1)_{ab}$, $(\delta a_2)_{ab}$,

$$\begin{array}{c} (\delta a_1)_{ab} \quad (\delta a_2)_{ab} \\ \begin{array}{c} (\delta a_1)_{ab} \\ (\delta a_2)_{ab} \end{array} \left[\begin{array}{cc} 1 - \sqrt{\rho_a \rho_b} & -\sqrt{\tau_a \tau_b} \\ -\sqrt{\tau_a \tau_b} & 1 - \sqrt{\rho_a \rho_b} \end{array} \right] \cdot 2, \end{array} \quad (142)$$

giving, altogether, a contribution

$$\left\{ \left[1 + \sqrt{(1-\tau_a)(1-\tau_b)} \right]^2 - \tau_a \tau_b \right\} \left\{ \left[1 - \sqrt{(1-\tau_a)(1-\tau_b)} \right]^2 - \tau_a \tau_b \right\}$$
$$= \left[2 - \tau_a - \tau_b + 2\sqrt{(1-\tau_a)(1-\tau_b)} \right] \left[2 - \tau_a - \tau_b - 2\sqrt{(1-\tau_a)(1-\tau_b)} \right]$$
$$= (\tau_a - \tau_b)^2. \quad (143)$$

Multiplying (140) and (143) and taking the square root, as required by Eq. (24), we find the result given in Eq. (27).

The case $\beta = 2$

We go back to Eq. (134). We can write the single summation in that equation (except for its last term) as

$$\sum_a \{ [(\delta s_1)_{aa} + (\delta s_3)_{aa}]^2 + [(\delta s_2)_{aa} + (\delta s_4)_{aa}]^2$$
$$- 2\tau_a [(\delta s_1)_{aa} - (\delta s_2)_{aa}] [(\delta s_3)_{aa} - (\delta s_4)_{aa}] \}$$
$$= \sum_a \left[(\delta x_a)^2 + (\delta y_a)^2 - 2\tau_a (\delta z_a)(\delta x_a - \delta y_a - \delta z_a) \right], \quad (144)$$

where we have defined the combinations

$$\delta x_a = (\delta s_1)_{aa} + (\delta s_3)_{aa} \quad (145)$$
$$\delta y_a = (\delta s_2)_{aa} + (\delta s_4)_{aa} \quad (146)$$
$$\delta z_a = (\delta s_1)_{aa} - (\delta s_2)_{aa}. \quad (147)$$

Notice that in ds^2 the $4N$ quantities $(\delta s_1)_{aa}$, $(\delta s_2)_{aa}$, $(\delta s_3)_{aa}$, $(\delta s_4)_{aa}$ appear only through the $3N$ combinations $\delta x_a, \delta y_a, \delta z_a$: these quantities, together with the $d\tau_a$ contribute $4N$ independent variations. The $(\delta s_i)_{ab}$ and the $(\delta a_i)_{ab}$ for $i = 1, \cdots, 4$ and

$a < b$ contribute $4 \cdot N(N-1)/2$ each, so that we have a total of $4N^2$ variations, which is the correct number of independent parameters for a $2N$-dimensional unitary matrix.

In terms of independent variations we can thus write the ds^2 of Eq. (134) as

$$ds^2 = \sum_a \left\{ \left[(\delta x_a)^2 + (\delta y_a)^2 - 2\tau_a(\delta z_a)(\delta x_a - \delta y_a - \delta z_a) \right] + \frac{(d\tau_a)^2}{2\tau_a \rho_a} \right\}$$
$$+ 2 \sum_{a<b} \left\{ \sum_{i=1}^{4} ((\delta a_i)_{ab})^2 + \sum_{i=1}^{4} ((\delta s_i)_{ab})^2 \right. \tag{148}$$
$$+ 2\sqrt{\rho_a \rho_b} \left[(\delta a_1)_{ab} (\delta a_3)_{ab} + (\delta a_2)_{ab} (\delta a_4)_{ab} + (\delta s_1)_{ab} (\delta s_3)_{ab} + (\delta s_2)_{ab} (\delta s_4)_{ab} \right]$$
$$\left. + 2\sqrt{\tau_a \tau_b} \left[(\delta a_2)_{ab} (\delta a_3)_{ab} + (\delta a_1)_{ab} (\delta a_4)_{ab} + (\delta s_2)_{ab} (\delta s_3)_{ab} + (\delta s_1)_{ab} (\delta s_4)_{ab} \right] \right\}.$$

The metric tensor appearing in Eq. (23) has a simple block structure, consisting of 1×1, 3×3 and 4×4 blocks along the diagonal, as follows. There are N 3-dimensional blocks with rows and columns labeled δx_a, δy_a, δz_a:

$$\begin{array}{c} \delta x_a \\ \delta y_a \\ \delta z_a \end{array} \begin{bmatrix} 1 & 0 & -\tau_a \\ 0 & 1 & \tau_a \\ -\tau_a & \tau_a & 2\tau_a \end{bmatrix} \tag{149}$$

and N 1-dimensional blocks $1/2\rho_a\tau_a$ labeled by $d\tau_a$, giving, altogether, the contribution

$$\prod_{a=1}^{N} \frac{2\tau_a(1-\tau_a)}{2\tau_a(1-\tau_a)} = 1 \tag{150}$$

to $\det g$. There are $N(N-1)/2$ 4-dimensional blocks with rows and columns labeled $(\delta s_1)_{ab}$, $(\delta s_2)_{ab}$, $(\delta s_3)_{ab}$, $(\delta s_4)_{ab}$ (for $a < b$)

$$\begin{array}{c} (\delta s_1)_{ab} \\ (\delta s_2)_{ab} \\ (\delta s_3)_{ab} \\ (\delta s_4)_{ab} \end{array} \begin{bmatrix} 1 & 0 & \sqrt{\rho_a\rho_b} & \sqrt{\tau_a\tau_b} \\ 0 & 1 & \sqrt{\tau_a\tau_b} & \sqrt{\rho_a\rho_b} \\ \sqrt{\rho_a\rho_b} & \sqrt{\tau_a\tau_b} & 1 & 0 \\ \sqrt{\tau_a\tau_b} & \sqrt{\rho_a\rho_b} & 0 & 1 \end{bmatrix} \tag{151}$$

and also $N(N-1)/2$ 4-dimensional blocks with rows and columns labeled $(\delta a_1)_{ab}$, $(\delta a_2)_{ab}$, $(\delta a_3)_{ab}$, $(\delta a_4)_{ab}$ and identical to the matrices of (151) giving, altogether, a contribution

$$(\tau_a - \tau_b)^4. \tag{152}$$

Multiplying (150) and (152) and taking the square root, as required by Eq. (24), we find the result given in Eq. (27).

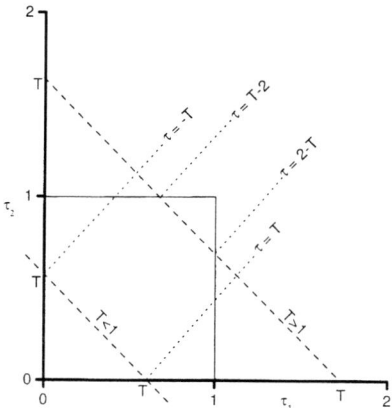

FIGURE 11. Region of integration for finding the distribution $w(T)$ in the $N = 2$ case.

THE DISTRIBUTION OF THE CONDUCTANCE IN THE ABSENCE OF DIRECT PROCESSES

We derive here the two-channel conductance distribution of Eqs. (79) and (80) and the behavior of that distribution for arbitrary N in the range $0 < T < 1$, as given by Eq. (82).

For the two-channel distribution we have to perform the integral (77) in the two-dimensional space τ_1, τ_2, inside the square $0 < \tau_1 < 1$, $0 < \tau_2 < 1$, indicated in Fig. 11. The change of variables

$$T = \tau_1 + \tau_2, \qquad \tau = \tau_1 - \tau_2 \tag{153}$$

will be found advantageous. As shown in Fig. 11, when $0 < T < 1$ the variable τ varies in the interval $-T < \tau < T$, whereas when $1 < T < 2$ we have $T - 2 < \tau < 2 - T$.

Derivation of Eq.(80).

From Eq. (27) for the invariant measure we find, for two channels

$$P^{(2)}(\tau_1, \tau_2) = C\left(\tau_1 - \tau_2\right)^2. \tag{154}$$

The normalization constant is found from

$$1 = C \int\int_0^1 (\tau_1 - \tau_2)^2 \, d\tau_1 d\tau_2$$
$$= \frac{C}{2} \left[\int_0^1 dT \int_{-T}^T d\tau \cdot \tau^2 + \int_1^2 dT \int_{T-2}^{2-T} d\tau \cdot \tau^2 \right] = \frac{C}{6}, \tag{155}$$

so that $C=6$.

The distribution of the spinless conductance is thus

$$w^{(2)}(T) = 3\left[\int_0^1 dT'\delta(T-T')\int_{-T'}^{T'} d\tau \cdot \tau^2 + \int_1^2 dT'\delta(T-T')\int_{T'-2}^{2-T'} d\tau \cdot \tau^2\right], \quad (156)$$

so that

$$w^{(2)}(T<1) = 2T^3, \qquad w^{(2)}(1<T<2) = 2(2-T)^3. \quad (157)$$

These two equations can be condensed in the result of Eq. (80).

Derivation of Eq. (79)

From Eq. (27) for the invariant measure we find, for two channels

$$P^{(1)}(\tau_1, \tau_2) = C\frac{|\tau_1 - \tau_2|}{\sqrt{\tau_1\tau_2}}. \quad (158)$$

The normalization constant is found from

$$1 = C\int\int_0^1 \frac{|\tau_1 - \tau_2|}{\sqrt{\tau_1\tau_2}}d\tau_1 d\tau_2 \quad (159)$$

$$= \frac{C}{2}\left[\int_0^1 dT \int_{-T}^{T} d\tau \frac{2|\tau|}{\sqrt{T^2-\tau^2}} + \int_1^2 dT \int_{T-2}^{2-T} d\tau \frac{2|\tau|}{\sqrt{T^2-\tau^2}}\right] = \frac{4}{3}C, \quad (160)$$

so that $C=3/4$.

The distribution of the spinless conductance is thus

$$w^{(1)}(T) = \frac{3}{8}[\int_0^1 dT'\delta(T-T')\int_{-T'}^{T'} d\tau \frac{2|\tau|}{\sqrt{(T')^2-\tau^2}}$$

$$+ \int_1^2 dT'\delta(T-T')\int_{T'-2}^{2-T'} d\tau \frac{2|\tau|}{\sqrt{(T')^2-\tau^2}}, \quad (161)$$

so that

$$w^{(1)}(T<1) = \frac{3}{2}T, \qquad w^{(1)}(1<T<2) = \frac{3}{2}\left[T - 2\sqrt{T-1}\right], \quad (162)$$

which are Eqs. (79) we wanted to prove.

Derivation of Eq. (82)

The integration region selected by the δ function in Eq. (77) is $N-1$ dimensional. In the particular case $T < 1$, each of the τ_a's ($a=1,\ldots,N$) varies, over that region, in the interval $0 < \tau_a < T$. This is clearly illustrated in Fig. 11 for $N=2$, where the relevant region is the segment extending from $(\tau_1, \tau_2) = (0, T)$ to $(\tau_1, \tau_2) = (T, 0)$. For $T < 1$ the corners of the N-dimensional hypercube are not relevant; they become relevant for $T > 1$. We can thus write, for $T < 1$

$$w_N^{(\beta)}(T<1) = C_N^{(\beta)} \int_0^T \cdots \int_0^T \delta\left(T - \sum_{a=1}^N \tau_a\right) \prod_{a<b} |\tau_a - \tau_b|^\beta \prod_c \tau_c^{\frac{\beta-2}{2}} \prod_i d\tau_i \quad (163)$$

Introducing the new variables $\sigma_a = \tau_a/T$ ($a = 1, \cdots N$) which vary in the interval $(0,1)$, we have

$$w_N^{(\beta)}(T<1)$$
$$= C_N^{(\beta)} \int_0^1 \cdots \int_0^1 \delta\left(T - T\sum_{a=1}^N \sigma_a\right) T^{\frac{N(N-1)}{2}\beta} \prod_{a<b}|\sigma_a - \sigma_b|^\beta T^{N\frac{\beta-2}{2}} \prod_c \sigma_c^{\frac{\beta-2}{2}} T^N \prod_i d\sigma_i$$
$$= C_N^{(\beta)} T^{\beta\frac{N^2}{2}-1} \int_0^1 \cdots \int_0^1 \delta\left(1 - \sum_{a=1}^N \sigma_a\right) \prod_{a<b}|\sigma_a - \sigma_b|^\beta \prod_c \sigma_c^{\frac{\beta-2}{2}} \prod_i d\sigma_i, \quad (164)$$

which behaves as the power of T indicated in Eq. (82).

REFERENCES

1. Newton, R. G., *Scattering Theory of Waves and Particles*, New York: Springer-Verlag, 1982.
2. van de Hulst, H. C., *Light Scattering from Small Particles*, New York, Dover, 1981.
3. Sheng, P., *Scattering and Localization of Classical Waves in Random Media*, Singapore: World Scientific, 1990.
4. Altshuler, B. L., Lee, P. A., and Webb, R. A., Eds., *Mesoscopic Phenomena in Solids*, Amsterdam: North-Holland, 1991.
5. Feshbach, H., *Theoretical Nuclear Physics— Nuclear Reactions* New York: Wiley, 1992.
6. Beenakker, C. W. J., and van Houten, H., in *Solid State Physics*, edited by Ehrenreich, H., and Turnbull, D., New York: Academic, 1991, Vol. 44, pp. 1-228.
7. Akkermans, E., Montambaux, G., Pichard, J.-L., and Zinn-Justin, J., Eds., *Mesoscopic Quantum Physics*, New York: Elsevier, 1995.
8. Sohn, L. L., Kouwenhoven, L. P., and Schön, G., Eds., *Mesoscopic Electron Transport*, Boston: Kluwer Academic, 1997.
9. Timp, G. L., editor, *Nanotechnology*, New York: Springer-Verlag, 1998.
10. Beenakker, C. W. J., *Rev. Mod. Phys.* **69**, 731 (1997).

11. Giannoni, M.-J., Voros, A., and Zinn-Justin, J., Eds., *Chaos and Quantum Physics*, New York: North-Holland, 1991.
12. Gutzwiller, M. C., *Chaos in Classical and Quantum Mechanics* New York: Springer Verlag, 1991.
13. Feshbach, H., Porter, C. E., and Weisskopf, V. F., *Phys. Rev.* **96**, 448 (1954).
14. Feshbach, H., "Topics in the Theory of Nuclear Reactions" in *Reaction Dynamics*, New York: Gordon and Breach, 1973.
15. Mehta, M. L., *Random Matrices*, New York: Academic, 1991.
16. Porter, C. E., *Statistical Theories of Spectra: Fluctuations*, New York: Academic, 1965.
17. French, J. B., Mello, P. A., and Pandey, A., *Phys. Lett. B* **80**, 17 (1978).
18. Agassi, D., Weidenmüller, H. A., and Mantzouranis, G., *Phys. Rep.* **22**, 145 (1975).
19. López, G., Mello, P. A., and Seligman, T. H., *Z. Phys. A* **302**, 351 (1981).
20. Mello, P. A., and Seligman, T. H., *Nucl. Phys. A* **344**, 489 (1980).
21. Levine, R. D., *Quantum Mechanics of Molecular Rate Processes*, Oxford: Oxford University Press, 1969, Ch. 3.5; Miller, W. J., *J. Chem. Phys.* **65**, 2216 (1976).
22. Ericson, T., and Mayer-Kuckuk, T., *Ann. Rev. Nucl. Sc.* **16**, 183 (1966).
23. Gutzwiller, M. C., *Physica D* **7**, 341 (1983).
24. Blümel, R., and Smilansky, U., *Phys. Rev. Lett.* **60**, 477 (1988); *Physica D* **36**, 111 (1989); *Phys. Rev. Lett.* **64**, 241 (1990).
25. For a review of classical and quantum chaotic scattering see Smilansky, U., in Ref. [11] pp. 371-441.
26. Doron, E., Smilansky, U., and Frenkel, A., *Phys. Rev. Lett.* **65**, 3072 (1990); *Physica D* **50**, 367 (1991).
27. Stein, J., Stöckmann, H.-J., and Stoffregen, U., *Phys. Rev. Lett.* **75**, 53 (1995) and references therein.
28. Kudrolli, A., Kidambi, V., and Sridhar, S., *Phys. Rev. Lett.* **75**, 822 (1995) and references therein.
29. Alt, H., Bäcker, A., Dembowski, C., Gräf, H.-D., Hofferbert, R., Rehfeld, H., and Richter, A., *Phys. Rev. E* **58**, 1737 (1998) and references therein.
30. Stoytchev, M., and Genack, A. Z., *Phys. Rev. Lett.* **79**, 309 (1997) and references therein.
31. Gertsenshtein, M. E., and Vasil'ev, V. B., *Theor. Probab. Appl.* **4**, 391 (1959); **5**, 340(e) (1960) [*Teor. Veroyatn. Primen.* **4**, 424 (1959); **5**, 3(E) (1960)].
32. Jalabert, R. A., Baranger, H. U., and Stone, A. D., *Phys. Rev. Lett.* **65**, 2442 (1990).
33. Marcus, C. M., Westervelt, R. M., Hopkins, P. F., and Gossard, A. C., *Phys. Rev. B* **48**, 2460 (1993).
34. Chang, A. M., Baranger, H. U., Pfeiffer, L. N., and West, K. W., *Phys. Rev. Lett.* **73**, 2111 (1994).
35. Berry, M. J., Katine, J. A., Westervelt, R. M., and Gossard, A. C., *Phys. Rev. B* **50**, 17721 (1994); Berry, M. J., Baskey, J. H., Westervelt, R. M., and Gossard, A. C., *Phys. Rev. B* **50**, 8857 (1994).
36. Bird, J. P., Ishibashi, K., Aoyagi, Y., Sugano, T., and Ochiai, Y., *Phys. Rev. B* **50**, 18678 (1994).
37. Keller, M. W., Mittal, A., Sleight, J. W., Wheeler, R. G., Prober, D. E., Sacks, R.

N., and Shtrikmann, H., *Phys. Rev. B* **53**, R1693 (1996).
38. Chan, I. H., Clarke, R. M., Marcus, C. M., Campman, K., and Gossard, A. C., *Phys. Rev. Lett.* **74**, 3876 (1995).
39. Lütjering, G., Richter, K., Weiss, D., Mao, J., Blick, R. H., von Klitzing, K., and Foxon, C. T., *Surf. Sci.* **361**, 709 (1996).
40. Zozoulenko, I. V., Schuster, R., Berggren, K.-F., and Ensslin, K., *Phys. Rev. B* **55**, R10209 (1997).
41. Lee, Y., Faini, G., and Mailly, D., *Phys. Rev. B* **56**, 9805 (1997).
42. Sachrajda, A. S., Ketzmerick, R., Gould, C., Feng, Y., Kelly, P. J., Delage, A., and Wasilewski, Z., *Phys. Rev. Lett.* **80**, 1948 (1998).
43. Christensson, L., Linke, H., Omling, P., Lindelof, P. E., Zozoulenko, I. V., and Berggren, K.-F., *Phys. Rev. B* **57**, 12306 (1998).
44. Huibers, A. G., Patel, S. R., Marcus, C. M., Brouwer, P. W., Duruöz, C. I., and Harris Jr., J. S., *Phys. Rev. Lett.* **81**, 1917 (1998) [cond-mat/9801174].
45. For reviews of the semiclassical treatment of ballistic cavities see Stone, A. D., in Ref. [7] and Baranger, H. U., in Ref. [9].
46. Baranger, H. U., and Mello, P. A., *Phys. Rev. Lett.* **73**, 142 (1994).
47. Baranger, H. U., and Mello, P. A., *Phys. Rev. B* **51**, 4703 (1995).
48. Baranger, H. U., and Mello, P. A., *Europhys. Lett.* **33**, 465 (1996).
49. Baranger, H. U., and Stone, A. D., *Phys. Rev. B* **40**, 8169 (1989).
50. McVoy, K. W., "Nuclear Resonance Reactions and S-Matrix Analyticity", in *Fundamentals in Nuclear Theory*, Vienna: International Atomic Energy Agency, 1967.
51. Lane, A. M., and Thomas, R. G., *Rev. Mod. Phys.* **30**, 257 (1958).
52. Merzbacher, E., *Quantum Mechanics*, 2nd. Ed., J. Wiley, 1970.
53. Dyson, F. J., *J. Math. Phys.* **3**, 140 (1962).
54. Hua, L. K.,*Harmonic Analysis of Functions of Several Complex Variables in the Classical Domains*, translated by Ebner, L., and Koranyi, A., Providence RI: Amer. Math. Soc., 1963.
55. Mello, P. A., and Pichard, J.-L., *J. Phys. I* **1**, 493 (1991).
56. Mello, P. A., Pereyra, P., and Kumar, N., *Ann. Phys. (N.Y.)* **181**, 290 (1988).
57. Büttiker, M., *IBM Jour. Res. Develop.* **32**, 317 (1988).
58. Levinson, Y. B., *Zh. Eksp. Teor. Fiz.* **95**, 2175 (1989) [*Sov. Phys. JETP* **68**, 1257 (1989)].
59. Levinson, Y. B., and Shapiro, B., *Phys. Rev. B* **46**, 15520 (1992).
60. Wigner, E. P., *Group Theory and its Applications to the Quantum Mechanics of Atomic Spectra*, New York: Academic Press, 1959.
61. Hamermesh, M., *Group Theory and Its Applications to Physical Problems* Reading MA: Addison-Wesley, 1962, p. 313.
62. Lass, H., *Vector and Tensor Analysis*, New York: McGraw-Hill, 1950, p. 277, Ex. 125; pp. 279, 280; p. 287, Pr. 7.
63. Jalabert, R. A., Pichard, J.-L., and Beenakker, C. W. J., *Europhys. Lett.* **27**, 255 (1994).
64. Yaglom, A. M., *An Introduction to the Theory of Stationary Random Functions* translated by Rilverman, A. S, New York: Prentice-Hall, 1962.
65. Gopar, V. A., Ph. D. Thesis, Universidad Nacional Autónoma de México, México,

1999.
66. Mello, P. A., Pereyra, P., and Seligman, T. H., *Ann. Phys. (N.Y.)* **161**, 276 (1985).
67. Brouwer, P. W., *Phys. Rev. B* **51**, 16878 (1995).
68. Brouwer, P. W., *On the Random-Matrix Theory of Quantum Transport*, Doctoral Thesis, University of Leiden, 1997.
69. Friedman, W., and Mello, P. A., *Ann. Phys. (N.Y.)* **161**, 276 (1985).
70. Nishioka, H., and Weidenmüller, H. A., *Phys. Lett. B* **157**, 101 (1985); Iida, S., Weidenmüller, H. A., and Zuk, J. A., *Ann. Phys. (N.Y.)* **200**, 219 (1990); Lewenkopf, C. H., and Weidenmüller, H. A., *Ann. Phys. (N.Y.)* **212**, 53 (1991).
71. Mello, P. A., *J. Phys. A* **23**, 4061 (1990).
72. Politzer, H. D., *Phys. Rev. B* **40**, 11917 (1989).
73. Abramowitz, M., and Stegun, I. E., *Handbook of Mathematical Functions*, Washington: National Bureau of Standards, 1964, Ch. 15.
74. Baranger, H. U., DiVincenzo, D. P., Jalabert, R. A., and Stone, A. D., *Phys. Rev. B* **44**, 10637 (1991).
75. Jensen, R., *Chaos* **1**, 101 (1991).
76. Baranger, H. U., Jalabert, R. A., and Stone, A. D., *Phys. Rev. B* **70**, 3876 (1993).
77. Baranger, H. U., Jalabert, R. A., and Stone, A. D., *Chaos* **3**, 665 (1993).
78. The same effect has been noticed in the context of random matrix Hamiltonians: one can have Gaussian ensemble eigenvalue statistics while not having Porter-Thomas wavefunction statistics (E. Mucciolo, private communication).
79. While $\langle S \rangle$ varies slightly over the full energy range used, the distribution of T implied by $\langle S \rangle$ does not vary. Thus, the theoretical curve shown in Fig. 7 is the average of the prediction of Poisson's kernel for each of the four intervals of 50 energies. The numerical result is the histogram of all of the data.
80. Büttiker, M., *Phys. Rev. B* **33** 3020 (1986).
81. Previous uses of the "fake lead" approach to dephasing include Datta, S., *Phys. Rev. B* **40**, 5830 (1989); Kirczenow, G., *Solid State Commun.* **74**, 1051 (1990); D'Amato, J. L., and Pastawski, H. M., *Phys. Rev.* **41**, 7441 (1990); Hershfield, S., *Phys. Rev. B* **43**, 11586 (1991); DiVincenzo, D. P.,*Phys. Rev. B* **48**, 1404 (1993).
82. Brouwer, P. W., and Beenakker, C. W. J., *Phys Rev. B* **51**, 7739 (1995).
83. Brouwer, P. W., and Beenakker, C. W. J., *Phys Rev. B* **55**, 4695 (1997).
84. To compensate for the thermal smearing in the experiment, we estimate the energy correlation length E_c from the data given and then multiply the experimental variance by $3.5kT/E_c$. For [38], this produces a factor of ~ 2.6-4.5.
85. Baranger, H. U., and Mello, P. A., *Phys. Rev. B* **54**, R14297 (1996).
86. Gopar, V. A., Martínez, M., Mello, P. A., and Baranger, H. U., *J. Phys. A* **29**, 881 (1996).
87. Dorokhov, O. N., *Pis'ma Zh. Eksp. Teor. Fiz.* **36**, 259 (1982) [*JETP Lett.* **36**, 318 (1982)].
88. Mello, P. A., Pereyra, P., and Kumar, N., *Ann. Phys. (N.Y.)* **181**, 290 (1988).
89. Mello, P. A., and Stone, A. D., *Phys. Rev. B* **44**, 3559 (1991).
90. For the simple structures of Sec. VI.A, in addition to comparing the numerical results to the invariant-measure theory, we also compared to Poisson's kernel. The agreement was also poor in this case because the statistics were not stationary so that an optical

S-matrix could not be unambiguously defined. In contrast, the structures of Sec. VI.B have smaller openings: this causes two widely separated time scales, and so Poisson's kernel is a good description.

91. Lewenkopf, C., and Weidenmüller, H. A., *Ann. Phys. (N.Y.)* **212**, 53 (1991).
92. Fyodorov, Y. V., and Sommers, H.-J., *J. Math. Phys.* **38**, 1918 (1997).
93. Gopar, V., Mello, P. A., and Büttiker, M., *Phys. Rev. Lett.* **77**, 3005 (1996).
94. Brouwer, P. W., Frahm, K. M., and Beenakker, C. W. J., *Phys. Rev. Lett.* **78**, 4737 (1997) and cond-mat/9809022.
95. Gopar, V. A., and Mello, P. A., *Europhys. Lett.* **42**, 131 (1998).
96. Tsallis, C., *J. Stat. Phys.* **52**, 479 (1988); *Physica A* **221**, 277 (1995).

Topics in Quantum Mechanics

Marcos Moshinsky*

Instituto de Física-UNAM.
Apartado Postal 20-364; 01000 México, D. F. México

Abstract. The present paper deals with three independent subjects.
I. We show how for classical canonical transformation we can pass, with the help of Wigner distribution functions, from their representation U in the configurational Hilbert space to a kernel K in phase space. The latter is a much more transparent way of looking at representations of canonical transformations, as the classical limit is reached when $\hbar \to 0$ and successive quantum corrections are related with powers of $\hbar^{2n}, n = 1, 2, \ldots$.
II. We discuss the coherent states solution for a charged particle in a constant magnetic field and show that it is the appropriate one for getting the classical limit of the problem, *i.e.*, motion in a circle around any point in the plane perpendicular to the field and with the square of the radius proportional to the energy of the particle.
III. We show that it is possible to have just one equation involving n α's and β matrices to get relativistic wave equations that can have spins with values up to $\frac{n}{2}$. We then decompose the α's and β's into direct products of ordinary spin matrices and a new type of them that we call sign spin. The problem reduces then to that of the generators of a SU(4) group, entirely similar to the one in the spin-isospin theory of nuclear physics. For a free particle of arbitrary spin the symmetry group is actually the unitary symplectic subgroup of SU(4), *i.e.*, Sp(4). As the latter is isomorphic to O(5), we can characterize our states by the canonical chain O(5)$\supset$ O(4) $\supset$ O(3) $\supset$ O(2), and from it obtain the spin and mass content of our relativistic equation.

---

*) Member of El Colegio Nacional

INTRODUCTION

My contribution to the 1998 Latin American School of Physics, that took place in El Colegio Nacional, July 17-August 14, was originally going to deal with my recent work on "Supermultiplicity and Relativistic Problems".

During the School I noticed that the interests of the speakers and participants were more in the direction of the interaction of particles and radiation in ion-traps and related subjects. In these fields the use of Wigner distributions and of coherent states was mentioned and thus I decided to divide my contribution into three chapters, the first two dealing with the subjects mentioned at the beginning of this phrase and the last one being the one mentioned in the first paragraph.

I THE REPRESENTATION OF CANONICAL TRANSFORMATIONS IN TERMS OF WIGNER DISTRIBUTION FUNCTIONS

One of the subjects mentioned at the school was the evolution in time of a quantum mechanical system. The corresponding classical problem is a canonical transformation as the initial state can be characterized by a point (q,p) when the phase space is two dimensional, and these observables satisfy the Poisson bracket $\{q,p\} = 1$.

Under the action of a Hamiltonian H, the q,p transform into $Q(q,p,t), P(q,p,t)$, where $Q(q,p,0) = q$, $P(q,p,0) = p$ but the Poisson bracket remains $\{Q,P\} = 1$, as can derived immediately from the Hamiltonian equations of motion $(dq/dt) = (\partial H/\partial p), (dp/dt) = -(\partial H/\partial q)$.

Thus the development in time is a canonical transformation, but the concept is even more general as it allows us to transform given Hamiltonians to simpler ones.

Our objective though to is not the study of the classical canonical transformations in phase space but in their representation in quantum mechanics.

Thus it is interesting to discuss the representation of canonical transformations in the phase space version of quantum mechanics that was developed originally by Wigner [1], with the help of the distribution functions that now bear his name. We shall do this here, illustrating the analysis with the representations of some simple examples of canonical transformations.

We begin by recalling the definition of Wigner's distribution function $f(q,p)$ for a given wavefunction $\psi(q)$, i.e.,

$$f(q,p) = (\pi\hbar)^{-1} \int_{-\infty}^{\infty} <\psi|q+y><q-y|\psi> \exp\left(\frac{2ipy}{\hbar}\right) dy, \qquad (1)$$

where we use Dirac's notation $<q|\psi> = \psi(q), <\psi|q> = \psi^*(q)$, and restrict ourselves to a single degree of freedom. As is well known [1], the integration of

$f(q,p)$ with respect to p or q gives the probability density for the state $|\psi>$ in configuration or momentum space respectively.

We consider now a canonical transformation

$$Q = Q(q,p), \qquad P = P(q,p); \qquad \{Q,P\} \equiv \frac{\partial Q}{\partial q}\frac{\partial P}{\partial p} - \frac{\partial Q}{\partial q}\frac{\partial P}{\partial p} = 1 \qquad (2)$$

under which a *classical* distribution function $f(q,p)$ would of course transform into $F(q,p)$ given by

$$F(q,p) = f[Q(q,p), P(q,p)] \qquad (3)$$

In quantum mechanics though, the state $|\psi>$ transforms into [2–4]

$$|\psi> \to |\Psi> = U|\psi>, \qquad (4)$$

where U is an unitary operator, and thus

$$F(q,p) = (\pi\hbar)^{-1} \int_{-\infty}^{\infty} <\Psi|q+y><q-y|\Psi> \exp\left(\frac{2ipy}{\hbar}\right) dy$$

$$= (\pi\hbar)^{-1} \int\int_{-\infty}^{\infty}\int dz_+ dy dz_- <\psi|z_+><z_+|U^\dagger|q+y>$$

$$\times <q-y|U|z_-><z_-|\psi> \exp\left(\frac{2ipy}{\hbar}\right). \qquad (5)$$

Writing $z_\pm = q' \pm y'$ when it is associated with ψ, and $z_\pm = \bar{q}' \pm \bar{y}'$ when it is associated with U, and integrating over $q', y', \bar{y}', y$, with the extra factor

$$\delta(y' - \bar{y}') = (\pi\hbar)^{-1} \int_{-\infty}^{\infty} \exp\left(\frac{2ip'(y' - \bar{y}')}{\hbar}\right) dp', \qquad (6)$$

we immediately arrive to the relation

$$F(q,p) = \int_{-\infty}^{\infty}\int_{-\infty}^{\infty} dq' dp' f(q',p') <q'p'|K|qp>, \qquad (7)$$

in which the kernel K is given by

$$<q'p'|K|qp> = 2(\pi\hbar)^{-1} \int_{-\infty}^{\infty}\int_{-\infty}^{\infty} dy dy' <q+y'|U^\dagger|q+y>$$

$$\times <q-y|U|q'-y'> \exp\left(\frac{i(2py - 2p'y')}{\hbar}\right), \qquad (8)$$

where from (3) we expect that

$$\lim_{\hbar \to 0} <q'p'|K|qp> = \delta[q' - Q(q,p)]\delta[p' - P(q,p)]. \qquad (9)$$

To obtain K we must known U which, *in principle* [3], is determined by the equations [2–4,6]

$$Q(q,p) = UqU^\dagger, \qquad P(q,p) = UpU^\dagger, \tag{10}$$

where q, p are now quantum mechanical operators. As $U^\dagger U = I$, we can pass $U^\dagger$ to the left-hand side, and taking matrix elements between a bra $<q'|$ and a ket $|q''>$ obtain the equations [2,6,4,7]

$$Q\left(q', \frac{\hbar}{i}\frac{\partial}{\partial q'}\right) <q'|U|q''> = q'' <q'|U|q''>, \tag{11}$$

$$P\left(q', \frac{\hbar}{i}\frac{\partial}{\partial q'}\right) <q'|U|q''> = -\frac{\hbar}{i}\frac{\partial}{\partial q''} <q'|U|q''>, \tag{12}$$

Of course these equations only make sense when Q, P are well defined operators; otherwise, more sophisticated procedures need to be used [4,7]

We shall now consider two simple examples of canonical transformations. The first will be the linear one

$$Q = aq + bp, \qquad P = cq + dp; \qquad ad - bc = 1, \quad b > 0, \tag{13}$$

where the constants are all real. We have then [8],

$$<q'|U|q''> = (2\pi b)^{-1/2} \exp[(-i/2b)(aq'^2 - 2q'q'' + dq''^2)], \tag{14}$$

which satisfies equations (11) and (12) if we note from (13) that $c = (ad-1)/b$. Introducing (14) in (8) and using the relation $<q'|U^\dagger|q''> = <q''|U|q'>^*$ we immediately obtain

$$<q'p'|K|qp> = \delta[q' - (aq + bp)]\delta[p' - (cq + dp)]. \tag{15}$$

Thus for the linear canonical transformation the kernel coincides with its classical limit (9), in agreement with the fact that for this type of transformation Poisson and Moyal [9] brackets coincide.

In the second example we take Q as the Hamiltonian of a linear potential [10], and thus we have the canonical transformation

$$Q = (p^2/2m) - F_0 q, \qquad P = -p/F_0, \tag{16}$$

where m is the mass, F_0 a constant of the dimension of force, and $\{Q, P\} = 1$. Equation (11) leads then to an Airy function [10] and we also satisfy (12) and get a normalized [10] unitary representation if we write

$$<q'|U|q''> = A\Phi(-\xi), \tag{17}$$

$$\xi = [q' + (q''/F_0)](2mF_0/\hbar^2)^{1/3}, \tag{18}$$

$$A = (2m)^{1/3}\pi^{-1/2}F_0^{-1/6}\hbar^{-2/3}, \tag{19}$$

$$\Phi(\xi) = (4\pi)^{-1/2}\int_{-\infty}^{\infty} \exp\{i[(u^3/3) + u\xi]\}du. \tag{20}$$

Substituting (17) into (8) and making use of (20) we can show straightforwardly that for canonical transformation (16) the kernel K becomes

$$<q'p'|K|qp> = \left\{2\left(\frac{m}{\hbar^2 F_0^2}\right)^{1/3} \pi^{-1/2} \Phi\left[2\left(\frac{m}{\hbar^2 F_0^2}\right)^{1/3}\left(\frac{p^2}{2m} - F_0 q - q'\right)\right]\right\} \delta\left(p' + \frac{p}{F_0}\right). \tag{21}$$

We note first that when $\hbar \to 0$ the function Φ becomes [10] either very small or very rapidly oscillating except when $q' \simeq (p^2/2m) - F_o q$. Furthermore, from (20) we easily see that $\pi^{-1/2} \int_{-\infty}^{\infty} \Phi(x) dx = 1$. Thus the expression in $\{\}$ in (20) tends to a δ function in the limit $\hbar \to 0$, so that the kernel K goes into the classical limit (9), where Q and P are given by (16).

To see what the quantum corrections are, it is best to apply the K of (21) to a smooth distribution function $f(q,p)$, rather than study it directly. We choose

$$f(q,p) = (\pi ab)^{-1} \exp[-(q^2/a^2) - (p^2/b^2)]. \tag{22}$$

where from (1) we will have the relation $b = \hbar/a$ if f is obtained from a Gaussian state in configuration space. Again using (20) we obtain for the new distribution function $F(q,p)$ the expression

$$F(q,p) = f(Q,P) \sum_{k=0}^{\infty} \frac{\hbar^{2k} F_0^{2k} m^{-k}}{(2a)^{3k}} \left(\sum_{\substack{t=0 \\ 3k-t \text{ even}}}^{3k} \frac{(-1)^{(3k-t)/2}(2Q)^t (3k)!}{a^t t! [(3k-t)/2]! k! 3^k}\right), \tag{23}$$

where Q, P are given by (16). As indicated in (3), $f(Q,P)$ is the classical change in the distribution function due to the canonical transformation, and it will be the only one remaining in (23) if $\hbar \to 0$. Thus the terms associated with the higher powers of $\hbar^2$ indicate the successive quantum corrections to the distribution function when we perform the canonical transformation.

The examples discussed here are very specialized, but they clearly indicate the procedure to be followed in general. Among the more interesting cases where this formalism can be applied are those of non-bijective [6,3,4,7] canonical transformations. The concepts of ambiguity group and ambiguity spin used in the derivation of the representation U can then give interesting insights into the structure of phase space as a carrier of canonical transformation.

II COHERENT STATE SOLUTION FOR A CHARGED PARTICLE IN A CONSTANT MAGNETIC FIELD

We proceed to discuss the coherent state of a charged particle in a constant magnetic field [11,12]. We start denoting by $\vec{r}'', \vec{p}''$ the vectors of position and momentum of the electron in the usual units ($c.\ g.\ s.$); by m_o, e its mass and charge; by $\vec{A}''$ the vector potential associated with a constant magnetic field $\vec{B}$, i.e.,

$$\vec{A}'' = \tfrac{1}{2}(\vec{B} \times \vec{r}''). \tag{24}$$

The Hamiltonian for the electron is then

$$H'' = (2m_0)^{-1}[\vec{p}'' + (e/2c)(\vec{B} \times \vec{r}'')]^2. \tag{25}$$

Introducing the dimensionless coordinates and momenta

$$x_i' = (m_0 e^2/\hbar^2) x_i'', \qquad p_i' = (\hbar/m_0 e^2) p_i'', \quad i = 1, 2, 3, \tag{26}$$

and choosing x_3' parallel to the field $\vec{B}$ we then have

$$H' \equiv (\hbar^2/m_0 e^4) H'' = \tfrac{1}{2}[(p_1' - b^2 x_2')^2 + (p_2' + b^2 x_1')^2 + p_3'^2] = H_\perp' + H_\parallel', \tag{27}$$

where [13]

$$b^2 = (\hbar^3 B / 2 m_0^2 c e^3), \tag{28}$$

is now a dimensionless parameter.

The $H_\perp', H_\parallel'$ in (27) are, respectively, the Hamiltonians associated with the coordinates and momenta perpendicular and parallel to the magnetic field. Since $H_\parallel'$ is just a one dimensional free particle we concentrate on $H_\perp'$. Carrying out the canonical transformation of dilation

$$x_i = b x_i', \qquad p_i = b^{-1} p_i', \quad i = 1, 2, \tag{29}$$

we obtain

$$\begin{aligned} H_\perp = b^{-2} H_\perp' &= \tfrac{1}{2}(p_1^2 + p_2^2 + x_1^2 + x_2^2) + (x_1 p_2 - x_2 p_1) \\ &= (\eta_1 \xi^1 + \eta_2 \xi^2 + 1) + (1/i)(\eta_1 \xi^2 - \eta_2 \xi^1) \\ &= (\eta_+ \xi^+ + \eta_- \xi^- + 1) + (\eta_+ \xi^+ - \eta_- \xi^-) \\ &= (2\eta_+ \xi^+ + 1), \end{aligned} \tag{30}$$

where

$$\eta_i = (1/\sqrt{2})(x_i - i p_i), \quad \xi^i = (1/\sqrt{2})(x_i + i p_i), \qquad i = 1, 2, \tag{31}$$
$$\eta_\pm = (1/\sqrt{2})(\eta_1 \pm i \eta_2), \tag{32}$$
$$\xi^\pm = (1/\sqrt{2})(\xi^1 \mp i \xi^2), \tag{33}$$

are creation and annihilation operators satisfying the commutation relations

$$[\eta_i, \eta_j] = [\xi^i, \xi^j] = 0, \tag{34}$$
$$[\xi^i, \eta_j] = \delta_j^i, \tag{35}$$

where i, j are either 1,2 or $+, -$. Then $\eta_\pm, \xi^\pm$ have the differential operator form

$$\eta_\pm = \left(\frac{1}{\sqrt{2}}\right)\left(x_\pm - \frac{\partial}{\partial x_\mp}\right), \tag{36}$$

$$\xi_\pm = \left(\frac{1}{\sqrt{2}}\right)\left(x_\mp - \frac{\partial}{\partial x_\pm}\right), \tag{37}$$

where

$$x_\pm = (1/\sqrt{2})(x_1 \pm ix_2). \tag{38}$$

The Hamiltonian $(2\eta_+\xi^+ + 1)$ of (7) does not contain η_-, ξ^-, so we are completely free in the way that we can characterize the eigenstate of $H_\perp$, in relation to these last operators. In this section we shall consider eigenstates of both $(2\eta_+\xi^+ + 1)$ and ξ^-. Since the latter is a non-Hermitian operator [12] we shall designate its eigenvalue by the complex number z_o, whose real and imaginary parts we denote by x_{o1}, x_{o2}, respectively, i.e.,

$$z_0 = x_{01} + ix_{02}. \tag{39}$$

Since $[\xi^+, \eta_+] = 1, \xi^=$ is a number operator and we shall designate its eigenvalues by $\nu = 0, 1, 2, \ldots$.

We are thus looking for kets $|\nu z_0>$ satisfying

$$(2\eta_+\xi^+ + 1)|\nu z_0> = (2\nu + 1)|\nu z_0>, \tag{40}$$

$$\xi^-|\nu z_0> = z_0|\nu z_0>, \tag{41}$$

which we shall call coherent states [12] since they are precisely characterized by Eq. (41).

To obtain the $|\nu z_0>$ we start with a ground state $|0>$ obeying

$$\xi^\pm|0> = 0, \tag{42}$$

which from (37) is then given by

$$|0> = \pi^{-1/2}\exp(-x_+x_-) = \pi^{-1/2}\exp\left[-\tfrac{1}{2}(x_1^2 + x_2^2)\right].) \tag{43}$$

¿From the commutation relations (34) and (35) we see that to obtain the eigenvalue ν for $\eta_+\xi^+$ we need to apply η_+^ν to $|0>$, while to obtain the eigenvalue z_o for ξ^- we need to apply $\exp(z_0\eta_-)$ to $|0>$. Thus the *normalized* coherent state $|\nu z_0>$ is given by

$$|\nu z_0> = \left[(\nu!)^{-1/2}\eta_+^\nu\right]\left[\exp(-\tfrac{1}{2}|z_0|^2)\exp(z_0\eta_-)\right]|0>, \tag{44}$$

where $|z_0|^2 = (x_{01}^2 + x_{02}^2)$ and $\exp(z_0\eta_-))$ should be understood as the exponential series in powers of $(z_0\eta_-)$.

From (36) and (43) we see that

$$\eta_-|0> = \sqrt{2}x_-|0> \tag{45}$$

and thus in (44) we can replace η_- by $\sqrt{2}x_-$ to obtain

$$|\nu z_0> = (\pi\nu!)^{1/2}\exp[-\tfrac{1}{2}(x_{01}^2 + x_{02}^2)]\eta_+^\nu \exp[(\sqrt{2}z_0 - x_+)x_-]|0>. \tag{46}$$

Finally, applying ν times η_+, we obtain

$$|\nu z_0> = (\pi\nu!)^{-1/2}[(x_1 - x_{01}) + i(x_2 - x_{02})]^\nu \exp\{-\tfrac{1}{2}[(x_1 - x_{01})^2 + (x_2 - x_{02})^2]\}$$
$$\times \exp[i(x_{02}x_1 - x_{01}x_2)]. \tag{47}$$

This state is normalized and besides, we obtain

$$<\nu z_0|x_i|\nu, z_0> = x_{0i}, \quad i = 1, 2, \tag{48}$$
$$<\nu z_0|(x_1 - x_{01})^2 + (x_2 - x_{02})^2|\nu z_0> = (\nu + 1). \tag{49}$$

Thus the coherent eigenstate $|\nu z_0>$ of the Hamiltonian $H_\perp$ is centered at the point x_{01}, x_{02} and its mean-square radius is given by $(\nu + 1)$.

All the states $|\nu z_0>$ of fixed ν and arbitrary z_0 correspond to the energy $2\nu + 1$ and thus for coherent states the accidental degeneracy is related to the arbitrariness in z_0.

To understand this result more fully we first consider the probability density for these states, which from (46) is given by

$$|<x_1, x_2|\nu z_0>|^2 = (\pi\nu!)^{-1}\rho^{2\nu}\exp(-\rho^2), \tag{50}$$

where

$$\rho^2 \equiv (x_1 - x_{01})^2 + (x_2 - x_{02})^2. \tag{51}$$

The maximum for this probability density occurs at $\rho^2 = \nu$ and there it takes the value

$$|<x_1, x_2|\nu z_0>|^2_{\max} = (\pi\nu!)^{-1}\nu^\nu e^{-\nu} \simeq \pi^{-1}(2\pi\nu)^{1/2}, \tag{52}$$

in which the right-hand side is valid for large ν when we use Stirling's formula [13].

In the classical limit, i.e., when $\nu \geq 1$, the probability density is concentrated in a ring of radius $\nu^{1/2}$ and width Δ where the latter can be estimated by considering that the product of $(2\pi\nu^{1/2}\Delta)$ with the maximum height (52) of the probability density should give 1, since the wave function (46) is normalized. For $\nu \geq 1$ we thus have

$$(2\pi\nu^{1/2}\Delta)[\pi^{1/2}] = (2/\pi)^{1/2}\Delta = 1. \tag{53}$$

The value of Δ in the units given by (26) and (29) is then $(\pi/2)^{1/2}$, which is of the order of 1, while in the classical limit, as $\nu \gg 1$, the radius of the ring is also $\nu^{1/2} \gg 1$ and thus is much larger than the width Δ. The coherent wavefunctions then represent the states associated with the classical circular orbits for particles of definite energy in a constant magnetic field. The accidental degeneracy of the quantum problem just reflects the classical fact that the centers of these circular orbits can be placed at any point in the plane perpendicular to the field.

The eigenstates and eigenvalues of the Hamiltonian $H_\perp$ of (30) could be found more directly if we used the polar coordinates r, φ as then

$$H_\perp = \tfrac{1}{2}(-\frac{1}{r}\frac{\partial}{\partial r}r\frac{\partial}{\partial r} + \frac{1}{r^2}\frac{\partial^2}{\partial \varphi^2}) + \frac{1}{i}\frac{\partial}{\partial \varphi} \qquad (54)$$

from which we see immediately that the eigenkets have the form

$$|nm> = R_{n|m|}(r)\exp(im\varphi), \quad n = 0, 1, 2, \ldots, \quad m = 0, \pm 1, \pm 2, \ldots \qquad (55)$$

with R being the radial part expressible in terms of Laguerre polynomials, while the eigenvalues have the form

$$E_{nm} = (2n + |m| + 1) + m, \qquad (56)$$

where the round bracket corresponds to the energy of a two dimensional harmonic oscillator

The states $|nm>$ of (55) give no indication of the character of the classical limit of the problem that was mentioned above. We note though that $|nm>$ is infinitely or finitely degenerate depending on wether m is negative or positive. Thus it is, in principle, possible to reconstruct the state $|\nu z_0>$ of (46) int terms of an infinite number of states $|nm>$, but this would be a very laborious way to recover a state from which we can obtain the classical limit.

Thus the importance of solving quantum mechanical problems in terms of coherent states, as this allows a more direct approach to the classical picture.

III SUPERMULTIPLICITY AND THE FREE RELATIVISTIC PARTICLE OF ARBITRARY SPIN

III.A Introduction and Summary

The correct equation for a free relativistic particle of spin $\tfrac{1}{2}$ was proposed by Dirac. Almost immediately there were efforts to obtain the corresponding equations for higher spins. One of the more successful ones was proposed by Bhabha [14,15] who considered the equation

$$(\Gamma^\nu p_\nu + M)\psi = 0 \qquad (57)$$

where M is a constant and Γ are appropriate matrices that may be associated with spins up to $\frac{n}{2}$, where n is an arbitrary integer. *In the following we shall refer to Eq. (57) as the Bhabha equation.*

We shall proceed to show in the next section an elementary way [16,17] in which the matrices Γ^ν can be derived with the property that they behave as a four vector under Lorentz transformation. As p_ν is also a four vector, and in (57) repeated indices are summed over the values $\nu = 0, 1, 2, 3$, we conclude that equation of the form (57) that we shall derive is Lorentz invariant.

In section III.C we rewrite the Γ^ν in the terms of matrices that are direct products of the ones associate with the ordinary spin and a new, but similar, concept we call sign spin [18]. Thus Eq. (57) can be expressed in terms of the generators of an SU(4), just as happens in nuclear physics with the spin-isospin states [19]. As the latter imply the presence of supermultiplicity that is the reason for the appearance of this word in the title. In fact the symmetry group of the relativistic problem turns out to be the unitary symplectic subgroup Sp(4) of SU(4).

In section III.D we use the fact that the groups Sp(4) and O(5) are isomorphic, to carry our analysis in terms of the latter, where can use the canonical chain O(5)⊃O(4)⊃O(3)⊃ O(2) to characterize our states.

In section III.E we discuss the matrix representation of the relevant generators of O(5) and write Eq. (57) in finite matrix form. From the latter we can obtain the energy eigenvalue E of our equation as function of the wave number k as well as the irrep $(n_1 n_2)$ of the O(5) group. It turns out to be much simpler to consider the case when $k = 0$, so E becomes the rest mass M, and the value of the latter as function of the irreps $(n_1 n_2)$ of O(5) and the spin s, are given in Table 1. The results of the table are still valid when $k \neq 0$, but then the mass M must be replaced by $(E^2 - k^2)^{1/2}$.

In conclusion we show that concepts of supermultiplicity allow a simple discussion of the relativistic equations involving spins higher than $\frac{1}{2}$.

III.B A linear equation in p_ν for a particle with arbitrary spins that is Lorentz invariant

We start with the well known proof of the Lorentz invariance of the ordinary Dirac equation so as to later extend it to problem we are interested in. Thus we have

$$(\gamma^\nu p_\nu + 1)\psi = 0 \tag{58}$$

where the index $\nu = 0, 1, 2, 3$ and when it is repeated it means a sum over the values indicated. Throughout we shall use units in which

$$\hbar = m = c = 1 \tag{59}$$

and the 4×4 matrices γ^μ are given by

$$\gamma^i = \begin{pmatrix} 0 & \sigma_i \\ -\sigma_i & 0 \end{pmatrix} \quad i = 1,2,3; \qquad \gamma^0 = \begin{pmatrix} I & 0 \\ 0 & -I \end{pmatrix}. \tag{60}$$

where $\sigma_i, i = 1, 2, 3$, are 2×2 Pauli spin matrices.

As p_ν is a four vector the Eq. (58) will be Lorentz invariant if the γ^ν, $\nu = 0, 1, 2, 3$ transform also as a four vector under this operation, which implies the existence of 4×4 matrix $\mathcal{U}$ such that

$$\gamma'^\nu = a^\nu_\mu \gamma^\mu = \mathcal{U}\gamma^\nu \mathcal{U}^{-1} \tag{61}$$

where $A \equiv \|a^\nu_\mu\|$ is a Lorentz transformation.

The existence of such a matrix $\mathcal{U}$ is given in many places [17] but for completness we derive it in the appendix.

We now consider n equations of the type (58) distinguished by the fact that we have $\gamma^\nu_r, p_{\nu r}, r = 1, 2, \ldots n$ and sum them making all four momenta equal to get the equation

$$\left(\Gamma^\nu p_\nu + n\right)\psi = 0 \tag{62}$$

where

$$\Gamma^\nu = \sum_{r=1}^n \gamma^\nu_r \tag{63}$$

with γ^ν_r being the direct product of 4×4 matrices

$$\gamma^\nu_r = I \otimes I \otimes \ldots \otimes I \otimes \gamma^\nu \otimes I \ldots \otimes I \tag{64}$$

with γ^ν in the r position where the σ_i in it is replaced by σ_{ir}. Because the $\sigma_{ir}, r = 1, \ldots, n$, this equation can represent particles with spin going from $\frac{n}{2}, \frac{n}{2} - 1, \ldots \frac{1}{2}$ or 0.

The Eq. (62) is Lorentz invariant because if we introduce the direct product matrix

$$U \equiv \mathcal{U}_1 \otimes \mathcal{U}_2 \otimes \ldots \otimes \mathcal{U}_r \otimes \ldots \otimes \mathcal{U}_n, \tag{65}$$

where $\mathcal{U}_r$ is given as in the appendix by replacing σ_i by σ_{ir} we immediately see that

$$\Gamma'^\nu = a^\nu_\mu \Gamma^\mu = U\Gamma^\nu U^{-1} \tag{66}$$

Thus (62) have the Lorentz invariance and spin properties of the Bhabha equation and we shall proceed to show, first going through its supermultiplet formulation, that it also invariant under an O(5) group.

III.C The supermultiplet form of the Bhabha equation

To achieve the object indicated in the title of this section we first have to review some results of reference [18], but now as applied to the γ^ν, $\nu = 0, 1, 2, 3$ matrices.

We start by introducing two types of spin vectors, the ordinary one and what we have called the sign spin, which have the same mathematical form but will be distinguish here by round and square brackets respectively [18]

$$\hat{I} = \begin{pmatrix} 1 & 0 \\ 0 & 1 \end{pmatrix}, \quad s_1 = \frac{1}{2}\begin{pmatrix} 0 & 1 \\ 1 & 0 \end{pmatrix}, \quad s_2 = \frac{1}{2}\begin{pmatrix} 0 & -i \\ i & 0 \end{pmatrix}, \quad s_3 = \frac{1}{2}\begin{pmatrix} 1 & 0 \\ 0 & -1 \end{pmatrix}, \quad (67)$$

$$\check{I} = \begin{bmatrix} 1 & 0 \\ 0 & 1 \end{bmatrix}, \quad t_1 = \frac{1}{2}\begin{bmatrix} 0 & 1 \\ 1 & 0 \end{bmatrix}, \quad t_2 = \frac{1}{2}\begin{bmatrix} 0 & -i \\ i & 0 \end{bmatrix}, \quad t_3 = \frac{1}{2}\begin{bmatrix} 1 & 0 \\ 0 & -1 \end{bmatrix}. \quad (68)$$

From (60) and (67,68) it is immediately clear that the γ^ν can be expressed as the direct products [18]

$$\gamma^j = i4 s_j \otimes t_2, \; j = 1, 2, 3; \quad \gamma^0 = 2\hat{I} \otimes t_3 \quad (69)$$

We can now add an index $r = 1, 2, \ldots n$, to all these matrices interpreting them in the direct product form (64) and we immediately see that Eq. (62) takes the form

$$\left\{ \sum_{r=1}^{n}[4i(s_{jr} \otimes t_{2r})p_j] + \sum_{r=1}^{n}[2(\hat{I} \otimes t_{3r})p_0] + n \right\}\psi = 0 \quad (70)$$

where repeated latin indices (i, j, k) are summed over their values 1,2,3.

Now we define

$$S_i = \sum_{r=1}^{n}(s_{ir} \otimes \check{I}), \quad R_{ij} = \sum_{r=1}^{n}(s_{ir} \otimes t_{jr}), \quad T_j = \sum_{r=1}^{n}(\hat{I} \otimes t_{jr}); \quad i, j = 1, 2, 3 \quad (71)$$

and, as we indicated in reference [18], the 15 operators close under commutation and correspond to the SU(4) Lie algebra.

Using the definitions (71) the Bhabha equation (70) can be written as

$$\{4i R_{j2} p_j + 2 T_3 p_0 + n\} \psi = 0. \quad (72)$$

As only R_{i2}, T_3 appear in the equation we may assume that it admits a smaller symmetry group than SU(4). In fact we see from the commutation relations given in reference [18], that the ten operators

$$S_i, R_{i1}, R_{i2}, T_3, \quad i = 1, 2, 3, \quad (73)$$

close under commutation as

$$\left[S_i, S_j\right] = i\epsilon_{ijk}S_k, \quad \left[S_i, R_{j1}\right] = i\epsilon_{ijk}R_{j1},$$

$$\left[S_i, R_{j2}\right] = i\epsilon_{ijk}R_{k2}, \quad \left[T_3, R_{j1}\right] = iR_{j2}, \quad \left[T_3, R_{j2}\right] = -iR_{j1}, \quad (74)$$

$$\left[R_{i1}, R_{j2}\right] = \frac{i}{4}T_3 \delta_{ij}.$$

Thus the ten operators of (73) form a Lie algebra which clearly is a subalgebra of SU(4) and in fact is the unitary symplectic algebra Sp(4) whose Casimir operators commute with the operators in (72) and thus is its symmetry Lie algebra.

As will be discussed in the next section Sp(4) is isomorphic to O(5) and thus we get the symmetry Lie algebra that Bhabha derived by a very different procedure.

III.D The O(5) symmetry algebra of the Bhabha equation

The generators of an orthogonal Lie algebra of dimension d are given by antisymmetric operators $\wedge_{m,m'} = -\wedge_{m'm}$ where $m, m' = 1, 2, \ldots d$, and thus there are $(d/2)(d-1)$ of them satisfying the commutations relations [21]

$$\left[\wedge_{mm'}, \wedge_{nn'}\right] = i\left[\delta_{m'n}\wedge_{n'm} + \delta_{mn'}\wedge_{nm'} + \delta_{mn}\wedge_{m'n'} + \delta_{m'n'}\wedge_{mn}\right] \tag{75}$$

Comparing them with the commutation relations (74) we easily see that when $d = 5$ the $\wedge_{mm'}$ with $m < m'$ (to avoid the repetition due to the antisymmetry) are correlated with $S_i, R_{i1}, R_{i2}, T_3; i = 1, 2, 3$ in the following way

$$\begin{aligned}
\wedge_{12} &= S_3, & \wedge_{14} &= 2R_{11}, & \wedge_{15} &= 2R_{12} \\
\wedge_{13} &= -S_2, & \wedge_{24} &= 2R_{21}, & \wedge_{25} &= 2R_{22}, & \wedge_{45} &= T_3 \\
\wedge_{23} &= S_1, & \wedge_{34} &= 2R_{31}, & \wedge_{35} &= 2R_{32}
\end{aligned} \tag{76}$$

The O(5) has the following chains of subgroups O(5)⊃ O(4)⊃ O(3)⊃ O(2) whose generators, in terms of the operators appearing in the commutation rules (74), can be selected as

$$\begin{array}{llll}
10 & S_i, R_{i1}, R_{i2}, T_3 \text{ or } & \wedge_{12}, \wedge_{13}, \wedge_{23}, \wedge_{i4}, \wedge_{i5}, \wedge_{45}, i=1,2,3 & O(5) \\
6 & S_i, R_{i2} \text{ or } & \wedge_{12}, \wedge_{13}, \wedge_{23}, \wedge_{i5}, i=1,2,3 & O(4) \\
3 & S_i \text{ or } & \wedge_{12}, \wedge_{13}, \wedge_{23} & O(3) \\
1 & S_3 \text{ or } & \wedge_{12} & O(2)
\end{array} \tag{77}$$

where on the left side we give the number of generators and on the right the group in question, with the generators expressed both in the supermultiplet notation $S_i, R_{ij}, T_3; i, j = 1, 2, 3$ and in the orthogonal one $\wedge_{mm'}, m < m', m = 1, 2, 3, 4, 5$.

We note now that in the supermultiplet notation the Bhabha equation is given by (72), so using the relations (76) we can write it also in the form

$$\left[2i\wedge_{i5} p_i + 2\wedge_{45} p_0 + n\right]\psi = 0 \tag{78}$$

It is with this equation we want to deal with but with a small modification that would allow us to make use of a simple form of the matrix representation of the generators of orthogonal groups that we require in (78). For this purpose we note that the transposition (79) can be represented by the 5×5 orthogonal matrix

$$\begin{pmatrix} 1 & 0 & 0 & 0 & 0 \\ 0 & 1 & 0 & 0 & 0 \\ 0 & 0 & 1 & 0 & 0 \\ 0 & 0 & 0 & 0 & 1 \\ 0 & 0 & 0 & 1 & 0 \end{pmatrix} \qquad (79)$$

and thus is an element of the group O(5).

As O(5) is the symmetry group of Eq. (78), we can apply to it (79) and get a completely equivalent equation that now has the form

$$H\psi \equiv \left[2i \sum_{q=-1}^{1} (-1)^q \wedge_{q4} p_{-q} - 2 \wedge_{45} p_0 + n \right] \psi = 0 \qquad (80)$$

where we also replaced the scalar product in cartesian coordinates $i = 1, 2, 3$ by the spherical ones $q = 1, 0, -1$, and denote the operator in the square bracket by H.

We shall proceed to discuss this equation by first getting the matrix elements, in an appropriate basis, of $\wedge_{q4}, q = 1, 0, -1$ and $\wedge_{45}$.

III.E Matrix elements of the generators $\wedge_{45}, \wedge_{q4}, q = 1, 0, -1$ in a basis of irreps in the chain O(5)⊃ O(4)⊃ O(3) ⊃ O(2)

As is well known [22,23] the irreps of $O(2k+1)$ and $O(2k)$ are characterized by partition involving k numbers that can be integer or seminteger and non-negative, except for the last one in the even case which sometimes can be negative.

Rather than discussing the general theory analyzed in references [22] and [23], we shall restrict our analysis to the chain of orthogonal groups that appear in the title of this section where the irreps will be denoted as follows:

$$\begin{aligned} &O(5) \ ; \ n_1, n_2 \\ &O(4) \ ; \ m_1, m_2 \\ &O(3) \ ; \ s \\ &O(2) \ ; \ \sigma \end{aligned} \qquad (81)$$

As O(5) is a symmetry group of the operator (80) the n_1, n_2 are integrals of motion of the problem and remain fixed. Turning now our attention to O(4), m_1, m_2 are restricted by the inequalities [22,23]

$$n_1 \geq m_1 \geq n_2 \geq |m_2|. \qquad (82)$$

For O(3) we have the single number s restricted by

$$m_1 \geq s \geq |m_2|. \qquad (83)$$

Finally σ of O(2) is restricted by $|\sigma| \leq s$ which implies that is given by [24]

$$\sigma = s, s-1, \ldots, -s+1, -s \tag{84}$$

as all the values indicated can only change by one unit at a time within the limits indicated in the inequalities. We note then that the integer or seminteger character of the representation (n_1, n_2) of O(5) propagates to all of its subgroups.

The kets for the spin part of O(5) $\supset$ O(4)$\supset$ O(3) $\supset$ O(2) chain of groups, can be denoted by

$$\left| \begin{array}{c} n_1 n_2 \\ m_1 m_2 \\ s \\ \sigma \end{array} \right\rangle \tag{85}$$

and the matrix elements of $\wedge_{45}, \wedge_{q4}$ with respect to them have been calculated in references [25,26]. Before giving them explicitly here, we note that $\wedge_{q4}$ is a Racah tensor of order 1 with respect to the O(3) group and, in particular, $\wedge_{04}$ corresponds to the component 0 of this tensor so we have by the Wigner–Eckart theorem that [24]

$$\left\langle \begin{array}{c} n_1 n_2 \\ m_1' m_2' \\ s' \\ \sigma' \end{array} \right| \wedge_{04} \left| \begin{array}{c} n_1 n_2 \\ m_1 m_2 \\ s \\ \sigma \end{array} \right\rangle = \langle s\sigma, 10 | s'\sigma' \rangle \left\langle \begin{array}{c} n_1 n_2 \\ m_1' m_2' \\ s' \end{array} \right\| \wedge_4 \left\| \begin{array}{c} n_1 n_2 \\ m_1 m_2 \\ s \end{array} \right\rangle, \tag{86}$$

where $\langle | \rangle$ is a standard O(3) Clebsch–Gordan coefficient. Thus for $\wedge_{q4}$ we need only the reduced matrix element on the right hand side of (86), and its explicit value, together with that of $\wedge_{45}$, is given below [25,26]

$$\left\langle \begin{array}{c} n_1 n_2 \\ m_1' m_2' \\ s \end{array} \right| \wedge_{45} \left| \begin{array}{c} n_1 n_2 \\ m_1 m_2 \\ s \end{array} \right\rangle =$$

$$-\frac{i}{2}\delta_{m_1',m_1+1}\delta_{m_2',m_2}A - \frac{i}{2}\delta_{m_1',m_1}\delta_{m_2',m_2+1}B$$

$$+\frac{i}{2}\delta_{m_1',m_1-1}\delta_{m_2',m_2}C + \frac{i}{2}\delta_{m_1',m_1}\delta_{m_2',m_2-1}D$$

$$A = \sqrt{\frac{(m_1-s+1)(m_1+s+2)(n_1-m_1)(n_1+m_1+3)(m_1-n_2+1)(m_1+n_2+2)}{(m_1+m_2+1)(m_1+m_2+2)(m_1-m_2+1)(m_1-m_2+2)}}$$

$$B = \sqrt{\frac{(s-m_2)(s+m_2+1)(n_2-m_2)(n_2+m_2+1)(n_1-m_2+1)(n_1+m_2+2)}{(m_1+m_2+2)(m_1+m_2+1)(m_1-m_2)(m_1-m_2+1)}}$$

$$C = \sqrt{\frac{(s+m_1+1)(m_1-s)(n_1-m_1+1)(n_1+m_1+2)(m_1-n_2)(m_1+n_2+1)}{(m_1+m_2)(m_1+m_2+1)(m_1-m_2)(m_1-m_2+1)}}$$

$$D = \sqrt{\frac{(s - m_2 + 1)(s + m_2)(n_2 - m_2 + 1)(n_2 + m_2)(n_1 - m_2 + 2)(m_2 + n_1 + 1)}{(m_1 + m_2)(m_1 + m_2 + 1)(m_1 - m_2 + 2)(m_1 - m_2 + 1)}} \tag{87}$$

$$\left\langle \begin{matrix} n_1 n_2 \\ m_1 m_2 \\ s' \end{matrix} \middle\| \wedge_4 \middle\| \begin{matrix} n_1 n_2 \\ m_1 m_2 \\ s \end{matrix} \right\rangle$$

$$= -i\sqrt{\frac{(m_1 - s)(m_1 + s + 2)(s - m_2 + 1)(s + m_2 + 1)}{(2s + 3)(s + 1)}} \delta_{s', s+1}$$

$$+ \frac{(m_1 + 1)m_2}{\sqrt{s(s+1)}} \delta_{s', s} + i\sqrt{\frac{(m_1 - s + 1)(m_1 + s + 1)(s - m_2)(s + m_2)}{(2s - 1)s}} \delta_{s', s-1} \tag{88}$$

We can now turn back to our equation (80) and see that the operator appearing it commutes with the components $p_q, q = 1, 0, -1$ of the momentum so they are integrals of motion that we can denote by the constants k_q. Furthermore, without loss of generality we can select our coordinate axis so the vector $\vec{k}$ is along the third of them so $k_0 \equiv k, k_{\pm 1} = 0$. The p_0 is also an integral of motion and we can replace it by a numerical constant we call E as in the units (59) it would be the energy. If we now consider the numerical finite and hermitian matrix

$$\left\| 2i\langle s\sigma, 10|s'\sigma\rangle k \left\langle \begin{matrix} n_1 n_2 \\ m'_1 m'_2 \\ s' \end{matrix} \middle\| \wedge_4 \middle\| \begin{matrix} n_1 n_2 \\ m_1 m_2 \\ s \end{matrix} \right\rangle - 2E \left\langle \begin{matrix} n_1 n_2 \\ m'_1 m'_2 \\ s' \end{matrix} \middle| \wedge_{45} \middle| \begin{matrix} n_1 n_2 \\ m_1 m_2 \\ s \end{matrix} \right\rangle \delta_{ss'} + \Delta \right\|$$

$$\Delta \equiv n \delta_{m'_1 m_1} \delta_{m'_2 m_2} \delta_{s's} \tag{89}$$

where the indices $m_1, m_2, s, m'_1, m'_2, s'$ vary according to the rules (82), (83), (84) and σ is diagonal, we see that if we equate its determinant to 0 we will get a secular equation that gives several expressions of E as function k, n_1, n_2, σ and n.

We can determine E in a much simpler way than the one indicated in the previous paragraph if we carry out a similarity transformation of the operator (80) along the lines used by Foldy and Wouthuysen [27]. We shall not present this here but rather limit ourselves to the analysis of (89) when $k = 0$, i.e., when E is replaced by the rest mass M, and (89) becomes the matrix

$$\left\| -2M \left\langle \begin{matrix} n_1 n_2 \\ m'_1 m'_2 \\ s \end{matrix} \middle| \wedge_{45} \middle| \begin{matrix} n_1 n_2 \\ m_1 m_2 \\ s \end{matrix} \right\rangle + n \delta_{m'_1 m_1} \delta_{m'_2 m_2} \delta_{s's} \right\|. \tag{90}$$

As $\wedge_{45}$ is an Hermitian operator its matrix can be diagonalized giving real eigenvalues. In fact as the permutation taking us from (12345) to (54321) is an element of the O(5) group, we see that the eigenvalues of $\wedge_{45}$ are the same as those of $\wedge_{12}$ and the latter, given by σ, are integers or semiintegers positive or negative restricted by the relations (82), (83) and (84). The eigenvalues of $2\Lambda_{45}$ are all integers and we shall denote them by 2λ. Thus the possible values of the masses are given by

$$M_\lambda = \frac{n}{2\lambda} \qquad (91)$$

It is of great interest to know the different values of spin and mass that the Bhabha particle can take. For the possibles values of the spin we have to use the inequalities (82) and (83). For the masses associated with a given irrep $(n_1 n_2)$ of $O(5)$ and definite spins, we have to find the eigenvalues λ of the matrix associated with Λ_{45}, whose elements are given by (87), and then use (91). In Table 1 we give these values up to $n = 4$ and, as we indicated in reference [27], they remain valid even if $k \neq 0$.

III.F Conclusion

We have shown the usefulness of the supermultiplicity approach to relativistic problems through the discussion of the Bhabha equation for a free particle with arbitrary spin.

If we have an external field, characterized by a four vector $A_\nu, \nu = 0, 1, 2, 3$, all we have to do is to replace p_ν by $p_\nu - A_\nu$ in Eq. (80) and all the following ones in which p_ν appears. The p_ν ceases then to be an integral of motion and we cannot arrive at a finite matrix expression such as Eq. (89). It is then necessary to follow a variational or perturbation procedure based on an appropriate complete set of states. One aspect of this procedure is discussed in another article [30].

APPENDIX

To determine the 4×4 matrix $\mathcal{U}$ satisfying (61) we first note that if the Lorentz transformation A is only a rotation, $\mathcal{U}$ will consist of two blocks in the diagonal associated with the spinor representation of the rotation, i.e., $D^{\frac{1}{2}}$.

We need then restrict ourselves only to boosts, and as they can be reduced by rotations to boosts in the $\nu = 3$ direction we only need to consider the $\mathcal{U}$ corresponding to the Lorentz transformation

$$A = \begin{pmatrix} c & 0 & 0 & s \\ 0 & 1 & 0 & 0 \\ 0 & 0 & 1 & 0 \\ s & 0 & 0 & c \end{pmatrix} \qquad (92)$$

where $c = cosh\delta, s = sinh\delta$ and δ being an arbitrary real parameter.

We now note that in terms of the γ's the spin operator [20] takes the form

$$S^{\mu\nu} = \tfrac{1}{2}(\gamma^\mu \gamma^\nu - \gamma^\nu \gamma^\mu) \qquad (93)$$

and thus for the particular Lorentz transformations in (92) we have that the matrix corresponding to an infinitesimal transformation in the γ space is given by

$$S^{03} = \frac{i}{2}(\gamma^0 \gamma^3 - \gamma^3 \gamma^0) = \frac{i}{2}\begin{pmatrix} 0 & \sigma_3 \\ \sigma_3 & 0 \end{pmatrix} \qquad (94)$$

and thus for the finite one corresponding to A in (92) we have

$$\mathcal{U} = \exp(i\delta S^{03}) = \begin{pmatrix} \bar{c}I & \bar{s}\sigma_3 \\ \bar{s}\sigma_3 & \bar{c}I \end{pmatrix} \qquad (95)$$

where we used the fact that $\sigma_3^2 = I$ is a 2×2 unit matrix and denote by $\bar{c}, \bar{s}$ the functions

$$\bar{c} = cosh(\delta/2); \quad \bar{s} = sinh(\delta/2) \qquad (96)$$

Thus we have shown the existence of a $\mathcal{U}$ related to A and, in particular, if the boost is in an arbitrary directions given by the unit vector $\vec{b}$ instead of $\nu = 3$, then obviously $\mathcal{U}$ becomes

$$\mathcal{U} = \begin{pmatrix} \bar{c}I & \bar{s}(\sigma \cdot \vec{b}) \\ \bar{s}(\sigma \cdot \vec{b}) & \bar{c}I \end{pmatrix} \qquad (97)$$

ACKNOWLEDGMENTS

Part of the material in this article was taken from the following publications Chapter I: García Calderón G. and Moshinsky M., *J: Phys. A: Math. Gen.* **13**, L185-188 (1980); Chapter II: Loyola G., Moshinsky M., Szczepaniak A., *Am. J. Phys.* **57**, 811 (1989); Chapter III: M. Moshinsky, Nikitin A.G., Sharma A. and Smirnov Yu. F., *J. Phys. A: Math. Gen.* **31**, 6095 (1998).

The author of the present article wishes to thank his collaborators in the elaboration of the material mentioned above.

REFERENCES

1. Wigner, E. P., *Phys. Rev.* **40**, 749 (1932).
2. Mello, P. A., and Moshinsky, M., *J. Math. Phys.* **16**, 2017 (1975).
3. Dirac, P. A. M., *The principles of Quantum Mechanics* Oxford: Oxford U.P., 1937.
4. Kramer, P., Moshinsky, M., and Seligman, T. H., *J. Math. Phys.* **19**, 683 (1978).
5. Moshinsky, M., and Seligman, T. H., *Ann. Phys.* (N. Y.) **120**, 402 (1979).
6. Moshinsky, M., and Seligman, T. H., *J. Phys. A: Math. Gen.* **114**, 243 (1978).
7. Moshinsky, M., and Seligman, T. H., *J. Phys. A: Math. Gen.* **12**, L135 (1979).
8. Moshinsky, M., and Quesne, C., *J. Math. Phys.* **12**, 1772 (1971).

9. Moyal, J. E., *Proc. Camb. Phil. Soc.* **45**, 99 (1949).
10. Landau, L. D., and Lifshitz, E. M., *Quantum Mechanics*, London: Pergamon, 1958, pp. 73-4.
11. Landau, L. D., and Lifshitz, E. M., *Quantum Mechanics* London: Pergamon, 1958, pp. 424-427.
12. Haken, H. A., *Light*, Amsterdam: North-Holland, 1981, Vol. 1, pp. 170-173.
13. Gradsteyn, I. S., and Ryzhik, I. M., *Tables of Integrals, Series and Products*, New York: Academic, 1965, pp. 937, formula 8.327.
14. Bhabha, H. J., *Rev. Mod. Phys.* **17**, 200 (1945).
15. Krajcik, R. A., and Nieto M. M. *Phys. Rev. D* **10**, 4049 (1974); **11**, 1442 (1975); **11**, 1459 (1975); **13**, 924 (1976); **14**, 418 (1977); **15**, 433 (1977).
    Loide, R. K., Ots, I., and Saar, R., *J. Phys. A: Math. Gen.* **39**, 4005 (1997).
16. Moshinsky, M., and del Sol Mesa, A., *J. Phys. A: Math. Gen.* **29**, 4217 (1996).
17. Moshinsky, M., Loyola, G., and Villegas, C., *J. Math. Phys.* **32**, 373 (1991).
18. Moshinsky, M., and Smirnov, Yu. F., *J. Phys. A: Math. Gen.* **29**, 6027 (1996).
19. Wigner, E. P., *Phys. Rev.* **51**, 106 (1937).
20. Kim, Y. S., and Noz, M. E., *Theory and applications of the Poincaré group*, Dordrecht: The Reider Publishing Co., 1986, pp. 69,70.
21. Moshinsky, M., *Group Theory and the many body problem*, New York: Gordon and Breach, 1968, pp. 36.
22. Gelfand, I. M., and Zetlin, M. L., *Dok. Akad. Nauk. USSR* **71**, 147 (1950) (In Russian).
23. Pang, S.C., and Hecht, K. T., *J. Math. Phys.* **8**, 1233 (1967).
24. Rose, M. E., *Elementary Theory of Angular Momentum*, New York: John Wiley and Sons, 1957, pp. 85-88.
25. Filippov, G. F., Ovcharenko, V. I., Smirnov, Yu. F., *Microscopic theory of collective excitations of atomic nuclei*, Kiev: Nauka Dumka, 1981, pp. 252-254 (in Russian).
26. Nikitin, A. G., and Tretynyk, V. V., *J. Phys. A: Math. Gen.* **28**, 1655 (1995).
27. Foldy, L. I., and Wouthuysen, S. A., *Phys. Rev.* **78**, 29 (1950).
28. Sharma, A., Smirnov, Yu. F., and Nikitin, A. G., "Mass and spin content of a free relativistic particle and the group reduction $Sp(4) \supset U(1) \otimes SU(2)$". To be published in *Rev. Mex. Fís.* **44**, (1998).
29. Moshinsky, M., Nikitin, A. G., Sharma, A., and Smirnov, Yu. F., "Analysis of relativistic particles through different chains of groups". To be published in *Rev. Mex. Fís.* **44**, (1998).
30. Moshinsky, M., and Sharma, A., *J. Phys. A. Math. Gen.* **31**, 397 (1998).

LIST OF PARTICIPANTS

AGUILAR PINEDA GABRIEL ELOY
UAM-Iztapalapa
Av. Michaoacán y Purísima s/n, Col. Vicentina

eloy@tonantzin.vam.mx

México

ALARCON CHAVEZ MARIANO
UMSNH Escuela Físico Matemáticas
U. Michoacán San Nicolas Hidalgo, Morelia

mach@itm1 ifm.umich.mx

México

ALARCON RONZON MARTIN
IF-UNAM

mar@graef.fciencias.unam.mx

México

ALBARRAN ARREGUIN JAIME
UAM-Iztapalapa

jalarr@sunserver.vaq.mx

México

ALCANTARA ONTIGOZE MARISOL
UNAM, Facultad de Ciencias

maricol@ce.ifisicam.unam.mx

México

ALBRECHT HERMANN
Universidad Simón Bolívar, Depto. de Química
A. P. 89000 Caracas

egalb@telcel.net.ve

Venezuela

ARRIETA CASTAÑEDA ALMA MIREYA
UAM-Iztapalapa

loc@xanum.uam.mx

México

AYALA RODRIGUEZ RAFAEL A.
UNAM, Instituto de Ciencias Nucleares

rayala@nuclecu.unam.mx

México

BASTIN THIERRY
University of Liège, Belgium
IPNE, B15 Sart Tilman b-4000, Liège

t.bastin@ulg.ac.be

Belgium

BENITEZ MARTINEZ FERNANDO
ICN-UNAM

benitez@nuclecu.unam.mx

México

BERNAL CASTAÑO URSULA
IF-UNAM

ursula@ce.ifisicam.unam.mx
México

BEREBICHEZ DEBORAH
IF-UNAM

deborah@fenix.ifisicacu.unam.mx
México

BERRONDO MANUEL
Brigham Young University
Dept. of Physics-N149, esc. .Provo Ut 84602

berrondo@byu.edu

USA

BUKOR ZELMIRA
UNAM-Instituto de Física

zelmira@fenix.ifisicam.unam.mx
México

CABRERA TRUJILLO REMIGIO
UAM-Iztapalapa
Av.Michoacan y Purisima s/n,D.F.

remigio@abaco.uam.mx

México

CAMACHO QUINTANA ABEL
UAM-Iztapalapa

abel@abaco.uam.mx
México

CANDIA SILVIA INES
Instituto Balseiro
Av. Dustillo km 5,300 Centro, Bariloche

candias@cab.cnea.edu.ar

Argentina

CASTELLANO OLGA
Universidad de Zulia, Maracaibo,
Grano de Oro, Depto de Química. Mod. 2.

olga@sinamaica.ciens.luz.ve
Venezuela

CASTRO BELTRAN HECTOR M
INAOE
Luis Erique Erro 1-72840, Tonantzintla, Pue.

hcastro@inaoep,.mx

México

CASTRO PEÑA JESÚS
UAM-Azcapotzalco

jdjcp@hp9000a1.uam.mx
México

CERVERA GOMORA PEDRO
UNAM, Facultad de Ciencias

cerverap@hotmail.com.mx
México

CETTO ANA MARIA
IF-UNAM

ana@fenix.ifisicacu.unam.mx
México

CORICHI ALEJANDRO
ICN-UNAM

corichi@nuclecu.unam.mx
México

CORTINA LOPEZ MANUEL ALFONSO cortina@fenix.ifisicacu.unam.mx
IF-UNAM México

DAGDUG LIMA LEONARDO dll@xanum.uam.mx
UAM-Iztapalapa México

DE FRANÇA SANTOS MARCELO P. mfranca@if.ufrj.br
Universidad Federal de Rio de Janeiro
Inst. de Fisica
Blocoa-Ct-Ilha do fundão, Rio de Janeiro Brasil

DE LA CRUZ GUTIERREZ MANUEL manuelastico@hotmail.com.mx
Universidad de Guadalajara, Jalisco México

DRIGO FILHO ELSO elso@df.ibille.unesp.br
Inst. de Biociencias, letras e ciencias exatas
UNESP
R. Cristovão Colombo, 2265, 15054-000.

DVOEGLAZOV VALERI V. valeri@cantera.reduaz.mx
Escuela de Fisica, UAZ México

DURDEVICH MICHO micho@matem.unam.mx
IM-UNAM México

ENTRALGO HERRERO ELIAS elias@isctn.edu.cu
Inst. Superior de Ciencia y Tecn.Nuclear
Plaza S. Allende Quinta de los Molinos Cuba

FERNANDEZ CABRERA DAVID JOSE david@fis.cinvestav.mx
CINVESTAV-Depto. de Fisica, IPN México

FUENTES GURIDI IVETTE ivette@ft.isicacu.unam.mx
IF-UNAM México

FUENTES Y MARTINEZ GILBERTO J.
UAM-Iztapalapa México

GARCIA CALDERON GASTON gaston@fenix.ifisicacu.unam.mx
IF-UNAM México

GARCIA GUTIERREZ JOSE A.
UNAM, Facultad de Ciencias México

GILLER MARIA
University of Lodz, Physics Department
Poland.

mgiller@zpk.u.lods.pl

GILLER STEFAN
University of Lodz, Physics Department

sgiller@krysia.uni.lodz.pl
Poland

GOIZ M. ALEJANDRO
Universidad Iberoamericana

México

GOMEZ GARCIA EDUARDO
UNAM, Centro de Instrumentos

cometa@mexred.net.mx
México

GONZALEZ GARCIA GERARDO
Escuela Superior de Física y Matemáticas
Inst. Politécnico Nacional s/n

México

GOMEZ-MONT CARLOS
CIMAT-Guanajuato

cgm@fractal.cimat.mx
México

GRABINSKY JAIME
UAM-Azcapotzalco

jaimeg@aztlan.uam.mx
México

GULZARI SEMIH
IF-UNAM

verda@internet.com.mx
México

GUTIERREZ MEDINA BRAULIO
U. of Texas, at Austin, Physics Dept.
RLM 5.208.

braulio@physics.utexas.edu

USA

HERNANDEZ CALDERON ISAAC
CINVESTAV-Depto. de Física, IPN

ihernand@fis.cinvestav.mx
México

HERNANDEZ DE LA PEÑA LEANDRO
ICN-UNAM

leandro@feynman.ifisicacu.unam.mx
México

HERNANDEZ DE LA PEÑA LISANDRO
ISCTN
Av. S. Allende Quinta de los Molinos
Plaza Cd. Habana

lisandro@isctn.edu.cu

Cuba

HERNANDEZ SALDAÑA HUGO
IF-UNAM

hugo@fenix.ifisicacu.unam.mx
México

HOJMAN SERGIO
Universidad de Chile, Santiago de Chile

shojman@abello.dic.uchile.cl
Chile

IZQUIERDO DE LA CRUZ ERICK
Escuela Superior de Física y Matemáticas
Unidad A. López Mateos, IPN

México

JIMENEZ ANGELES FELIPE
UAM–Iztapalapa

México

JIMENEZ HERRERA URIEL
UNAM, Facultad de Ciencias

jimenhu@servidor.unam.mx
México

KLIMOV ANDREI
Universidad de Guadalajara, Jal.

klimov@cencar.udg.mx
México

KOUZNETSOV DMITRI
UNAM, Centro de Instrumentos

kusnecov@aleph.cinstrum.unam.mx
México

LEMUS CASILLAS RENATO
ICN-UNAM

lemus@servidor.unam.mx
México

LEY KOO EUGENIO
IF-UNAM

eleykoo@fenix.ifisicacu.unam.mx
México

LICONA IBARRA MA. L
UAM-Iztapalapa

lrli@xanum.uam.mx
México

LOPEZ GONZALEZ JOSE LUIS
ICN-UNAM

lopezg@nuclecu.unam.mx
México

LOPEZ SANCHEZ ERICK JAVIER
FC-UNAM

lsej@minervaux.fciencias.unam.mx
México

LOPEZ SUAREZ ALEJANDRA
IF-UNAM

chipi@fenix.ifisicacu.unam.mx
México

LOYOLA GERARDO
Ohio University

loyola@ouhelios.edu
USA

LOZANO OCHOA ENRIQUE
ICN-UNAM

lozano@nuclecu.unam.mx
México

LUTTERBACK LUIZ GUILLERME
Universidad Federal de Rio de Janeiro
Inst. de Fisica
Blocoa-Ct-Ilha do fundão, Rio de Janeiro

lutter@if.ufrj.br

Brasil

MAN'KO MARGARITA
Lebedev Physical Inst.
117333, Moscow Lenisnky prospect 53, 117333

manko@lebedev.ru

Rusia

MARTINEZ GALICIA RICARDO
FC-UNAM

México

MENDEZ PEREZ SANTIAGO ANTONIA
IF-UNAM

México

MENDEZ VILLUENDAS EDUARDO
UAM-Azcapoltzalco

eduardo@fismol.uam.mx
México

MERA OLGUIN SILVIA
UAM-Iztapalapa

loc@xanum.uam.mx
México

MILMAN PEROLA R.V.
Universidade Federal do Rio de Janeiro

perola@if.ufrj.br
Brasil

MORALES GUZMAN D. DR.
UAM-Azcapotzalco

dmorales@fis.cinvestav.mx
México

MORALES RUIZ B. ALEJANDRO
Universidad de Sonora

bamr@fisica.uson.mx
México

MOYA CESSA HECTOR MANUEL
INAOE
AP. 51 y 216, 72000 Puebla, Pue.

México

NERI FAJARDO JAVIER
Escuela Normal Superior
Manuel Salazar 201 Col. El Rosario

hmmc@inaoep.mx

México

NORMAN BENJAMIN PABLO
IF-UNAM

benpablo@ft.ifisicacu.unam.mx

México

OCHOA ENRIQUE LOZANO
ICN-UNAM

lozano@nuclecu.unam.mx

México

OLIVEIRA TERRA CUNHA MARCELO
UFMG-Univ. Fed. de Minas Gerais
Dept. de Matematica–Campus Pampucha

tcunha@mat.ufmg.br

ORTEGA JIMENEZ ROBERTO
UAM-Azcapotzalco

México

ORTEGA MARTINEZ ROBERTO
UNAM, Centro de Instrumentos

roberto@aleph.cinstrum.unam.mx

México

ORTIZ CASTRO ANTONIO
CINVESTAV-IPN

aortiz@fis. cinvestav.mx

México

OZIEWICZ ZBIGNIEW
UNAM-FES-Cuautitlán

oziewicz@servidor.unam.mx

México

PADILLA RODAL ELIZABETH
ICN-UNAM

padilla@xochitl.nuclear.unam.mx

México

PATIÑO J. ERICK LEONARDO
IF-UNAM

leonardo@ft.ifisicacu.unm.mx

México

PONCE JUAREZ ROBERTO
UNAM, Facultad de Ciencias

pasivo@grex.cyberspace.org

México

PRADA ROJAS INGMAR AUGUSTO
IF-UNAM

ingmar@ft.ifisicam.unam.mx

México

QUEIPO RUIZ JOEL
Inst. Superior de Ciencia y Tecn.Nuclear
Plaza S. Allende, Quinta de los Molinos

jqueipo@isctn.edu.cu

Cuba

QUINTO SU PEDRO A.   qspa@minervaux.fciencias.unam.mx
FC-UNAM   México

RAMIREZ BOLAÑOS EDGAR HUGO   cverap@hotmail.com.mx
UNAM, Facultad de Ciencias   México

RAMIREZ MARTINEZ FERNANDO   feram@servidor.unam.mx
UNAM, Facultad de Ciencias   México

REBOIRO MARTA   reboiro@venus.fisica.unlp.edu.ar
UNLP-CONICET
Calle 47 y 115. La Plata, CC 67 (1900)   Argentina

RESENDIS ANTONIO OSBALDO   rean@xanum.uam.mx
UAM-Iztapalapa   México

RIBEIRO DE CARVALHO ANDRE R.   andre@if.ufrj.br
Universidad Federal de Rio de Janeiro
Depto. Fisica Matematica
Rio de Janeiro   Brasil

RIQUER VERONICA IVETTE   chloe@fenix.ifisicacu.unam.mx
IF-UNAM   México

ROA NERI JOSE ANTONIO E.   rnjae@h9000al.vam.mx
UAM-Azcapotzalco   México

ROMAN MORENO CARLOS JESUS   crm@labvis.unam.mx
UNAM, Centro Instrumentos   México

ROMERO IBARRA JOSE LUIS
Universidad de Guadalajara, Jal.   México

ROMERO ROCHIN VICTOR   romero@fenix.ifisicacu.unam.mx
IF-UNAM   México

ROSAS ORTIZ JOSE OSCAR   orosas@klander.fam.cie.uva.es
Universidad de Valladolid, Fac. de Ciencias   España

RUIZ VICENT ORELLANA LUIS E.   vicent@ce.ifisicam.unam.mx
IIMAS-Cuernavaca, Mor.   México

SANABRIA GOMEZ JOSE DAVID
CINVESTAV-IPN

sanabria@fis.cinvestav.mx
México

SANCHEZ ARELLANO ENRIQUE
UAM-Iztapalapa

enrique@tepetl.uam.mx
México

SANCHEZ PINEDA ANGELICA E.
UAM-Iztapalapa

clf@xanum.uam.mx
México

SANCHEZ TORRES YANET NORMA
UNAM, Facultad de Ciencias

lsej@minervaux.fciencias.unam.mx
México

SANCHEZ VILLICAÑA VICENTE
INAOE
Luis Enrique Erro 1, Tonanzintla, Pue.

vsanchez@inaoep.mx

México

SANDOVAL R. MA. DE LOS ANGELES
UNAM, Facultad de Ciencias

México

SANTOS RODRIGUEZ ELI
CINVESTAV-IPN

eli@fis.cinvestav.mx
México

SAPIRES FILHO PAULO AUGUSTO
Universidad Federal de Rio Janeiro
Inst. de Fisica-UFRJ
CxP 68528, 21945-970

sapires@novanet.com.br

Brasil

SAULES ESTRADA GUSTAVO
UAM-Iztapalapa

gsau@xanum.uam.mx
México

SCOTTI ANTONIO
Physics Defortunist Counfus Universitari
Universitá di PARMA

scott@vaxpr.pr.ingn.it

Italy

SELIGMAN H. THOMAS
IF-UNAM, Lab. de Cuernavaca, Mor.

seligman@fenix.ifisicacu.unam.mx
México

SILVA PEREYRA HECTOR GABRIEL
FI-UNAM

México

SOCOLOVSKY MIGUEL
ICN-UNAM

socolovs@nuclecu.unam.mx
México

SOLANO ENRIQUE
UFRJ (Rio-Brasil)
Isla del Fundão, Cd. Universitaria Rio

solano@if.ufrj.br

Brasil

SOSA FONSECA REBECA
UAM-Iztapalapa

rebe@xanum.uam.mx
México

SOSCUN HUMBERTO
Universidad de Zulia
Fac. Exp. de Ciencias
Av. Universidad, AP. 556
Grano de Oro, Maracaibo

humberto@sinamaica.ciens.luz.ve

Venezuela

TORRES VEGA GABINO
CINVESTAV-IPN

gabino@fis.cinvestav.mx
México

TEJEDA YEOMANS MA. ELENA
Universidad de Sonora

elenaty@fisica.uson.mx
México

URRUTIA LUIS F.
ICN-UNAM

México

VALDES FERNANDEZ MA. TERESA
UNAM, Instituto de Física

maria@ft.ifisicacu.unam.mx
México

VALDEZ BALDERAS DANIEL
Universidad de Sonora

dvaldezb@fisica.uson.mx
México

VARGAS MADRAZO CARLOS ERNESTO
CINVESTAV-IPN

cvargas@rosa.fis.cinvestav.mx
México

VARGAS MARTINEZ JOSE MANUEL
INAOE
Tonantzintla, Pue.

vargas@licuadora.inaoep.mx

México

VAZQUEZ COUTIÑO GUILLERMO
FESC-UNAM y UAM-I

garc@xanum.uam.mx
México

VAZQUEZ LIMA SAMUEL
UAM-Iztapalapa · México

VELAZQUEZ AGUILAR VICTOR M. · vvelaz@linda.fis.cinvestav.mx
CINVESTAV-IPN · México

VILLA TORRES GUADALUPE · villa@ifisicacu.unam.mx
UNAM · México

VILLARREAL LUJAN CARLOS · villarreal@ft.ifisicacu.unam.mx
UNAM · México

VILLEGAS LELOVSKY LEONARDO · lvl@fis.cinvestav.mx
CINVESTAV-IPN · México

VUCETICH HECTOR · pipi@natura.fcaglp.unep.edu.ar
UNLP-CONICET, Observatorio Astronómico
Paseo del Bosque s/n, 1900 La Plata · Argentina

ZAKHARIEV BORIS · zakharev@thsun1.jinr.ru
Joint Institute for Nuclear Research
Russia 141980 Dubna, Moscow District · Russia

AUTHOR INDEX

B

Baranger, H. U., 281

C

Cetto, A. M., 151

D

Davidovich, L., 3
de la Peña, L., 151

G

Gómez, E., 221

H

Haroche, S., 45

L

Lee, H., 221
Leyvraz, F., 253

M

Man'ko, V. I., 191
Mello, P. A., 281
Moshinsky, M., 335

O

Orozco, L. A., 67
Ortega-Martínez, R., 221

R

Rolston, S. L., 91

S

Scully, M. O., 221

T

Thompson, R. C., 111